T0211894

Studies in Classification, Data Analysis, and Knowledge Organization

Studies in Classification, Data Analysis, and Knowledge Organization is a book series which offers constant and up-to-date information on the most recent developments and methods in the fields of statistical data analysis, exploratory statistics, classification and clustering, handling of information and ordering of knowledge. It covers a broad scope of theoretical, methodological as well as application-oriented articles, surveys and discussions from an international authorship and includes fields like computational statistics, pattern recognition, biological taxonomy, DNA and genome analysis, marketing, finance and other areas in economics, databases and the internet. A major purpose is to show the intimate interplay between various, seemingly unrelated domains and to foster the cooperation between mathematicians, statisticians, computer scientists and practitioners by offering well-based and innovative solutions to urgent problems of practice.

Paula Brito · José G. Dias · Berthold Lausen ·
Angela Montanari · Rebecca Nugent
Editors

Classification and Data Science in the Digital Age

 Springer

Editors
Paula Brito
Faculty of Economics
University of Porto
Porto, Portugal

INESC TEC, Centre for Artificial
Intelligence and Decision Support
(LIAAD)
Porto, Portugal

Berthold Lausen
Department of Mathematical Sciences
University of Essex
Colchester, UK

Rebecca Nugent
Department of Statistics & Data Science
Carnegie Mellon University
Pittsburgh, PA, USA

José G. Dias
Business Research Unit
University Institute of Lisbon
Lisbon, Portugal

Angela Montanari
Department of Statistical Sciences
"Paolo Fortunati"
University of Bologna
Bologna, Italy

ISSN 1431-8814   ISSN 2198-3321   (electronic)
Studies in Classification, Data Analysis, and Knowledge Organization
ISBN 978-3-031-09033-2   ISBN 978-3-031-09034-9   (eBook)
https://doi.org/10.1007/978-3-031-09034-9

Mathematics Subject Classification: 62H30, 62H25, 62R07, 68T09, 62H86, 68T10, 94A16, 68T30

This Springer imprint is published by the registered company Springer Nature Switzerland AG
The registered company address is: Gewerbestrasse 11, 6330 Cham, Switzerland

Preface

"Classification and Data Science in the Digital Age", the 17th Conference of the International Federation of Classification Societies (IFCS), is held in Porto, Portugal, from July 19th to July 23rd 2022, locally organised by the Faculty of Economics of the University of Porto and the Portuguese Association for Classification and Data Analysis, CLAD.

The International Federation of Classification Societies (IFCS), founded in 1985, is an international scientific organization with non-profit and non-political motives. Its purpose is to promote mutual communication, co-operation and interchange of views among all those interested in scientific principles, numerical methods, theory and practice of data science, data analysis, and classification in a broad sense and in as wide a range of applications as possible; to serve as an agency for the dissemination of scientific information related to these areas of interest; to prepare international conferences; to publish a newsletter and other publications. The scientific activities of the Federation are intended for all people interested in theory of classification and data analysis, and related methods and applications. IFCS 2022 – originally scheduled for August 2021, and postponed due to the Covid-19 pandemic – will be its 17th edition; previous editions were held in Thessaloniki (2019), Tokyo (2017) and Bologna (2015).

Keynote lectures are addressed by Genevera Allen (Rice University, USA), Charles Bouveyron (Université Côte d'Azur, Nice, France), Dianne Cook (Monash University, Melbourne, Australia), and João Gama (Faculty of Economics, University of Porto & LIAAD INESC TEC, Portugal). The conference program includes two tutorials: "Analysis of Data Streams" by João Gama (Faculty of Economics, University of Porto & LIAAD INESC TEC, Portugal) and "Categorical Data Analysis of Visualization" by Rosaria Lombardo (Università degli Studi della Campania Luigi Vanvitelli, Italy) and Eric Beh (University of Newcastle, Australia). IFCS 2022 has highlighted topics, which lead to Semi-Plenary Invited Sessions. The Conference program also includes Thematic Tracks on specific areas, as well as free contributed sessions in different topics (both oral communications and posters).

The Conference Scientific Program Committee is co-chaired by Paula Brito, José G. Dias, Berthold Lausen, and Angela Montanari, and includes representatives of the IFCS member societies: Adalbert Wilhelm – GfKl, Ahmed Moussa – MCS, Arthur White – IPRCS, Brian Franczak – CS, Eva Boj del Val – SEIO, Fionn Murtagh – BCS, Francesco Mola – CLADAG, Hyunjoong Kim – KCS, Javier Trejos Zelaya – SoCCCAD, Koji Kurihara – JCS, Krzysztof Jajuga – SKAD, Mark de Rooij – VOC, Mohamed Nadif – SFC, Niel le Roux – MDAG, Simona Korenjak Černe – SSS, Theodore Chadjipadelis – GSDA, who were responsible for the Conference Scientific Program, and whom the organisers wish to thank for their precious cooperation. Special thanks are also due to the chairs of the Thematic Tracks, for their invaluable collaboration.

The papers included in this volume present new developments in relevant topics of Data Science and Classification, constituting a valuable collection of methodological and applied papers that represent the current research in highly developing areas. Combining new methodological advances with a wide variety of real applications, this volume is certainly of great value for Data Science researchers and practitioners alike.

First of all, the organisers of the Conference and the editors would like to thank all authors, for their cooperation and commitment. We are specially grateful to all colleagues who served as reviewers, and whose work was decisive to the scientific quality of these proceedings. We also thank all those who have contributed to the design and production of this Book of Proceedings at Springer, in particular Veronika Rosteck, for her help concerning all aspects of publication.

The organisers would like to express their gratitude to the Portuguese Association for Classification and Data Analysis, CLAD, as well as to the Faculty of Economics of the University of Porto (FEP–UP), who enthusiastically supported the Conference from the very start, and contributed to its success. We cordially thank all members of the Local Organising Committee – Adelaide Figueiredo, Carlos Ferreira, Carlos Marcelo, Conceição Rocha, Fernanda Figueiredo, Fernanda Sousa, Jorge Pereira, M. Eduarda Silva, Paulo Teles, Pedro Campos, Pedro Duarte Silva, and Sónia Dias – and all people at FEP–UP who worked actively for the conference organisation, and whose work is much appreciated. We are very grateful to all our sponsors, for their generous support. Finally, we thank all authors and participants, who made the conference possible.

Porto,   *Paula Brito*
July 2022   *José G. Dias*
                                                                    Berthold Lausen
                                                                  Angela Montanari
                                                                   Rebecca Nugent

Acknowledgements

The Editors are extremely grateful to the reviewers, whose work was determinant for the scientific quality of these proceedings. They were, in alphabetical order:

Adalbert Wilhelm
Agustín Mayo-Iscar
Alípio Jorge
André C. P. L. F. de Carvalho
Ann Maharaj
Anuška Ferligoj
Arthur White
Berthold Lausen
Brian Franczak
Carlos Soares
Christian Hennig
Conceição Amado
Eva Boj del Val
Francesco Mola
Francisco de Carvalho
Geoff McLAchlan
Gilbert Saporta
Glòria Mateu-Figueras
Hans Kestler
Hélder Oliveira
Hyunjoong Kim
Jaime Cardoso
Javier Trejos
Jean Diatta
José A. Lozano
José A. Vilar
José Matos

Koji Kurihara
Krzysztof Jajuga
Laura Palagi
Laura Sangalli
Lazhar Labiod
Luis Angel García-Escudero
Luis Teixeira
M. Rosário Oliveira
Margarida G. M. S. Cardoso
Mark de Rooij
Michelangelo Ceci
Mohamed Nadif
Niel Le Roux
Paolo Mignone
Patrice Bertrand
Pedro Campos
Pedro Duarte Silva
Pedro Ribeiro
Peter Filzmoser
Rosanna Verde
Rosaria Lombardo
Salvatore Ingrassia
Satish Singh
Simona Korenjak-Černe
Theodore Chadjipadelis
Veronica Piccialli
Vladimir Batagelj

Partners & Sponsors

We are extremely grateful to the following institutions whose support contributes to the success of IFCS 2022:

Sponsors

Banco de Portugal

Berd

Comissão de Viticultura da Região dos Vinhos Verdes

Indie Campers

INESC/TEC

Luso-American Development Foundation

PSE

Sociedade Portuguesa de Estatística

Instituto Nacional de Estatística/Statistics Portugal

Unilabs

Universidade do Porto

Partners

Associação Portuguesa para a Investigação Operacional

Associação Portuguesa de Reconhecimento de Padrões

Associação de Turismo do Porto e Norte

Centro Internacional de Matemática

Faculdade de Engenharia da Universidade do Porto

International Association of Statistical Computing

International Association of Statistical Education

Sociedade Portuguesa de Matemática

Springer

Organisation

CLAD - Associação Portuguesa de Classificação e Análise de Dados

Faculdade de Economia da Universidade do Porto

Contents

A Topological Clustering of Individuals 1
Rafik Abdesselam

Model Based Clustering of Functional Data with Mild Outliers 11
Cristina Anton and Iain Smith

**A Trivariate Geometric Classification of Decision Boundaries for
Mixtures of Regressions** 21
Filippo Antonazzo and Salvatore Ingrassia

Generalized Spatio-temporal Regression with PDE Penalization 29
Eleonora Arnone, Elia Cunial, and Laura M. Sangalli

A New Regression Model for the Analysis of Microbiome Data 35
Roberto Ascari and Sonia Migliorati

**Stability of Mixed-type Cluster Partitions for Determination of the
Number of Clusters** 43
Rabea Aschenbruck, Gero Szepannek, and Adalbert F. X. Wilhelm

**A Review on Official Survey Item Classification for Mixed-Mode Effects
Adjustment** 53
Afshin Ashofteh and Pedro Campos

**Clustering and Blockmodeling Temporal Networks – Two Indirect
Approaches** 63
Vladimir Batagelj

Latent Block Regression Model 73
Rafika Boutalbi, Lazhar Labiod, and Mohamed Nadif

**Using Clustering and Machine Learning Methods to Provide Intelligent
Grocery Shopping Recommendations** 83
Nail Chabane, Mohamed Achraf Bouaoune, Reda Amir Sofiane Tighilt,
Bogdan Mazoure, Nadia Tahiri, and Vladimir Makarenkov

**COVID-19 Pandemic: a Methodological Model for the Analysis of
Government's Preventing Measures and Health Data Records** 93
Theodore Chadjipadelis and Sofia Magopoulou

**pcTVI: Parallel MDP Solver Using a Decomposition into Independent
Chains** 101
Jaël Champagne Gareau, Éric Beaudry, and Vladimir Makarenkov

Three-way Spectral Clustering 111
Cinzia Di Nuzzo and Salvatore Ingrassia

**Improving Classification of Documents by Semi-supervised Clustering
in a Semantic Space** 121
Jasminka Dobša and Henk A. L. Kiers

Trends in Data Stream Mining 131
João Gama

Old and New Constraints in Model Based Clustering 139
Luis A. García-Escudero, Agustín Mayo-Iscar, Gianluca Morelli, and Marco
Riani

Clustering Student Mobility Data in 3-way Networks 147
Vincenzo Giuseppe Genova, Giuseppe Giordano, Giancarlo Ragozini, and
Maria Prosperina Vitale

Clustering Brain Connectomes Through a Density-peak Approach 155
Riccardo Giubilei

Similarity Forest for Time Series Classification 165
Tomasz Górecki, Maciej Łuczak, and Paweł Piasecki

**Detection of the Biliary Atresia Using Deep Convolutional Neural
Networks Based on Statistical Learning Weights via Optimal Similarity
and Resampling Methods** 175
Kuniyoshi Hayashi, Eri Hoshino, Mitsuyoshi Suzuki, Erika Nakanishi,
Kotomi Sakai, and Masayuki Obatake

Some Issues in Robust Clustering 183
Christian Hennig

Robustness Aspects of Optimized Centroids 193
Jan Kalina and Patrik Janáček

Data Clustering and Representation Learning Based on Networked Data 203
Lazhar Labiod and Mohamed Nadif

Towards a Bi-stochastic Matrix Approximation of k-means and Some Variants 213
Lazhar Labiod and Mohamed Nadif

Clustering Adolescent Female Physical Activity Levels with an Infinite Mixture Model on Random Effects 223
Amy LaLonde, Tanzy Love, Deborah R. Young, and Tongtong Wu

Unsupervised Classification of Categorical Time Series Through Innovative Distances 233
Ángel López-Oriona, José A. Vilar, and Pierpaolo D'Urso

Fuzzy Clustering by Hyperbolic Smoothing 243
David Masís, Esteban Segura, Javier Trejos, and Adilson Xavier

Stochastic Collapsed Variational Inference for Structured Gaussian Process Regression Networks 253
Rui Meng, Herbert K. H. Lee, and Kristofer Bouchard

An Online Minorization-Maximization Algorithm 263
Hien Duy Nguyen, Florence Forbes, Gersende Fort, and Olivier Cappé

Detecting Differences in Italian Regional Health Services During Two Covid-19 Waves 273
Lucio Palazzo and Riccardo Ievoli

Political and Religion Attitudes in Greece: Behavioral Discourses 283
Georgia Panagiotidou and Theodore Chadjipadelis

Supervised Classification via Neural Networks for Replicated Point Patterns 293
Kateřina Pawlasová, Iva Karafiátová, and Jiří Dvořák

Parsimonious Mixtures of Seemingly Unrelated Contaminated Normal Regression Models 303
Gabriele Perrone and Gabriele Soffritti

Penalized Model-based Functional Clustering: a Regularization Approach via Shrinkage Methods 313
Nicola Pronello, Rosaria Ignaccolo, Luigi Ippoliti, and Sara Fontanella

Emotion Classification Based on Single Electrode Brain Data: Applications for Assistive Technology 323
Duarte Rodrigues, Luis Paulo Reis, and Brígida Mónica Faria

The Death Process in Italy Before and During the Covid-19 Pandemic: a Functional Compositional Approach 333
Riccardo Scimone, Alessandra Menafoglio, Laura M. Sangalli, and Piercesare Secchi

Clustering Validation in the Context of Hierarchical Cluster Analysis: an Empirical Study 343
Osvaldo Silva, Áurea Sousa, and Helena Bacelar-Nicolau

An MML Embedded Approach for Estimating the Number of Clusters .. 353
Cláudia Silvestre, Margarida G. M. S. Cardoso, and Mário Figueiredo

Typology of Motivation Factors for Employees in the Banking Sector: An Empirical Study Using Multivariate Data Analysis Methods 363
Áurea Sousa, Osvaldo Silva, M. Graça Batista, Sara Cabral, and Helena Bacelar-Nicolau

A Proposal for Formalization and Definition of Anomalies in Dynamical Systems 373
Jan Michael Spoor, Jens Weber, and Jivka Ovtcharova

New Metrics for Classifying Phylogenetic Trees Using K-means and the Symmetric Difference Metric 383
Nadia Tahiri and Aleksandr Koshkarov

On Parsimonious Modelling via Matrix-variate t Mixtures 393
Salvatore D. Tomarchio

Evolution of Media Coverage on Climate Change and Environmental Awareness: an Analysis of Tweets from UK and US Newspapers 403
Gianpaolo Zammarchi, Maurizio Romano, and Claudio Conversano

A Topological Clustering of Individuals

Rafik Abdesselam

Abstract The clustering of objects-individuals is one of the most widely used approaches to exploring multidimensional data. The two common unsupervised clustering strategies are Hierarchical Ascending Clustering (HAC) and k-means partitioning used to identify groups of similar objects in a dataset to divide it into homogeneous groups. The proposed Topological Clustering of Individuals, or TCI, studies a homogeneous set of individual rows of a data table, based on the notion of neighborhood graphs; the columns-variables are more-or-less correlated or linked according to whether the variable is of a quantitative or qualitative type. It enables topological analysis of the clustering of individual variables which can be quantitative, qualitative or a mixture of the two. It first analyzes the correlations or associations observed between the variables in a topological context of principal component analysis (PCA) or multiple correspondence analysis (MCA), depending on the type of variable, then classifies individuals into homogeneous group, relative to the structure of the variables considered. The proposed TCI method is presented and illustrated here using a real dataset with quantitative variables, but it can also be applied with qualitative or mixed variables.

Keywords: hierarchical clustering, proximity measure, neighborhood graph, adjacency matrix, multivariate data analysis

1 Introduction

The objective of this article is to propose a topological method of data analysis in the context of clustering. The proposed approach, Topological Clustering of Individuals

Rafik Abdesselam (✉)
University of Lyon, Lyon 2, ERIC - COACTIS Laboratories
Department of Economics and Management, 69365 Lyon, France,
e-mail: `rafik.abdesselam@univ-lyon2.fr`

© The Author(s) 2023
P. Brito et al. (eds.), *Classification and Data Science in the Digital Age*,
Studies in Classification, Data Analysis, and Knowledge Organization,
https://doi.org/10.1007/978-3-031-09034-9_1

1

(TCI) is different from those that already exist and with which it is compared. There are approaches specifically devoted to the clustering of individuals, for example, the Cluster procedure implemented in SAS software, but as far as we know, none of these approaches has been proposed in a topological context.

Proximity measures play an important role in many areas of data analysis [16, 5, 9]. The results of any operation involving structuring, clustering or classifying objects are strongly dependent on the proximity measure chosen.

This study proposes a method for the topological clustering of individuals whatever type of variable is being considered: quantitative, qualitative or a mixture of both. The eventual associations or correlations between the variables partly depends on the database being used and the results can change according to the selected proximity measure. A proximity measure is a function which measures the similarity or dissimilarity between two objects or variables within a set.

Several topological data analysis studies have been proposed both in the context of factorial analyses (discriminant analysis [4], simple and multiple correspondence analyses [3], principal component analysis [2]) and in the context of clustering of variables [1], clustering of individuals [10] and this proposed TCI approach.

This paper is organized as follows. In Section 2, we briefly recall the basic notion of neighborhood graphs, we define and show how to construct an adjacency matrix associated with a proximity measure within the framework of the analysis of the correlation structure of a set of quantitative variables, and we present the principles of TCI according to continuous data. This is illustrated in Section 3 using an example based on real data. The TCI results are compared with those of the well-known classical clustering of individuals. Finally, Section 4 presents the concluding remarks on this work.

2 Topological Context

Topological data analysis is an approach based on the concept of the neighborhood graph. The basic idea is actually quite simple: for a given proximity measure for continuous or binary data and for a chosen topological structure, we can match a topological graph induced on the set of objects.

In the case of continuous data, we consider $E = \{x^1, \cdots, x^j, \cdots, x^P\}$, a set of p quantitative variables. We can see in [1] cases of qualitative or even mixed variables.

We can, by means of a proximity measure u, define a neighborhood relationship, V_u, to be a binary relationship based on $E \times E$. There are many possibilities for building this neighborhood binary relationship.

Thus, for a given proximity measure u, we can build a neighborhood graph on E, where the vertices are the variables and the edges are defined by a property of the neighborhood relationship.

Many definitions are possible to build this binary neighborhood relationship. One can choose the Minimal Spanning Tree (MST) [7], the Gabriel Graph (GG) [11] or, as is the case here, the Relative Neighborhood Graph (RNG) [14].

For any given proximity measure u, we can construct the associated adjacency binary symmetric matrix V_u of order p, where, all pairs of neighboring variables in E satisfy the following RNG property:

$$V_u(x^k, x^l) = \begin{cases} 1 & \text{if } u(x^k, x^l) \leq \max[u(x^k, x^t), u(x^t, x^l)] ; \\ & \forall x^k, x^l, x^t \in E, \; x^t \neq x^k \text{ and } x^t \neq x^l \\ 0 & \text{otherwise.} \end{cases}$$

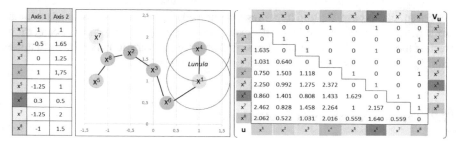

Fig. 1 Data - RNG structure - Euclidean distance - Associated adjacency matrix.

Figure 1 shows a simple illustrative example in $\mathbb{R}^2$ of a set of quantitative variables that verify the structure of the RNG graph with Euclidean distance as proximity measure: $u(x^k, x^l) = \sqrt{\sum_{j=1}^{2}(x_j^k - x_j^l)^2}$.

This generates a topological structure based on the objects in E which are completely described by the adjacency binary matrix V_u.

2.1 Reference Adjacency Matrices

Three topological factorial approaches are described in [1] according to the type of variables considered: quantitative, qualitative or a mixture of both. We consider here the case of a set of quantitative variables.

We assume that we have at our disposal a set $E = \{x^j; j = 1, \cdots, p\}$ of p quantitative variables and n individuals-objects. The objective here is to analyze in a topological way, the structure of the correlations of the variables considered [2], from which the clustering of individuals will then be established.

We construct the reference adjacency matrix named $V_{u_\star}$ from the correlation matrix. Expressions of suitable adjacency reference matrices for cases involving qualitative variables or mixed variables are given in [1].

To examine the correlation structure between the variables, we look at the significance of their linear correlation. The reference adjacency matrix $V_{u_\star}$ associated with reference measure $u_\star$, can be written using the Student's t-test of the linear correlation coefficient ρ of Bravais-Pearson:

Definition 1 For quantitative variables, $V_{u_\star}$ is defined as:

$$V_{u_\star}(x^k, x^l) = \begin{cases} 1 \text{ if } p\text{-value} = P[\ |T_{n-2}| > \text{t-value} \ ] \leq \alpha \ ; \ \forall k, l = 1, p \\ 0 \text{ otherwise.} \end{cases}$$

where the p-value is the significance test of the linear correlation coefficient for the two-sided test of the null and alternative hypotheses, $H_0 : \rho(x^k, x^l) = 0$ vs. $H_1 : \rho(x^k, x^l) \neq 0$.

Let T_{n-2} be a t-distributed random variable of Student with $\nu = n - 2$ degrees of freedom. In this case, the null hypothesis is rejected if the p-value is less than or equal to a chosen α significance level, for example, $\alpha = 5\%$. Using a linear correlation test, if the p-value is very small, it means that there is a very low likelihood that the null hypothesis is correct, and consequently we can reject it.

2.2 Topological Analysis - Selective Review

Whatever the type of variable set being considered, the built reference adjacency matrix $V_{u_\star}$ is associated with an unknown reference proximity measure $u_\star$.

The robustness depends on the α error risk chosen for the null hypothesis: no linear correlation in the case of quantitative variables, or positive deviation from independence in the case of qualitative variables, can be studied by setting a minimum threshold in order to analyze the sensitivity of the results. Certainly the numerical results will change, but probably not their interpretation.

We assume that we have at our disposal $\{x^k; k = 1, .., p\}$ a set of p homogeneous quantitative variables measured on n individuals. We will use the following notations:

- $X_{(n,p)}$ is the data matrix with n rows-individuals and p columns-variables,
- $V_{u_\star}$ is the symmetric adjacency matrix of order p, associated with the reference measure $u_\star$ which best structures the correlations of the variables,
- $\widehat{X}_{(n,p)} = XV_{u_\star}$ is the projected data matrix with n individuals and p variables,
- M_p is the matrix of distances of order p in the space of individuals,
- $D_n = \frac{1}{n}I_n$ is the diagonal matrix of weights of order n in the space of variables.

We first analyze, in a topological way, the correlation structure of the variables using a Topological PCA, which consists of carrying out the standardized PCA [6, 8] triplet $(\widehat{X}, M_p, D_n)$ of the projected data matrix $\widehat{X} = XV_{u_\star}$ and, for comparison, the duality diagram of the Classical standardized PCA triplet (X, M_p, D_n) of the initial data matrix X. We then proceed with a clustering of individuals based on the significant principal components of the previous topological PCA.

Definition 2 TCI consist of performing a HAC, based on the Ward criterion[1] [15], on the significant factors of the standardized PCA of the triplet $(\widehat{X}, M_p, D_n)$.

---

[1] Aggregation based on the criterion of the loss of minimal inertia.

3 Illustrative Example

The data used [13] to illustrate the TCI approach describe the renewable electricity (RE) of the 13 French regions in 2017, described by 7 quantitative variables relating to RE. The growth of renewable energy in France is significant. Some French regions have expertise in this area; however, the regions' profiles appear to differ.

The objective is to specify regional disparities in terms of RE by applying topological clustering to the French regions in order to identify which were the country's greenest regions in 2017. Statistics relating to the variables are displayed in Table 1.

Table 1 Summary statistics of renewable energy variables.

Variable	Frequency	Mean	Standard Deviation (N)	Coefficient of variation (%)	Min	Max
Total RE production (TWH)	13	6.84	6.58	96.19	0.59	2.34
Total RE consumption (TWH)	13	3.70	1.87	50.67	2.18	7.06
Coverage RE consumption (%)	13	0.18	0.11	59.01	0.02	0.36
Hydroelectricity (%)	13	0.34	0.30	87.47	0.01	0.89
Solar electricity (%)	13	0.13	0.09	72.57	0.02	0.31
Wind electricity (%)	13	0.39	0.29	76.12	0.01	0.86
Biomass electricity (%)	13	0.15	0.19	130.54	0.01	0.79

Table 2 Correlation matrix (p-value) - Reference adjacency matrix $V_{u_\star}$.

Production	1.000						
Consumption	**0.575**	1.000					
	(0.040)						
Coverage	**0.798**	0.090	1.000				
	(0.001)	(0.771)					
Hydroelectricity	**0.720**	0.138	**0.872**	1.000			
	(0.006)	(0.653)	**(0.000)**				
Solar	-0.272	-0.477	0.105	0.168	1.000		
	(0.369)	(0.099)	(0.734)	(0.582)			
Wind	-0.408	-0.305	-0.524	**-0.772**	-0.395	1.000	
	(0.167)	(0.311)	(0.066)	**(0.002)**	(0.181)		
Biomass	-0.365	0.489	**-0.609**	-0.459	-0.149	-0.135	1.000
	(0.220)	(0.090)	**(0.027)**	(0.114)	(0.627)	(0.660)	

$$V_{u_\star} = \begin{pmatrix} 1 & 1 & 1 & 1 & 0 & 0 & 0 \\ 1 & 1 & 0 & 0 & 0 & 0 & 0 \\ 1 & 0 & 1 & 1 & 0 & 0 & -1 \\ 1 & 0 & 1 & 1 & 0 & -1 & 0 \\ 0 & 0 & 0 & 0 & 1 & 0 & 0 \\ 0 & 0 & 0 & -1 & 0 & 1 & 0 \\ 0 & 0 & -1 & 0 & 0 & 0 & 1 \end{pmatrix}$$

Significance level: p–value $\leq \alpha = 5\%$

The adjacency matrix $V_{u_\star}$, associated with the proximity measure $u_\star$, adapted to the data considered, is built from the correlations matrix Table 2 according to Definition 1. Note that in this case, which uses quantitative variables, it is considered that two positively correlated variables are related and that two negatively correlated variables are related but remote. We will therefore take into account any sign of correlation between variables in the adjacency matrix.

We first carry out a Topological PCA to identify the correlation structure of the variables. A HAC, according to Ward's criterion, is then applied to the significant principal components of the PCA of the projected data. We then compare the results of a topological and a classical PCA.

Figure 2 presents, for comparison on the first factorial plane, the correlations between principal components-factors and the original variables.

We can see that these correlations are slightly different, as are the percentages of the inertias explained on the first principal planes of Topological and Classic PCA.

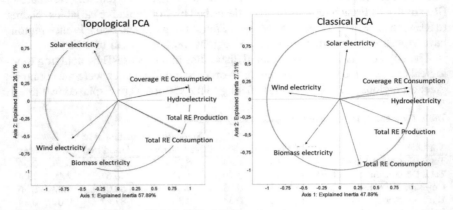

Fig. 2 Topological & Classical PCA of RE of the French regions.

The two first factors of the Topological PCA explain 57.89% and 26.11%, respectively, accounting for 83.99% of the total variation in the data set; however, the two first factors of the Classical PCA add up to 75.20%. Thus, the first two factors provide an adequate synthesis of the data, that is, of RE in the French regions. We restrict the comparison to the first significant factorial axes.

For comparison, Figure 3 shows dendrograms of the Topological and Classical clustering of the French regions according to their RE. Note that the partitions chosen in 5 clusters are appreciably different, as much by composition as by characterization. The percentage variance produced by the TCI approach, $R^2 = 86.42\%$, is higher than that of the classic approach, $R^2 = 84.15\%$, indicating that the clusters produced via the TCI approach are more homogeneous than those generated by the Classical one.

Based on the TCI analysis, the Corse region alone constitutes the fourth cluster, and the Nouvelle-Acquitaine region is found in the second cluster with the Grand-Est, Occitanie and Provence-Alpes-Côte-d'Azur (PACA) regions; however, in the Classical clustering, these two regions - Corse and Nouvelle-Aquitaine - together constitute the third cluster.

Figure 4 summarizes the significant profiles (+) and anti-profiles (-) of the two typologies; with a risk of error less than or equal to 5%, they are quite different.

The first cluster produced via the TCI approach, consisting of a single region, Auvergne-Rhônes-Alpes (AURA), is characterized by high share of hydroelectricity, a high level of coverage of regional consumption, and high RE production and consumption. The second cluster - which groups together the four regions of Grand-Est, Occitanie, Provence-Alpes-Côte-d'Azur (PACA) and Nouvelle-Aquitaine - is considered a homogeneous cluster, which means that none of the seven RE characteristics differ significantly from the average of these characteristics across all regions. This cluster can therefore be considered to reflect the typical picture of RE in France.

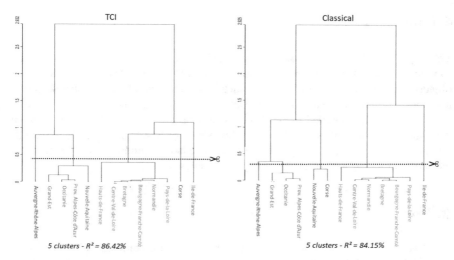

Fig. 3 Topological and Classical dendrograms of the French regions.

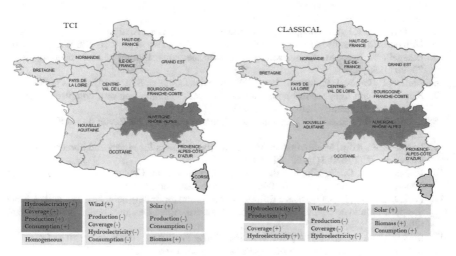

Fig. 4 Typologies - Characterization of TCI & Classical clusters

Cluster 3, which consists of six regions, is characterized by a high degree of wind energy, a low degree of hydroelectricity, low coverage of regional consumption, and low production and consumption of RE compared to the national average. Cluster 4, represented by the Corse region, is characterized by a high share of solar energy and low production and consumption of RE. The last class, represented by the Ile-de-France region, is characterized by a high share of biomass energy. Regarding the other types of RE, their share is close to the national average.

4 Conclusion

This paper proposes a new topological approach to the clustering of individuals which can enrich classical data analysis methods within the framework of the clustering of objects. The results of the topological clustering approach, based on the notion of a neighborhood graph, are as good - or even better, according to the R-squared results - than the existing classical method. The TCI approach is be easily programmable from the PCA and HAC procedures of SAS, SPAD or R software. Future work will involve extending this topological approach to other methods of data analysis, in particular in the context of evolutionary data analysis.

References

1. Abdesselam, R.: A topological clustering of variables. Journal of Mathematics and System Science. Accepted (2022)
2. Abdesselam, R.: A topological approach of Principal Component Analysis. International Journal of Data Science and Analysis. **77**(2), 20–31 (2021)
3. Abdesselam, R.: A topological Multiple Correspondence Analysis. Journal of Mathematics and Statistical Science, ISSN 2411-2518, **5**(8), 175–192 (2019)
4. Abdesselam, R.: A topological Discriminant Analysis. Data Analysis and Applications 2, Utilization of Results in Europe and Other Topics, Vol.3, Part 4. pp. 167–178 Wiley, (2019)
5. Batagelj, V., Bren, M.: Comparing resemblance measures. Journal of Classification, **12**(1), 73–90 (1995)
6. Caillez, F., Pagès, J. P.: Introduction à l'Analyse des Données. S.M.A.S.H., Paris (1976)
7. Kim, J. H. and Lee, S.: Tail bound for the minimal spanning tree of a complete graph. In Statistics & Probability Letters, **4**(64), 425–430 (2003)
8. Lebart, L.: Stratégies du traitement des données d'enquêtes. La Revue de MODULAD, **3**, 21–30 (1989)
9. Lesot, M. J., Rifqi, M., Benhadda, H.: Similarity measures for binary and numerical data: a survey. In: IJKESDP, **1**(1), 63–84 (2009)
10. Panagopoulos, D.: Topological data analysis and clustering. Chapter for a book, Algebraic Topology (math.AT) arXiv:2201.09054, Machine Learning (2022)
11. Park, J. C., Shin, H., Choi, B. K.: Elliptic Gabriel graph for finding neighbors in a point set and its application to normal vector estimation. Computer-Aided Design Elsevier, **38**(6), 619–626 (2006)
12. SAS Institute Inc. SAS/STAT Software, the Cluster Procedure, Available via DIALOG. https://support.sas.com/documentation/onlinedoc/stat/142/cluster.pdf
13. Selectra: Electricité renouvelable: quelles sont les régions les plus vertes de France ? http://selectra.info/energie/actualites/expert/electricite-renouvelable-regions-plus-vertes-france (2020)
14. Toussaint, G. T.: The relative neighbourhood graph of a finite planar set. Pattern Recognition, **12**(4) 261–268 (1980)
15. Ward, J. R.: Hierarchical grouping to optimize an objective function. Journal of the American Statistical Association, **58**(301), 236–244 (1963)
16. Zighed, D., Abdesselam, R., Hadgu, A.: Topological comparisons of proximity measures. In: Tan et al. (Eds). In Proc. 16th PAKDD 2012 Conference, pp. 379–391. Springer, (2012)

Model Based Clustering of Functional Data with Mild Outliers

Cristina Anton and Iain Smith

Abstract We propose a procedure, called CFunHDDC, for clustering functional data with mild outliers which combines two existing clustering methods: the functional high dimensional data clustering (FunHDDC) [1] and the contaminated normal mixture (CNmixt) [3] method for multivariate data. We adapt the FunHDDC approach to data with mild outliers by considering a mixture of multivariate contaminated normal distributions. To fit the functional data in group-specific functional subspaces we extend the parsimonious models considered in FunHDDC, and we estimate the model parameters using an expectation-conditional maximization algorithm (ECM). The performance of the proposed method is illustrated for simulated and real-world functional data, and CFunHDDC outperforms FunHDDC when applied to functional data with outliers.

Keywords: functional data, model-based clustering, contaminated normal distribution, EM algorithm

1 Introduction

Recently, model-based clustering for functional data has received a lot of attention. Real data are often contaminated by outliers that affect the estimations of the model parameters. Here we propose a method for clustering functional data with mild outliers. Mild outliers are usually sampled from a population different from the

Cristina Anton (✉)
MacEwan University, 10700 – 104 Avenue Edmonton, AB, T5J 4S2, Canada,
e-mail: popescuc@macewan.ca

Iain Smith
MacEwan University, 10700 – 104 Avenue Edmonton, AB, T5J 4S2, Canada,
e-mail: smithi23@mymacewan.ca

© The Author(s) 2023
P. Brito et al. (eds.), *Classification and Data Science in the Digital Age*,
Studies in Classification, Data Analysis, and Knowledge Organization,
https://doi.org/10.1007/978-3-031-09034-9_2

assumed model, so we need to choose a model flexible enough to accommodate them.

Functional data live in an infinite dimensional space and model-based methods for clustering are not directly available because the notion of probability density function generally does not exist for such data. A first approach is to use a two-step method and first do a discretization or a decomposition of the functional data in a basis of functions (such as Fourier series, B-splines, etc.), and then directly apply multivariate clustering methods to the discretization or the basis coefficients. A second approach, which allows the interaction between the discretization and the clustering steps, is based on a probabilistic model for the basis coefficients [1, 2].

We follow the second approach, and we propose a method, called CFunHDDC, which extends the functional high dimensional data clustering (FunHDDC) [1] to clustering functional data with mild outliers. There are several methods to detect outliers of functional data and a robust clustering methodology based on trimming is presented in [4]. Our approach does not involve trimming the outliers and it is inspired by the method CNmixt [3] for clustering multivariate data with mild outliers. We propose a model for the basis coefficients based on a mixture of contaminated multivariate normal distributions. A multivariate contaminated normal distribution is a two-component normal mixture in which the bad observations (outliers) are represented by a component with a small prior probability and an inflated covariance matrix.

In the next section we present the model and its parsimonious variants. Parameter estimation is included in Section 3. In Section 4 we present applications to simulated and real-world data. The last section includes the conclusions.

2 The Model

We suppose that we observe n curves $\{x_1, \ldots, x_n\}$ and we want to cluster them in K homogeneous groups. For each curve x_i we have access to a finite set of values $x_{ij} = x_i(t_{ij})$, where $0 \leq t_{i1} < t_{i2} < \ldots < t_{im_i} \leq T$. We assume that the observed curves are independent realizations of a L^2- continuous stochastic process $X = \{X(t)\}_{t \in [0,T]}$ for which the sample paths are in $L^2[0,T]$. To reconstruct the functional form of the data we assume that the curves belong to a finite dimensional space spanned by a basis of functions $\{\xi_1, \ldots, \xi_p\}$, so we have the expansion for each curve

$$x_i(t) = \sum_{j=1}^{p} \gamma_{ij} \xi_j(t).$$

Here we assume that the dimension p is fixed and known. We consider a model based on a mixture of multivariate contaminated normal distributions for the coefficients vectors $\{\gamma_1, \ldots, \gamma_n\} \subset \mathbb{R}^p$, $\gamma_i = (\gamma_{i1}, \ldots, \gamma_{ip})^\top \in \mathbb{R}^p$, $i = 1, \ldots, n$.

We suppose that there exists two unobserved random variables $Z = (Z_1, \ldots, Z_K)$, $\Upsilon = (\Upsilon_1, \ldots, \Upsilon_K) \in \{0, 1\}^K$ where Z indicates the cluster membership and Υ

whether an observation is good or bad (outlier). $Z_k = 1$ if $X \in k$th cluster and $Z_k = 0$ otherwise, and $Y_k = 1$ if $X \in k$th cluster and it is a good observation, and $Y_k = 0$ otherwise. For clustering we need to predict the value $z_i = (z_{i1}, \ldots, z_{iK})$ of Z, and to determine the bad observations we need to predict the value $v_i = (v_{i1}, \ldots, v_{iK})$ of Y for each observed curve x_i, $i = 1, \ldots, n$.

We consider a set of n_k observed curves of the kth cluster with the coefficients $\{\gamma_1, \ldots, \gamma_{n_k}\} \subset \mathbb{R}^p$. We assume that $\{\gamma_1, \ldots, \gamma_{n_k}\}$ are independent realizations of a random vector $\Gamma \in \mathbb{R}^p$, and that the stochastic process associated with the kth cluster can be described in a lower dimensional subspace $\mathbb{E}^k[0, T] \subset L^2[0, T]$ with dimension $d_k \leq p$ and spanned by the first d_k elements of a group specific basis of functions $\{\phi_{kj}\}_{j=1,\ldots,d_k}$ that can be obtained from $\{\xi_j\}_{j=1,\ldots,p}$ by a linear transformation

$$\phi_{kj} = \sum_{l=1}^{p} q_{k,jl} \xi_l,$$

with an $p \times p$ orthogonal matrix $Q_k = (q_{k,jl})$. In [1] for FunHDDC the assumption is that the distribution of Γ for the kth cluster is $\Gamma \sim N(\mu_k, \Sigma_k)$, $\Sigma_k = Q_k \Delta_k Q_k^\top$, where

$$\Delta_k = \left. \left(\begin{array}{cccc|cccc} a_{k1} & & 0 & & & & & \\ & \ddots & & & & \mathbf{0} & & \\ 0 & & a_{kd_k} & & & & & \\ \hline & & & & b_k & & 0 & \\ & \mathbf{0} & & & & \ddots & & \\ & & & & 0 & & b_k & \end{array} \right) \right\} p$$

with $a_{ki} > b_k$, $i = 1, \ldots, d_k$. We can say that the variance of the actual data in the kth cluster is modeled by $a_{k1}, \ldots, a_{kd_k}$ and the parameter b_k models the variance of the noise [1].

We follow the approach in [3] and we assume that Γ for the kth cluster has the multivariate contaminated normal distribution with density

$$f(\gamma_i; \theta_k) = \alpha_k \phi(\gamma_i; \mu_k, \Sigma_k) + (1 - \alpha_k)\phi(\gamma_i; \mu_k, \eta_k \Sigma_k), \tag{1}$$

where $\alpha_k \in (0.5, 1)$, $\eta_k > 1$, $\theta_k = \{\alpha_k, \mu_k, \Sigma_k, \eta_k\}$, and $\phi(\gamma_i; \mu_k, \Sigma_k)$ is the density for the $p-$variate normal distribution $N(\mu_k, \Sigma_k)$:

$$\phi(\gamma_i; \mu_k, \Sigma_k) = (2\pi)^{-p/2} |\Sigma_k|^{-1/2} \exp\left(-\frac{1}{2}(\gamma_i - \mu_k)^\top \Sigma_k^{-1} (\gamma_i - \mu_k)\right) \tag{2}$$

Here α_k defines the proportion of uncontaminated data in the $k$the cluster and η_k represents the degree of contamination. We can see η_k as an inflation parameter that measures the increase in variability due to the bad observations.

Each curve x_i has a basis expansion with coefficient γ_i such that γ_i is a random vector whose distributions is a mixture of contaminated Gaussians with density

$$p(\gamma; \theta) = \sum_{k=1}^{K} \pi_k f(\gamma; \theta_k) \tag{3}$$

where $\pi_k = P(Z_k = 1)$ is the prior probability of the kth the cluster and $\theta = \bigcup_{k=1}^{k} (\theta_k \cup \{\pi_k\})$ is the set formed by all the parameters. We refer to this model as FCLM$[a_{kj}, b_k, Q_k, d_k]$ (functional contaminated latent mixture). As in [1] we consider the parsimonious sub-models: FCLM$[a_{kj}, b, Q_k, d_k]$, FCLM$[a_k, b_k, Q_k, d_k]$, FCLM$[a, b_k, Q_k, d_k]$, FCLM$[a_k, b, Q_k, d_k]$, FCLM$[a, b, Q_k, d_k]$.

3 Model Inference

To fit the models we use the ECM algorithm [3], which is a variant of the EM algorithm. In the ECM algorithm we replace the M-step in the EM algorithm by two simpler CM-steps given by the partition of the set with the parameters $\theta = \{\Psi_1, \Psi_2\}$, where $\Psi_1 = \{\pi_k, \alpha_k, \mu_k, a_{kj}, b_k, q_{kj}, k = 1, \ldots, K, j = 1, \ldots, d_k\}$, $\Psi_2 = \{\eta_k, k = 1, \ldots, K\}$, and q_{kj} is the jth column of Q_k.

We have two sources of missing data: the clusters' labels and the type of observation (good or bad). Thus the complete data are given by $S = \{\gamma_i, z_i, v_i\}_{i=1,\ldots,n}$, and the complete-data likelihood is

$$L_c(\theta; S) = \prod_{i=1}^{N} \prod_{k=1}^{K} \left\{ \pi_k \left[\alpha_k \phi(\gamma_i; \mu_k, \Sigma_k) \right]^{v_{ik}} \left[(1 - \alpha_k) \phi(\gamma_i; \mu_k, \eta_k \Sigma_k) \right]^{1-v_{ik}} \right\}^{z_{ik}}$$

We denote the complete-data log-likelihood by $l_c(\theta; S) = \log(L_c(\theta; S))$.

Next we present the ECM algorithm for the model FCLM$[a_{kj}, b_k, Q_k, d_k]$. At the q iteration of the ECM algorithm in the E-step we calculate $E[l_c(\theta^{(q-1)}; S)|\gamma_1, \ldots, \gamma_n, \theta^{(q-1)}]$, given the current values of the parameters $\theta^{(q-1)}$. This reduces to the calculation of $z_{ik}^{(q)} := E[Z_{ik}|\gamma_i, \theta^{(q-1)}]$, $v_{ik}^{(q)} := E[\Upsilon_{ik}|\gamma_i, z_i, \theta^{(q-1)}]$.

In the first CM step in the q iteration of the ECM algorithm we calculate $\Psi_1^{(q)}$ as the value of Ψ_1 that maximize $l_c^{(q-1)}$ with Ψ_2 fixed at $\Psi_2^{(q-1)}$. We obtain

$$\pi_k^{(q)} = \frac{\sum_{i=1}^{n} z_{ik}^{(q)}}{n}, \quad \alpha_k^{(q)} = \frac{\sum_{i=1}^{n} z_{ik}^{(q)} v_{ik}^{(q)}}{\sum_{i=1}^{n} z_{ik}^{(q)}}, \quad \mu_k^{(q)} = \frac{\sum_{i=1}^{n} z_{ik}^{(q)} \left(v_{ik}^{(q)} + \frac{1-v_{ik}^{(q)}}{\eta_k^{(q-1)}} \right) \gamma_i}{\sum_{i=1}^{n} z_{ik}^{(q)} \left(v_{ik}^{(q)} + \frac{1-v_{ik}^{(q)}}{\eta_k^{(q-1)}} \right)} \tag{4}$$

$$\Sigma_k^{(q)} = \frac{1}{\sum_{i=1}^{n} z_{ik}^{(q)}} \sum_{i=1}^{n} z_{ik}^{(q)} \left(v_{ik}^{(q)} + \frac{1-v_{ik}^{(q)}}{\eta_k^{(q-1)}} \right) (\gamma_i - \mu_k^{(q)})(\gamma_i - \mu_k^{(q)})^{\top} \tag{5}$$

We introduce a value α^* and we constrain $\alpha_k \in (\alpha^*, 1)$. If the estimation $\alpha_k^{(q)}$ in (4) is less than α^*, we use the *optimize()* function in the *stats* package in R to do a numerical search for $\alpha_k^{(q)}$.

As in [1] we get the updated values $a_{kj}^{(q)}, b_k^{(q)}, q_{kj}^{(q)}, k = 1, \ldots, K, j = 1, \ldots, d_k$ from the sample covariance matrix $\Sigma_k^{(q)}$ of cluster k, using also the matrix of inner products between the basis functions $W = (w_{jl})_{1 \le j, l \le p}$, where $w_{jl} = \int_0^T \xi_j(t) \xi_l(t) dt$.

In the second CM step in the q iteration of the ECM algorithm we calculate $\eta_k^{(q)}$ as the value that maximize $l_c^{(q-1)}$ with Ψ_1 fixed at $\Psi_1^{(q)}$.

At the end of the ECM algorithm, we do a two-step classification to provide the expected clustering. If q_f is the last iteration of the algorithm before convergence, an observation $\gamma_i \in \mathbb{R}^p$ is assigned to the cluster $k_0 \in \{1, \ldots, K\}$ with the largest $z_{ik}^{(q_f)}$. Next, an observation γ_i that was assigned to the cluster k_0 is considered good if $v_{ik_0}^{(q_f)} > 0.5$, and it is considered bad otherwise. After the classification step we can eliminate the bad observations and run FunHDDC to re-cluster the remaining observations.

The class specific dimension d_k is selected through the scree-test of Cattell by comparison of the difference between eigenvalues with a given threshold [1]. The number of clusters K as well as the parsimonious model are selected using the BIC criterion.

4 Applications

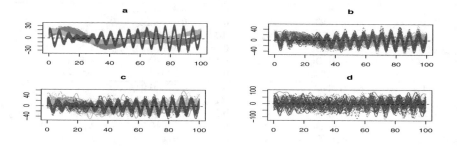

Fig. 1 Smooth data simulated without oultiers (a), according to scenario A (b), scenario B (c), and scenario C (d), coloured by group for one simulation.

We simulate 1000 curves based on the model $FCLM[a_k, b_k, Q_k, d_k]$. The number of clusters is fixed to $K = 3$ and the mixing proportions are equal $\pi_1 = \pi_2 = \pi_3 = 1/3$. We consider the following values of the parameters

   Group 1: $d = 5, a = 150, b = 5, \mu = (1, 0, 50, 100, 0, \ldots, 0)$
   Group 2: $d = 20, a = 15, b = 8, \mu = (0, 0, 80, 0, 40, 2, 0, \ldots, 0)$

Group 3: $d = 10$, $a = 30$, $b = 10$, $\mu = (0, \ldots, 0, 20, 0, 80, 0, 0, 100)$,

where d is the intrinsic dimension of the subgroups, μ is the mean vector of size 70, a is the value of the d-first diagonal elements of Δ, and b the value of the $70 - d$- last ones. Curves are smoothed using 35 Fourier basis functions. We repeat the simulation 100 times. A sample of theses data is plotted in Figure 1 a. We consider the following contamination schemes where the scores are simulated from contaminated normal distributions with the previous parameters and

A: $\alpha_i = 0.9$, $i = 1, \ldots, 3$, and $\eta_1 = 7$, $\eta_2 = 10$, $\eta_3 = 17$.
B: $\alpha_i = 0.9$, $i = 1, \ldots, 3$, and $\eta_1 = 5$, $\eta_2 = 50$, $\eta_3 = 15$.
C: $\alpha_i = 0.9$, $i = 1, \ldots, 3$, and $\eta_1 = 100$, $\eta_2 = 70$, $\eta_3 = 170$.

Samples for data generated according to scenarios A, B, C are plotted in Figure 1 b, c, d, respectively. We notice that there is more overlapping between the 3 groups when we increase the values of η.

Table 1 Mean (and standard deviation) of ARI for BIC best model on 100 simulations. Bold values indicates the highest value for each method.

Scenario	Method	α^*	ϵ	ARI	ARI Outliers
A	FunHDDC	-	0.05	**0.519 (0.11)**	-
A	FunHDDC	-	0.1	0.499(0.05)	-
A	FunHDDC	-	0.2	0.494 (0.01)	-
A	CFunHDDC	0.75	0.05	0.769 (0.23)	0.959(0.04)
A	CFunHDDC	0.75	0.1	0.986(0.08)	0.998(0.01)
A	CFunHDDC	0.75	0.2	**0.9995 (0.001)**	**1 (0)**
B	FunHDDC	-	0.05	**0.861 (0.23)**	-
B	FunHDDC	-	0.1	0.754(0.25)	-
B	FunHDDC	-	0.2	0.52 (0.09)	-
B	CFunHDDC	0.75	0.05	0.807 (0.22)	0.961(0.05)
B	CFunHDDC	0.75	0.1	0.948 (0.14)	0.99(0.03)
B	CFunHDDC	0.75	0.2	**0.990 (0.062)**	**0.971 (0.149)**
C	FunHDDC	-	0.05	0.490 (0.02)	-
C	FunHDDC	-	0.1	0.491(0.02)	-
C	FunHDDC	-	0.2	**0.494 (0.01)**	-
C	CFunHDDC	0.75	0.05	0.736 (0.23)	0.928(0.10)
C	CFunHDDC	0.75	0.1	0.911 (0.18)	0.958(0.15)
C	CFunHDDC	0.75	0.2	**0.965 (0.11)**	**0.994 (0.03)**

The quality of the estimated partitions obtained using FunHDDC and CFunHDDC is evaluated using the Adjusted Rand Index (ARI) [3], and the results are included in Table 1. For FunHDDC we use the library *funHDDC* in R. We run both algorithms for $K = 3$ with all 6 sub-models and the best solution in terms of the highest BIC value for all those submodels is returned. The initialization is done with the k-means

Table 2 Correct classification rates for each method.

Method	ϵ	CCR	Method	α^*	ϵ	CCR	Method	α^*	CCR
FunHDDC	0.01	0.68	CFunHDDC	0.85	0.01	0.67	CNmixt	0.5	0.67
FunHDDC	0.05	0.64	CFunHDDC	0.85	0.05	0.70	CNmixt	0.75	0.66
FunHDDC	0.1	0.59	CFunHDDC	0.85	0.1	0.70	CNmixt	0.85	0.67
FunHDDC	0.2	0.57	CFunHDDC	0.85	0.2	0.6	CNmixt	0.9	0.66

strategy with 50 repetitions, and the maximum number of iterations is 200 for the stopping criterion. We use $\epsilon \in \{0.05, 0.1, 0.2\}$ in the Cattell test.

We notice that CFunHDDC outperforms FunHDDC, and it gives excellent results even in Scenario C. For CFunHDDC the best results are obtained for $\epsilon = 0.2$ in the Catell test, and the values of the ARI are close to 1.

Next, we consider the NOx data available in the *fda.usc* library in R and representing daily curves of Nitrogen Oxides (NOx) emissions in the neighborhood of the industrial area of Poblenou, Barcelona (Spain). The measurements of NOx (in $\mu g/m^3$) were taken hourly resulting in 76 curves for "working days" and 39 curves for "non-working days" (see Figure 2 a). Since NOx is a contaminant agent, the detection of outlying emission is useful for environmental protection. This data set has been used for testing methods for the detection of outliers and to illustrate robust clustering based on trimming for functional data [4].

We apply CFunHDDC, FunHDDC, and CNmixt to the NOx data. Curves are smoothed using a basis of 8 Fourier functions, and we run the algorithms for $K = 2$ clusters. For CFunHDDC, FunHDDC we use $\epsilon \in \{0.001, 0.05, 0.1, 0.2\}$ in the Cattell test and the rest of the settings are the same as in the simulation study. We run CNmixt for all 14 models from the *ContaminatedMixt* R library, based on the coefficients in the Fourier basis, with 1000 iterations for the stopping criteria, and initialization done with the k-means method. The correct classification rates (CCR) are reported in Table 2.

The CCR for CFunHDDC are slightly better than the ones for FunHDDC and CNmixt, and are comparable with the ones reported in Table 1 in [4] for Funclust,

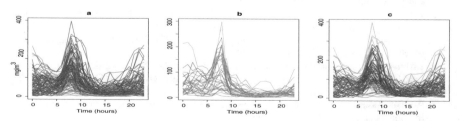

Fig. 2 a.Daily NOx curves for 115 days; b. c. Clustering obtained with CFunHDDC, $\epsilon = 0, 05, \alpha^* = 0.85$; Non-working days (blue), working days (red), outliers (green).

RFC, and TrimK. In Figure 2 b, c we present the clusters and the detected outliers for $\epsilon = 0.05$ and $\alpha^* = 0.85$. The curves that are detected as outliers (green lines) exhibit different patterns from the rest of the curves.

One of the advantages of extending the FunHDDC to CFunHDDC is the outlier detection. For $\alpha^* = 0.85$ and $\epsilon = 0.05$, CFunHDDC detects 16 outliers, which are the same with the outliers mentioned in [4]. For the data without outliers, CFunHDDC becomes equivalent to FunHDDC, and for the trimmed data the CCR increases to 0.79.

5 Conclusion

We propose a new method, CFunHDDC, that extends the FunHDDC functional clustering method to data with mild outliers. Unlike other robust functional clustering algorithms, CFunHDDC does not involve trimming the data. CFunHDDC is based on a model formed by a mixture of contaminated multivariate normal distributions, which makes parameter estimation more difficult than for FunHDDC, so we use an ECM instead of an EM algorithm. The clustering and outlier detection performance of CFunHDDC is tested for simulated data and the NOx data and it always outperforms FunHDDC. Moreover, CFunHDDC has a comparable performance with robust functional clustering methods based on trimming, such as RFC and TrimK, and it has similar or better performance when compared to a two-step method based on CNmixt. Although there are several model-based methods for multivariate data with outliers that can be used to construct two-step methods for functional data, as observed in [1], these two-step methods always suffers from the difficulty to choose the best discretization. CFunHDDC can be extended to multivariate functional data, and recently, independently of our work, a similar approach was followed in [5], but without considering the parsimonious models and the value α^*.

References

1. Bouveyron, C., Jacques, J.: Model-based clustering of time series in group-specific functional subspaces. Adv. Data. Anal. Classif. **5**(4), 281–300 (2011)
2. Jacques, J., Preda, C.: Funclust: a curves clustering method using functional random variables density approximation. Neurocomputing **112**, 164–171 (2013)
3. Punzo, A., McNicholas, P. D.: Parsimonious mixtures of multivariate contaminated normal distributions. Biom. J. **58**, 1506–1537 (2016)
4. Rivera-Garcia, D., Garcia-Escudero, L. A., Mayo-Iscar, A., Ortega, J.: Robust clustering for functional data based on trimming and constraints. Adv. Data Anal. Classif. **13**, 201–225 (2019)
5. Amovin-Assagba, M., Gannaz, I., Jacques, J.: Outlier detection in multivariate functional data through a contaminated mixture model. (2021)
   https://doi.org/10.48550/arXiv.2106.07222

A Trivariate Geometric Classification of Decision Boundaries for Mixtures of Regressions

Filippo Antonazzo and Salvatore Ingrassia

Abstract Mixtures of regressions play a prominent role in regression analysis when it is known the population of interest is divided into homogeneous and disjoint groups. This typically consists in partitioning the observational space into several regions through particular hypersurfaces called decision boundaries. A geometrical analysis of these surfaces allows to highlight properties of the used classifier. In particular, a geometrical classification of decision boundaries for the three most used mixtures of regressions (with fixed covariates, with concomitant variables and random covariates) was provided in case of one and two covariates, under Gaussian assumptions and in presence of only one real response variable. This work aims to extend these results to a more complex setting where three independent variables are considered.

Keywords: mixtures of regressions, decision boundaries, hyperquadrics, model-based clustering

1 Introduction

Linear regression is commonly employed to model the relationship between a d-dimensional real vector of covariates $\mathbf{X}$ and a real response variable Y. It is well suited if we can assume that regression coefficients are fixed over all possible realizations $(\mathbf{x}, y) \in \mathbb{R}^{d+1}$ of the couple $(\mathbf{X}, Y)$. This assumption falls if it is a-priori known that realizations come from a population Ω which can be partitioned into G disjoint

Filippo Antonazzo (✉)
Inria, Université de Lille, CNRS, Laboratoire de mathématiques Painlevé 59650 Villeneuve d'Ascq, France, e-mail: filippo.antonazzo@inria.fr

Salvatore Ingrassia
Dipartimento di Economia e Impresa, Università di Catania, Corso Italia 55, 95129 Catania, Italy, e-mail: salvatore.ingrassia@unict.it

© The Author(s) 2023
P. Brito et al. (eds.), *Classification and Data Science in the Digital Age*,
Studies in Classification, Data Analysis, and Knowledge Organization,
https://doi.org/10.1007/978-3-031-09034-9_3

homogeneous groups $\Omega_g, g = 1, \ldots, G$. In this case, a mixture of linear regressions (or clusterwise regression) is a more appropriate statistical tool. According to their degree of flexibility and generality, we can distinguish three types of mixtures of regressions: mixtures of regressions with fixed covariates (MRFC) [3]; mixtures of regressions with concomitant variables (MRCV) [6] and mixtures of regressions with random covariates (MRRC), also referred to in literature as cluster-weighted models [3, 4].

Mixtures of regressions can also be employed from a classification point of view to identify the group membership of each observation. In this case, the generated classifier divides the real space into G regions through particular $\mathbb{R}^{d+1}$ surfaces called decision boundaries. In [5], the decision boundaries generated by each type of mixture are analyzed from a geometrical point of view, especially in those cases where $d = 1, 2$ and $G = 2$. The aim of the present work is to extend the results presented in the aforementioned paper to a higher dimensional case where $d = 3$, giving more insight into the properties of these classifiers. The rest of the paper is organized as follows. In Section 2 we summarize the main ideas about mixtures of regressions. In Section 3 decision boundaries will be defined, finally proposing a geometrical classification in Section 4 when $d = 3$ and $G = 2$. In Section 5, we will conclude investigating with practical example the shape of three-dimensional decision boundaries in presence of variables following heavy-tailed t-distributions.

2 Mixtures of Regressions

Below we briefly define three types of mixtures of regressions, ordered according to their generality and flexibility, given by an increasing number of parameters.

MRFC. Mixtures of regressions with fixed covariates have the following density:

$$p(y|\mathbf{x}; \psi) = \sum_{g=1}^{G} \pi_g f(y|\mathbf{x}; \theta_g). \tag{1}$$

The density $f(y|\mathbf{x}; \theta_g)$ is indexed by a parameter vector θ_g belonging to an Euclidean parametric space Θ_g. Moreover, every π_g is positive and $\sum_{g=1}^{G} \pi_g = 1$. The vector $\psi = (\pi_1, \ldots, \pi_G, \theta_1, \ldots, \theta_G)$ denotes the set of all the parameters of the model.

MRCV. The density of a mixture of regressions with concomitant variables is:

$$p(y|\mathbf{x}; \psi) = \sum_{g=1}^{G} f(y|\mathbf{x}; \theta_g) p(\Omega_g|\mathbf{x}; \alpha), \tag{2}$$

where the vector $\psi = (\theta_1, \ldots, \theta_G, \alpha)$ contains all parameters indexing the model. More specifically, $p(\Omega_g|\mathbf{x}; \alpha)$ is a function depending on $\mathbf{x}$ according to a vector of real parameters α. Typically, the probability $p(\Omega_g|\mathbf{x}; \alpha)$ is a multinomial logistic density with $\alpha = (\alpha_1^t, \ldots, \alpha_G^t)^t$ and $\alpha_g = (\alpha_{g0}, \alpha_{g1}^t)^t \in \mathbb{R}^{d+1}$, i.e.:

$$p(\Omega_g | \mathbf{x}; \alpha) = \frac{\exp(\alpha_{g0} + \alpha_{g1}^t \mathbf{x})}{\sum_{g=1}^{G} \exp(\alpha_{g0} + \alpha_{g1}^t \mathbf{x})}.$$

Due to identifiability reasons, it is necessary to add the constraint $\alpha_1 = \mathbf{0}$, see [2].

MRRC. Mixtures of regressions with random covariates propose the following decomposition for the conjoint density $p(\mathbf{x}, y; \psi)$:

$$p(\mathbf{x}, y; \psi) = \sum_{g=1}^{G} f(y|\mathbf{x}, \theta_g) p(\mathbf{x}; \xi_g) \pi_g, \tag{3}$$

where $\pi_g > 0$ and $\sum_{g=1}^{G} \pi_g = 1$. Furthermore, the model is totally parametrized by the vector $\psi = (\pi_1, \ldots, \pi_G, \theta_1, \ldots, \theta_G, \xi_1, \ldots, \xi_G)$, where each θ_g indexes the conditional density $f(y|\mathbf{x}, \theta_g)$, while each ξ_g refers to the density of $\mathbf{X}$ in the group Ω_g, denoted with $p(\mathbf{x}; \xi_g)$.

In particular, under Gaussian assumptions it results $Y|\mathbf{x}, \Omega_g \sim N(\beta_{g0} + \beta_{g1}^t \mathbf{x}, \sigma_g^2)$, where each $\beta_g = (\beta_{g0}, \beta_{g1})$ is a vector of real parameters. Only for MRRC model, we will further assume $\mathbf{X}|\Omega_g \sim N(\mu_g, \Sigma_g)$ for all $g = 1, \ldots, G$, where μ_g denotes the mean of the Gaussian distribution, while Σ_g is its covariance matrix. Denoting with $\phi(\cdot)$ the Gaussian density function, equations (1)-(3) can be, respectively, rewritten as

$$p(y|\mathbf{x}; \psi) = \sum_{g=1}^{G} \phi(y; \beta_{g0} + \beta_{g1}^t \mathbf{x}; \sigma_g^2) \pi_g, \tag{4}$$

$$p(y|\mathbf{x}; \psi) = \sum_{g=1}^{G} \phi(y; \beta_{g0} + \beta_{g1}^t \mathbf{x}; \sigma_g^2) p(\Omega_g | \mathbf{x}; \alpha), \tag{5}$$

$$p(\mathbf{x}, y; \psi) = \sum_{g=1}^{G} \phi(y, \beta_{g0} + \beta_{g1}^t \mathbf{x}; \sigma_g^2) \phi(\mathbf{x}; \mu_g, \Sigma_g) \pi_g. \tag{6}$$

Maximum likelihood estimate for ψ are usually obtained with the Expectation-Maximization (EM) algorithm. Then, the final estimate is used to build classifiers which group observations into G disjoint classes.

3 Decision Boundaries: Generality

There are different ways to build classifiers. One of the best known is the method of discriminant functions. The aim of this procedure is to define G functions $D_g(\mathbf{x}, y; \psi)$ and a decision rule to divide the real space $\mathbb{R}^{d+1}$ into G decision regions, named

$\mathcal{R}_1, \dots, \mathcal{R}_G$. The decision regions have a one-to-one relationship with the subgroups Ω_g, i.e., if an observation $(\mathbf{x}, y) \in \mathbb{R}^{d+1}$ is assigned to $\mathcal{R}_g$, it is classified as part of Ω_g. Among all possible decision rules, the most used one consists in assigning $(\mathbf{x}, y)$ to $\mathcal{R}_g$ if:

$$D_g(\mathbf{x}, y; \psi) > D_j(\mathbf{x}, y; \psi) \quad \forall j \neq g. \tag{7}$$

Then, decision boundaries are defined as the surfaces in $\mathbb{R}^{d+1}$ separating the decision regions $\mathcal{R}_g$, where observations cannot be uniquely classified. Formally, each decision boundary is a hypersurface represented by the mathematical equation $D_j(\mathbf{x}, y; \psi) - D_k(\mathbf{x}, y; \psi) = 0$, $j \neq k$.

Different choices for discriminant functions are possible: under Gaussian assumptions it is convenient to define $D_g(\cdot)$ as the logarithm of the g-th component mixture density, as it conveys useful computational simplification [5]. So, we can define, for all the three models, these discriminant functions:

$$MRFC: \quad D_g(\mathbf{x}, y; \psi) = \ln[\phi(y; \beta_{g0} + \beta_{g1}^t \mathbf{x}, \sigma_g^2)\pi_g] \tag{8}$$

$$MRCV: \quad D_g(\mathbf{x}, y; \psi) = \ln[\phi(y; \beta_{g0} + \beta_{g1}^t \mathbf{x}, \sigma_g^2) \exp(\alpha_{g0} + \alpha_{g1}^t \mathbf{x})] \tag{9}$$

$$MRRC: \quad D_g(\mathbf{x}, y; \psi) = \ln[\phi(y; \beta_{g0} + \beta_{g1}^t \mathbf{x}, \sigma_g^2)\phi(\mathbf{x}; \mu_g, \Sigma_g)\pi_g] \tag{10}$$

3.1 The Case with $G = 2$

In the case of interest where $G = 2$, there is a single decision boundary defined by the equation $D(\mathbf{x}, y; \psi) = D_2(\mathbf{x}, y; \psi) - D_1(\mathbf{x}, y; \psi) = 0$. Thus, the assignment rule for every point $(\mathbf{x}, y) \in \mathbb{R}^{d+1}$ is based on the sign of $D(\mathbf{x}, y; \psi)$. It assigns $(\mathbf{x}, y)$ to Ω_2 if $D(\mathbf{x}, y; \psi) > 0$; to Ω_1, otherwise.

In [5] the geometrical properties of the hypersurfaces, defined by the equation $D(\mathbf{x}, y; \psi) = 0$, have been investigated up to dimension $d = 2$, providing the following propositions for quadrics.

Proposition 1 (MRFC quadrics) *The decision boundary between Ω_1 and Ω_2 is always a degenarate quadric.*

Proposition 2 (MRCV quadrics) *If $\alpha^t(\beta_{21} - \beta_{11}) \neq 0$, then the decision boundary between Ω_1 and Ω_2 is a paraboloid; otherwise it is a degenarate quadric.*

Proposition 3 (MRRC quadrics) *Under convenient conditions, the decision boundary between Ω_1 and Ω_2 can be a degenerate quadric but it can be also assume any of the general quadric forms.*

These results show that models with more flexibility, i.e. with more parameters, can generate more varieties of decision boundaries. In the following section, we will extend these statements to dimension $d = 3$.

4 Geometrical Classification of Decision Boundaries with $G = 2$ and $D = 3$

In this section we extend previous results for mixtures of regression in presence of two classes and $d = 3$, where decision boundaries reveal to be hyperquadrics in $\mathbb{R}^4$. Mathematical proofs of results for MRFC and MRCV models are based on an algebraic analysis of the matrices representing these hyperquadrics.

MRFC. Mixtures of regressions with fixed covariates are characterized by a low degree of flexibility. Indeed, all decision boundaries are degenerate hyperquadrics as the following result shows.

Proposition 4 (MRFC hyperquadrics) *The decision boundary between Ω_1 and Ω_2 is a degenerate hyperquadric of rank at most equal to 3. The rank is less than 3 if $\beta_{11} = \beta_{21}$ or $\frac{1-\pi_1}{\pi_1} = \frac{\sigma_2}{\sigma_1}$.*

MRCV. A MRCV allows more degrees of freedom than a MRFC. A consequence is that the obtained decision boundaries are higher rank hyperquadrics, as the following result states.

Proposition 5 (MRCV hyperquadrics) *The decision boundary between Ω_1 and Ω_2 is a degenerate hyperquadric with rank at most equal to 4. In particular, rank is equal to 4 if $\alpha^t(\beta_{21} - \beta_{11}) \neq 0$. In addition, if $\alpha^t(\beta_{21} - \beta_{11}) = 0$ and $\sigma_1^2 = \sigma_2^2$; the matrix has rank at most equal to 2, therefore the hyperquadric is reducible.*

MRRC. Proposition 3 shows MRRC exhibit a high number of possible types of conics and quadrics [5]. This fact is confirmed in dimension $d = 3$, even if a strong theoretical result is difficult to obtain with simple algebra due to the mathematical complexity of the MRRC hyperquadric matrix. Indeed, it is possible to show such flexibility by building several practical examples (not displayed here), where hyperquadrics of various shapes arise.

Analyzing the provided results, we can note that they perfectly match the hierarchy established in dimension $d = 2$. Indeed, a MRFC can generate only degenerate hyperquadrics of rank 3; the surfaces generated by a MRCV, which has more parameters, are still degenerate, but with a higher rank (equal to 4) depending on the same mathematical condition of Proposition 2; finally a MRRC, the most flexible model in terms of number of parameters, can give rise to various hyperquadrics, as in $d = 2$.

5 Beyond Gaussian Assumptions: t-distribution in $d = 2$

In [5], Gaussian assumptions were crossed by illustrating the case of a simple linear regression ($G = 2$ and $d = 1$) where more general t-distributions were required for robustness reasons. It is shown that the generated decision boundaries are more flexible than their Gaussian counterparts, as they can assume more various shapes, although these surfaces can be calculated only numerically. In this section, we continue the exploration of the t-distribution case adding one more variable, thus $d = 2$. Under these more general assumptions, discriminant functions (8) – (10) become:

$$MRFC\text{-}t: \quad D_g(\mathbf{x}, y; \psi) = \ln[q(y; \beta_{g0} + \beta_{g1}^t \mathbf{x}, \sigma_g^2, \eta_g) \pi_g], \tag{11}$$

$$MRCV\text{-}t: \quad D_g(\mathbf{x}, y; \psi) = \ln[q(y; \beta_{g0} + \beta_{g1}^t \mathbf{x}, \sigma_g^2, \eta_g) \exp(\alpha_{g0} + \alpha_{g1}^t \mathbf{x})], \tag{12}$$

$$MRRC\text{-}t: \quad D_g(\mathbf{x}, y; \psi) = \ln[q(y; \beta_{g0} + \beta_{g1}^t \mathbf{x}, \sigma_g^2, \eta_g) q(\mathbf{x}; \mu_g, \Sigma_g, \nu_g) \pi_g], \tag{13}$$

where $q(y; \beta_{g0} + \beta_{g1}^t \mathbf{x}, \sigma_g^2, \eta_g)$ denotes a generalized t-distribution density, with non-centrality parameter equal to $\beta_{g0} + \beta_{g1}^t \mathbf{x}$, scaling parameter equal to σ_g^2 and degrees of freedom given by η_g. Similarly, $q(\mathbf{x}; \mu_g, \Sigma_g, \nu_g)$ is a multivariate generalized t-distribution density, where μ_g is the non-centrality parameter, Σ_g denotes the scaling and ν_g represents the degrees of freedom. Figure 1-2 display the decision boundaries for the three considered models whose parameters are presented in Table 1: they clearly show the gain in flexibility given by the more general distributional assumptions. Moreover, t-boundaries with $\eta_1 = \eta_2 = 10$ (Figure 2; red curves) seem to be closer to Gaussian ones (blue curves) than those with $\eta_1 = \eta_2 = 3$ (Figure 1; orange curves): this is coherent with standard probabilistic theory.

Table 1 Parameters used in Figure 1-2. MRRC: covariance matrices Σ_1 and Σ_2 are equal to the identity matrix $\mathbf{I}_2$.

Model	Group	π_g	β_{g0}	β_{g1}	σ_g^2	α_{g0}	α_{g1}	μ_g	ν_g
MRFC	1	0.3	1	(2,-3)	0.5				
	2	0.7	1	(-4,3)	0.5				
MRCV	1	0.3	1	(2,-3)	0.5				
	2	0.7	1	(-4,3)	0.5	1	(-1,0.5)		
MRRC	1	0.3	1	(2,-3)	0.5			(1,2)	5
	2	0.7	1	(-4,3)	0.5	1	(-1,0.5)	(-1,-2)	5

6 Conclusions

This work has provided a trivariate multivariate geometrical classification for the decision boundaries generated by mixtures of regressions in presence of two classes. Under Gaussian assumptions, our results confirmed the same hierarchy that was shown in $d = 2$, as MRRC turns out to exhibits a huge variety of decision boundaries, while other models generate only degenerate surfaces. This is coherent with its high degree of flexibility given by its very general parametrization. The provided results

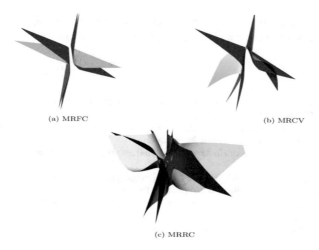

(a) MRFC   (b) MRCV

(c) MRRC

Fig. 1 Decision boundaries under assumptions of Gaussian (in blue) and t-distributed variables with $\eta_1 = \eta_2 = 3$ (in orange) for the three considered mixtures of regressions.

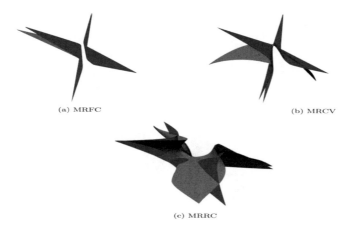

(a) MRFC   (b) MRCV

(c) MRRC

Fig. 2 Decision boundaries under assumptions of Gaussian (in blue) and t-distributed variables with $\eta_1 = \eta_2 = 10$ (in red) for the three considered mixtures of regressions.

could help to select the right model depending on the shape of data. For example, if in a descriptive analysis data turn out to be approximately separated by a simple degenerate hyperquadric, it will be better to estimate a MRFC or a MRCV instead of a complex MRRC. On the contrary, if the separation surface seems to be non-degenerate, then it will be preferable to fit a general MRRC. Moreover, this work also showed that the degree of flexibility (thus, the variety of possible decision boundaries) can be enhanced by go further Gaussianity, assuming, for example, t-distributed variables. This encourage additional extensions where more general distributions can be included, allowing a better comprehension of mixtures and possible applications to generalized linear models where categorical variables are considered.

References

1. DeSarbo, W. S., Cron, W. L.: A maximum likelihood methodology for clusterwise linear regression. J. Classif. **5**, 249–282 (1988)
2. Grun, B., Leisch, F.: FlexMix version 2: finite mixtures with concomitant variables and varying and constant parameters. J. Stat. Softw. **28**, 1–35 (2008)
3. Hennig, C.: Identifiablity of models for clusterwise linear regression. J. Classif. **17**, 273–296 (2000)
4. Ingrassia, S., Minotti, S. C., Vittadini, G.: Local Statistical Modeling via a Cluster-Weighted Approach with Elliptical Distributions. J. Classif. **29**, 363-401 (2012)
5. Ingrassia, S., Punzo, A.: Decision boundaries for mixtures of regressions. J. Korean Stat. Soc. **45**, 295-306 (2016)
6. Wedel, M.: Concomitant variables in finite mixture models. Stat. Neerl. **56**, 362–375 (2002)

Generalized Spatio-temporal Regression with PDE Penalization

Eleonora Arnone, Elia Cunial, and Laura M. Sangalli

Abstract We develop a novel generalised linear model for the analysis of data distributed over space and time. The model involves a nonparametric term f, a smooth function over space and time. The estimation is carried out by the minimization of an appropriate penalized negative log-likelihood functional, with a roughness penalty on f that involves space and time differential operators, in a separable fashion, or an evolution partial differential equation. The model can include covariate information in a semi-parametric setting. The functional is discretized by means of finite elements in space, and B-splines or finite differences in time. Thanks to the use of finite elements, the proposed method is able to efficiently model data sampled over irregularly shaped spatial domains, with complicated boundaries. To illustrate the proposed model we present an application to study the criminality in the city of Portland, from 2015 to 2020.

Keywords: functional data analysis, spatial data analysis, semiparametric regression with roughness penalty

Eleonora Arnone (✉)
Dipartimento di Scienze Statistiche, Università di Padova, Via Cesare Battisti, 241, 35121 Padova, Italy, e-mail: `eleonora.arnone@unipd.it`

Elia Cunial
Dipartimento di Matematica, Politecnico di Milano, Piazza Leonardo da Vinci 32, 20133 Milano, Italy, e-mail: `elia.cunial@mail.polimi.it`

Laura M. Sangalli
Dipartimento di Matematica, Politecnico di Milano, Piazza Leonardo da Vinci 32, 20133 Milano, Italy, e-mail: `laura.sangalli@polimi.it`

© The Author(s) 2023
P. Brito et al. (eds.), *Classification and Data Science in the Digital Age*,
Studies in Classification, Data Analysis, and Knowledge Organization,
https://doi.org/10.1007/978-3-031-09034-9_4

1 Introduction

In this work we develop a novel generalised linear model for the analysis of data distributed over space and time. Let Y be a real-valued variable of interest, and $\mathbf{W}$ a vector of q covariates, observed in n spatio-temporal locations $\{\mathbf{p}_i, t_i\}_{i=1,\ldots,n} \in \Omega \times T$, where $\Omega \subset \mathbb{R}^2$ is a bounded spatial domain, and $T \subset \mathbb{R}$ a temporal interval. We assume that the expected value of Y, conditional on the covariates and the location of observation, can be modeled as:

$$g(\mathbb{E}[Y|W, \mathbf{p}, t]) = \mathbf{W}^\top \beta + f(\mathbf{p}, t)$$

where g is a known monotone link function, chosen on the basis of the stochastic nature of Y, $\beta \in \mathbb{R}^q$ is an unknown vector of regression coefficients, and $f : \Omega \times T \to \mathbb{R}$ is an unknown deterministic function, which captures the spatio-temporal variation of the phenomenon under study. Starting from the values $\{y_i, \mathbf{w}_i\}_{i=1,\ldots,n}$, of the observed response variable and covariates, we estimates β and f in a semiparametric fashion. In particular, following the approach in [9], that consider a similar problem for data scattered over space only, we minimize the functional

$$\ell\left(\{y_i, \mathbf{w}_i, \mathbf{p}_i, t_i\}_{i=1,\ldots,n}; \beta, f\right) + \mathcal{P}(f)$$

where ℓ is the appropriate negative log-likelihood, and $\mathcal{P}(f)$ is a penalty that enforces f to be a regular function.

Similarly to the regression methods in [1, 2, 3, 4, 5, 7, 8], the roughness penalty on f, $\mathcal{P}(f)$, involves some partial differential operators. In particular, our aim is to extend the Spatial-Temporal regression with partial differential equations regularization (ST-PDE), developed in [2, 3, 4], to generalized linear model settings, further broadening the class of regression models with PDE regularization reviewed in [6]. Hence, likewise ST-PDE, the proposed generalized linear model has a roughness penalty that involves a second order linear differential operator L applied to f. Specifically, as in [4], we may consider the penalty

$$\mathcal{P}(f) = \lambda_T \int_\Omega \int_0^T \left(\frac{\partial^2 f}{\partial t^2}\right)^2 + \lambda_S \int_\Omega \int_0^T (Lf)^2,$$

where the first term accounts for the regularity of the function in time, while the second accounts for the regularity of the function in space; the importance of each term is controlled by two smoothing parameters λ_T and λ_S. Alternatively, as in [2], we may consider a single penalty which accounts for the spatial and temporal regularity:

$$\mathcal{P}(f) = \lambda \int_\Omega \int_0^T \left(\frac{\partial f}{\partial t} + Lf - u\right)^2.$$

Differently from the models in [2, 3, 4], the estimation functional to be minimized is not quadratic. This poses increased difficulties from the computational point of view. The minimization is performed via a functional version of the penalized iterative reweighted least square algorithm.

The estimation problem is appropriately discretized. In particular, in time, the discretization involves either cubic B-splines, for the two-penalties case, or finite differences, when the single penalty is employed. The discretization in space is performed via finite elements, on a triangulation of the spatial domain of interest. This enables to appropriately considered spatial domains with complicated boundaries, such as the one considered in the following section, concerning the study of criminality data over the city of Portland.

2 Application to Criminality Data

This section describes the Portland criminality data, that will be used to illustrate the proposed methodology. We will present a Poisson model to count the crimes in the city, and study their evolution from April 2015 to November 2020. In addition, we shall consider as a covariate the population of the city neighborhoods. The crime data are publicly available on the website of the Police Bureau of the city[1].

The crimes counts are aggregated by trimesters and at a neighborhoods level. Figure 1 shows the city neighborhoods, each neighborhood colored according to its total population. The bottom part of the same figure shows the temporal evolution of the crimes in each neighborhood. Each curve corresponds to a neighborhood and is colored according to the neighborhood population. In both panels, the three neighborhoods with the highest number of crimes are indicated by numbers 1, 2 and 3. The figure highlights the presence of some correlation between neighborhood population and the number of crimes. However, criminality is not fully explained by population. For instance, neighborhoods 1 and 3 present an high number of crimes with a moderate population. This raises the interest towards a semiparametric generalized linear model, as the one introduced in Section 1, with a nonparametric term accounting for the spatio-temporal variability in the phenomenon, that cannot be explained by population or other census quantities. Figure 2 shows the same data for four different trimesters on the Portland map. As already pointed out, the three area with the highest number of crimes are in the city center, and in the Hazelwood neighborhood, in the east part of the city.

From Figures 1 and 2 we can see that the shape of the domain is complicated; the city is indeed crossed by a river, with few bridges connecting the two parts, most of them placed downtown. Therefore, neighborhoods at opposite side of the river and far from the center, where most bridges are located, are close in euclidean distance, but far apart in reality. This particular morphology influences the phenomenon under study, for example, in the north of the city, the east side of the river is characterized by an higher number of crimes with respect to the west part. Due to this characteristics of the data and the domain, is is of crucial importance to take into account the shape

---

[1] Police Bureau crime data: https://www.portlandoregon.gov/police/71978

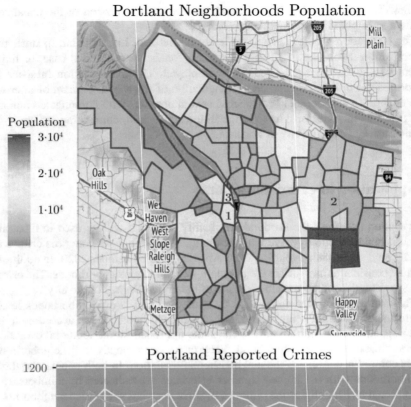

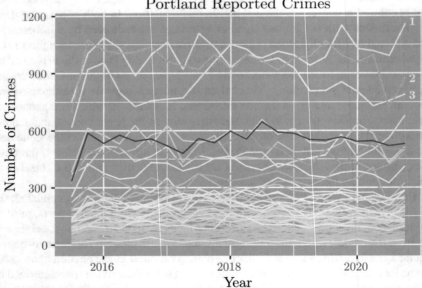

Fig. 1 Top: the city of Portland divided into neighborhoods, each neighborhood colored according to the total population. Bottom: the total crimes over time for each neighborhood; each curve corresponds to a neighborhood and is colored according to the neighborhood's population. The three neighborhoods with the highest number of crimes are indicated by numbers 1, 2 and 3.

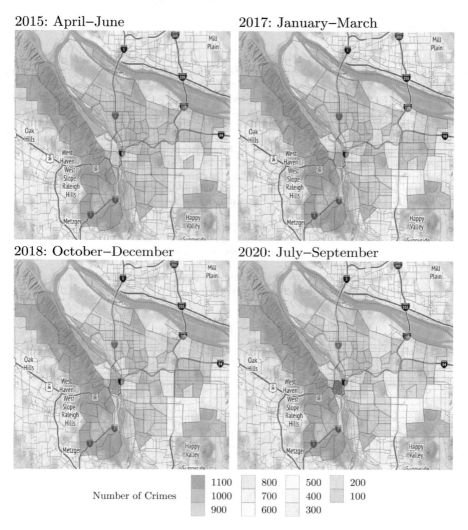

Fig. 2 Total crime counts per neighborhood per trimester; green indicates lower number of crimes, red indicates a higher number of crimes.

of the domain during the estimation process. For this reason, estimation based on classical semiparametric models, such as those based on thin-plate splines, would give poor results, while the proposed method is particularly well suited, being able to complying the nontrivial form of the domain.

References

1. Aguilera-Morillo, M. C., Durbán, M., Aguilera, A. M.: Prediction of functional data with spatial dependence: a penalized approach. Stoch. Environ. Res. Risk Assess. **31**, 7–22 (2017)
2. Arnone, E., Azzimonti, L., Nobile, F., Sangalli, L. M.: Modeling spatially dependent functional data via regression with differential regularization. J. Multivariate Anal. **170**, 275–295 (2019)
3. Arnone, E., Sangalli, L. M., Vicini, A.: Smoothing spatio-temporal data with complex missing data patterns. Stat. Model. Int. J. (2021)
4. Bernardi, M. S., Sangalli, L. M., Mazza, G., Ramsay, J. O.: A penalized regression model for spatial functional data with application to the analysis of the production of waste in Venice province. Stoch. Environ. Res. Risk Assess. **31**, 23–38 (2017)
5. Marra, G., Miller, D. L., Zanin, L.: Modelling the spatiotemporal distribution of the incidence of resident foreign population. Statistica Neerlandica **66**(2) 133–160 (2012)
6. Sangalli, L. M.: Spatial regression with partial differential equation regularization. Int. Stat. Rev. **89**(3), 505–531 (2021)
7. Ugarte, M. D., Goicoa, T., Militino, A. F., Durbán, M.: Spline smoothing in small area trend estimation and forecasting. Comput. Stat. Data Anal. **53**(10), 3616–3629 (2009)
8. Ugarte, M. D., Goicoa, T., Militino, A. F.: Spatio-temporal modeling of mortality risks using penalized splines. Environmetrics **21**, 270–289 (2010)
9. Wilhelm M., Sangalli L. M.: Generalized spatial regression with differential regularization. J. Stat. Comput. Simulat. **86**(13), 2497–2518 (2016)

A New Regression Model for the Analysis of Microbiome Data

Roberto Ascari and Sonia Migliorati

Abstract Human microbiome data are becoming extremely common in biomedical research due to the relevant connections with different types of diseases. A widespread discrete distribution to analyze this kind of data is the Dirichlet-multinomial. Despite its popularity, this distribution often fails in modeling microbiome data due to the strict parameterization imposed on its covariance matrix. The aim of this work is to propose a new distribution for analyzing microbiome data and to define a regression model based on it. The new distribution can be expressed as a structured finite mixture model with Dirichlet-multinomial components. We illustrate how this mixture structure can improve a microbiome data analysis to cluster patients into "enterotypes", which are a classification based on the bacteriological composition of gut microbiota. The comparison between the two models is performed through an application to a real gut microbiome dataset.

Keywords: count data, Bayesian inference, mixture model, multivariate regression

1 Introduction

The human microbiome is defined as the set of genes associated with the microbiota, i.e. the microbial community living in the human body, including bacteria, viruses and some unicellular eukaryotes [1, 8]. The mutualistic relationship between microbiota and human beings is often beneficial, though it can sometimes

Roberto Ascari (✉)
Department of Economics, Management and Statistics (DEMS), University of Milano-Bicocca, Milan, Italy, e-mail: `roberto.ascari@unimib.it`

Sonia Migliorati
Department of Economics, Management and Statistics (DEMS), University of Milano-Bicocca, Milan, Italy, e-mail: `sonia.migliorati@unimib.it`

© The Author(s) 2023
P. Brito et al. (eds.), *Classification and Data Science in the Digital Age*,
Studies in Classification, Data Analysis, and Knowledge Organization,
https://doi.org/10.1007/978-3-031-09034-9_5

become detrimental for several health outcomes. For example, changes in the gut microbiome composition can be associated with diabetes, cardiovascular disesase, obesity, autoimmune disease, anxiety and many other factors impacting on human health [1, 5, 12, 14]. Moreover, the development of next-generation sequencing technologies allows nowadays to survey the microbiome composition using direct DNA sequencing of either marker genes or the whole metagenomics, without the need for isolation and culturing. These are the two main reasons for the recent explosion of research on microbiome, and highlight the importance of understanding the association between microbiome composition and biological and environmental covariates.

A widespread distribution for handling microbiome data is the Dirichlet-multinomial (DM) (e.g., see [4, 16]), a generalization of the multinomial distribution obtained by assuming that, instead of being fixed, the underlying taxa proportions come from a Dirichlet distribution. This allows to model overdispersed data counts, that is data showing a variance much larger than that predicted by the multinomial model. Despite its popularity, the DM distribution is often inadequate to model real microbiome datasets due to the strict covariance structure imposed by its parameterization, which hinders the description of co-occurrence and co-exclusion relationships between microbial taxa.

The aim of this work is to propose a new distribution that generalizes the DM, namely the flexible Dirichlet-multinomial (FDM), and a regression model based on it. The new model provides a better fit to real microbiome data, still preserving a clear interpretation of its parameters. Moreover, being a finite mixture with DM components, it enables to account for the data latent group structure, and thus to identify clusters sharing similar biota compositions.

2 Statistical Models for Microbiome Data

In this section, we define a new distribution for multivariate counts and a regression model based on it, that allows to link microbiome abundances with covariates. Note that, once the DNA sequence reads have been aligned to the reference microbial genomes, the abundances of microbial taxa can be quantified. Thus, microbiome data represent the count composition of D bacterial taxa in a specific biological sample, and a microbiome dataset is a sequence of D-dimensional vectors $\mathbf{Y}_1, \mathbf{Y}_2, \ldots, \mathbf{Y}_N$, where Y_{ir} counts the number of occurrences of taxon r in the i-th sample ($i = 1, \ldots, N$ and $r = 1, \ldots, D$). Since the i-th sample contains a number n_i of bacteria, microbiome observations are subject to a fixed-sum constraint, that is $\sum_{r=1}^{D} Y_{ir} = n_i$.

2.1 Count Distributions

Following a compound approach, we assume that $\mathbf{Y}|\mathbf{\Pi} = \boldsymbol{\pi} \sim \text{Multinomial}(n, \boldsymbol{\pi})$, and we consider suitable distributions for the vector of probabilities $\mathbf{\Pi} \in \mathcal{S}^D$. The set $\mathcal{S}^D = \{\boldsymbol{\pi} = (\pi_1, \ldots, \pi_D)^\mathsf{T} : \pi_r > 0, \sum_{r=1}^D \pi_r = 1\}$ is the D-part simplex and it is the proper support of continuous compositional vectors. A distribution for $\mathbf{Y}$ is obtained by marginalizing the joint distribution of $(\mathbf{Y}, \mathbf{\Pi})^\mathsf{T}$. A common choice for this distribution is the mean-precision parameterized Dirichlet, whose probability density function (p.d.f.) is

$$f_{\text{Dir}}(\boldsymbol{\pi}; \boldsymbol{\mu}, \alpha^+) = \frac{\Gamma(\alpha^+)}{\prod_{r=1}^D \Gamma(\alpha^+ \mu_r)} \prod_{r=1}^D \pi_r^{(\alpha^+ \mu_r)-1},$$

where $\boldsymbol{\mu} = \mathbb{E}[\mathbf{\Pi}] \in \mathcal{S}^D$, and $\alpha^+ > 0$ is a precision parameter. Compounding the multinomial distribution with the Dirichlet one leads to the DM distribution, widely used in microbiome data analysis, whose probability mass function (p.m.f.) is

$$f_{\text{DM}}(\mathbf{y}; n, \boldsymbol{\mu}, \alpha^+) = \frac{n! \Gamma(\alpha^+)}{\Gamma(\alpha^+ + n)} \prod_{r=1}^D \frac{\Gamma(\alpha^+ \mu_r + y_r)}{(y_r!) \Gamma(\alpha^+ \mu_r)}.$$

The mean vector of a DM distribution is $\mathbb{E}[\mathbf{Y}] = n\boldsymbol{\mu}$, so that the parameter $\boldsymbol{\mu} = \mathbb{E}[\mathbf{Y}]/n$ can be thought of as a scaled mean vector. Moreover, its covariance matrix is

$$\mathbb{V}[\mathbf{Y}] = n\mathbf{M}\left[1 + \frac{n-1}{\alpha^+ + 1}\right], \tag{1}$$

where $\mathbf{M} = (\text{Diag}(\boldsymbol{\mu}) - \boldsymbol{\mu}\boldsymbol{\mu}^\mathsf{T})$. Equation (1) highlights how the additional parameter α^+ allows to increase flexibility in the variability structure with respect to the standard multinomial distribution.

We propose to take advantage of an alternative sound distribution defined on $\mathcal{S}^D$, namely the flexible Dirichlet (FD) [7, 9]. The latter is a structured finite mixture with Dirichlet components, entailing some constraints among the components' parameters to ensure model identifiability. Thanks to its mixture structure, the p.d.f. of a FD-distributed random vector can be expressed as

$$f_{\text{FD}}(\boldsymbol{\pi}; \boldsymbol{\mu}, \alpha^+, w, \mathbf{p}) = \sum_{j=1}^D p_j f_{\text{Dir}}\left(\boldsymbol{\pi}; \lambda_j, \frac{\alpha^+}{1-w}\right), \tag{2}$$

where

$$\lambda_j = \boldsymbol{\mu} - w\mathbf{p} + w\mathbf{e}_j \tag{3}$$

is the mean vector of the j-th component, $\boldsymbol{\mu} = \mathbb{E}[\mathbf{\Pi}] \in \mathcal{S}^D$, $\alpha^+ > 0$, $\mathbf{p} \in \mathcal{S}^D$, $0 < w < \min\left\{1, \min_{r \in \{1, \ldots, D\}}\left\{\frac{\mu_r}{p_r}\right\}\right\}$, and $\mathbf{e}_j$ is a vector with all elements equal to zero except for the j-th which is equal to one.

Equation (2) points that the Dirichlet components have different mean vectors and a common precision parameter, the latter being determined by α^+ and w. In particular, inspecting Equation (3), it is easy to observe that any two vectors λ_r and λ_h, $r \neq h$, coincide in all the elements except for the r-th and the h-th.

If Π is supposed to be FD distributed, a new discrete distribution for count vectors can be defined (we shall call flexible Dirichlet-multinomial (FDM)). The p.m.f. of the FDM can be expressed as

$$f_{\mathrm{FDM}}(\mathbf{y}; n, \boldsymbol{\mu}, \alpha^+, \mathbf{p}, w) = \sum_{j=1}^{D} p_j f_{\mathrm{DM}}\left(\mathbf{y}; n, \lambda_j, \frac{\alpha^+}{1-w}\right) \tag{4}$$

$$= \sum_{j=1}^{D} p_j \frac{n!\,\Gamma(\frac{\alpha^+}{1-w})}{\Gamma(\frac{\alpha^+}{1-w} + n)} \prod_{r=1}^{D} \frac{\Gamma(\frac{\alpha^+}{1-w}\lambda_{jr} + y_r)}{(y_r!)\Gamma(\frac{\alpha^+}{1-w}\lambda_{jr})},$$

where λ_j is defined in Equation (3). Interestingly, it is possible to recognize the flexible beta-binomial (FBB) [3] distribution as a special case of the FDM. The FBB is a generalization of the binomial distribution successful in dealing with overdispersion. Moreover, note that when $\mathbf{p} = \boldsymbol{\mu}$ and $w = 1/(\alpha^+ + 1)$ the DM distribution is recovered.

Equation (4) shows that the FDM is a finite mixture with DM components displaying a common precision parameter and different scaled mean vectors λ_j, $j = 1, \ldots, D$. The overall mean vector and the covariance matrix of the FDM can be expressed as

$$\mathbb{E}[\mathbf{Y}] = n\boldsymbol{\mu},$$
$$\mathbb{V}[\mathbf{Y}] = n\mathbf{M}\left[1 + \frac{n-1}{\phi+1}\right] + n\frac{(n-1)\phi w^2}{\phi+1}\mathbf{P}, \tag{5}$$

where $\mathbf{M} = (\mathrm{Diag}(\boldsymbol{\mu}) - \boldsymbol{\mu}\boldsymbol{\mu}^{\mathsf{T}})$, $\mathbf{P} = (\mathrm{Diag}(\mathbf{p}) - \mathbf{p}\mathbf{p}^{\mathsf{T}})$, and $\phi = \alpha^+/(1-w)$ is the common precision parameter of the DM components. A comparison between Equations (5) and (1) points out that the covariance matrix of the FDM distribution is a very easily interpretable extension of the DM's covariance matrix. Indeed, it is composed of two terms, the first one coinciding with the DM's covariance matrix, whereas the second one depends on the mixture structure of the FDM model. In particular, the FDM covariance matrix has D additional parameters with respect to the DM, namely $D-1$ distinct elements in the vector of mixing weights $\mathbf{p}$, and the parameter w which controls the distance among the components' barycenters [7]. This is the key element explaining the better ability of the FDM in modeling a wide range of scenarios.

2.2 Regression Models

With the aim of performing a regression analysis, let $\mathbf{Y} = (\mathbf{Y}_1, \ldots, \mathbf{Y}_N)^\mathsf{T}$ be a set of independent multivariate responses collected on a sample of N subjects/units. For the i-th subject, $\mathbf{Y}_i$ counts the number of times that each of D possible taxa occurred among n_i trials, and $\mathbf{x}_i$ is a $(K + 1)$-dimensional vector of covariates.

A parameterization of the FDM useful in a regression perspective is the one based on $\boldsymbol{\mu}$, $\mathbf{p}$, α^+, and $\tilde{w}$, where

$$\tilde{w} = \frac{w}{\min\left\{1, \min_r \left\{\frac{\mu_r}{p_r}\right\}\right\}} \in (0, 1). \tag{6}$$

We can define the FDM regression (FDMReg) and the DM regression (DMReg) models assuming that $\mathbf{Y}_i$ follows an FDM$(n_i, \boldsymbol{\mu}_i, \alpha^+, \mathbf{p}, \tilde{w})$ or a DM$(n_i, \boldsymbol{\mu}_i, \alpha^+)$ distribution, respectively. Even if the FDM and DM distributions do not belong to the dispersion-exponential family, we can follow a GLM-type approach, [6] by linking the parameter $\boldsymbol{\mu}_i$ to the linear predictor through a proper link function such as the multinomial logit link function, that is

$$g(\mu_{ir}) = \log\left(\frac{\mu_{ir}}{\mu_{iD}}\right) = \mathbf{x}_i^\mathsf{T} \boldsymbol{\beta}_r, \qquad r = 1, \ldots, D - 1, \tag{7}$$

where $\boldsymbol{\beta}_r = (\beta_{r0}, \beta_{r1}, \ldots, \beta_{rK})^\mathsf{T}$ is a vector of regression coefficients for the r-th element of $\boldsymbol{\mu}_i$. Note that the last category has been conventionally chosen as baseline category, thus $\boldsymbol{\beta}_D = \mathbf{0}$.

The parameterization of the FDMReg based on $\boldsymbol{\mu}$, $\mathbf{p}$, α^+, and $\tilde{w}$ defines a variation independent parameter space, meaning that no constraints exist among parameters. In a Bayesian framework, this allows to assume prior independence, and, consequently, we can specify a prior distribution for each parameter separately. In order to induce minimum impact on the posterior distribution, we select weakly-informative priors: (i) $\boldsymbol{\beta}_r \sim N_{K+1}(\mathbf{0}, \Sigma)$, where $\mathbf{0}$ is the $(K + 1)$-vector with zero elements, and Σ is a diagonal matrix with 'large' variance values, (ii) $\alpha^+ \sim Gamma(g_1, g_2)$ for small values of g_1 and g_2, (iii) $\tilde{w} \sim Unif(0, 1)$, and (iv) a uniform prior on the simplex for $\mathbf{p}$.

Inferential issues are dealt with by a Bayesian approach through a Hamiltonian Monte Carlo (HMC) algorithm [10], which is a popular generalization of the Metropolis-Hastings algorithm. The Stan modeling language [13] allows implementing an HMC method to obtain a simulated sample from the posterior distribution.

To compare the fit of the models we use the Watanabe-Akaike information criterion (WAIC) [15, 17], a fully Bayesian criterion that balances between goodness-of-fit and complexity of a model: lower values of WAIC indicate a better fit.

3 A Gut Microbiome Application

In this section, we fit the DM and the FDM regression models to a microbiome dataset analyzed by Xia et al. [19] and previously proposed by Wu et al. [18]. They collected gut microbiome data on 98 healthy volunteers. In particular, the counts of three bacteria genera were recorded, namely Bacteroides, Prevotella, and Ruminococcus. Arumugam et al. [2] used these three bacteria to define three groups they called enterotypes. These enterotypes provide information about the human's body ability to produce vitamins.

Wu et al. analyzed the same dataset conducting a cluster analysis via the 'partitioning around medoids' (PAM) approach. They detected only two of the three enterotypes defined in the work by Arumugam et al. Moreover, these two clusters are characterized by different frequencies: 86 out of the 98 samples were allocated to the first enterotype, whereas only 12 samples were clustered into enterotype 2. This is due to the small number of subjects with a high abundance of Prevotella (i.e., only 36 samples showed a Prevotella count greater than 0).

Besides the bacterial data, we consider also $K = 9$ covariates, representing information on micro-nutrients in the habitual long-term diet collected using a food frequency questionnaire. These 9 additional variables have been selected by Xia et al. using a l_1 penalized regression approach.

Table 1 shows the posterior mean and 95% credible set (CS) of each parameter involved in the DMReg and the FDMReg models. Though the significant covariates are the same across the models, the FDMReg shows a lower WAIC, thus being the best model in terms of fit. This is due to the additional set of parameters involved in the mixture structure that help in providing information on this dataset.

The mixture structure of the FDMReg model can be exploited to cluster observations into groups through a model-based approach. More specifically, each observation can be allocated to the mixture component that most likely generated it. Indeed, note that the mixing weights estimates (0.637, 0.357 and 0.006, from Table 1) confirm the presence of two out of the three enterotypes defined by Arumugam et al. [2]. To further illustrate the benefits of the FDReg model in a microbiome data analysis, we compare the clustering profile obtained by the FDMReg model and the one obtained with the PAM approach used by Wu et al. In particular, Table 2 summarizes this comparison in a confusion matrix. Despite the clustering generated by the FDMReg being based on some distributional assumptions (i.e., the response is FDM distributed), it highly agrees with the one obtained by the PAM algorithm for 84% of the observations. This percentage is obtained using the covariates selected by Xia et al. in a logistic normal multinomial regression model context. Clearly, the results could be improved by developing an ad hoc variable selection procedure for the FDMReg model. The main advantage to considering the FDMReg (that is a model-based clustering approach) is that, besides the clustering of the data points, it provides also some information on the detected clusters (e.g., their size and a measure of their distance) and the relationship between the response and the set of covariates. This additional information may increase the insight we can gain from

Table 1 Posterior mean and 95% CS for the parameters of the DMReg and FDMReg models. Regression coefficients in bold are related to 95% CS's not containing the zero value.

		DM		FDM	
		Post. Mean	95% CS	Post. Mean	95% CS
Bacteroides	Intercept	**2.197**	**(1.844, 2.546)**	**2.642**	**(2.215, 3.034)**
	Proline	-0.039	(-0.344, 0.273)	-0.036	(-0.325, 0.261)
	Sucrose	-0.257	(-0.555, 0.039)	-0.208	(-0.471, 0.064)
	Vitamin E, food fortification	-0.016	(-0.351, 0.336)	-0.043	(-0.351, 0.299)
	Beta cryptoxanthin	-0.073	(-0.357, 0.237)	-0.059	(-0.334, 0.214)
	Added germa from wheats	-0.147	(-0.477, 0.196)	-0.042	(-0.411, 0.271)
	Vitamin C	0.300	(-0.031, 0.771)	0.267	(-0.035, 0.673)
	Maltose	-0.031	(-0.311, 0.260)	0.034	(-0.237, 0.302)
	Palmitelaidic trans fatty acid	0.019	(-0.292, 0.328)	-0.044	(-0.336, 0.251)
	Acrylamide	0.133	(-0.167, 0.455)	0.184	(-0.094, 0.474)
Prevotella	Intercept	**-1.196**	**(-1.715, -0.699)**	-0.402	(-1.094, 0.245)
	Proline	-0.053	(-0.571, 0.443)	-0.018	(-0.663, 0.546)
	Sucrose	0.029	(-0.437, 0.476)	0.126	(-0.335, 0.591)
	Vitamin E, food fortification	0.109	(-0.355, 0.548)	0.113	(-0.473, 0.574)
	Beta cryptoxanthin	0.263	(-0.230, 0.762)	0.349	(-0.386, 0.812)
	Added germa from wheats	0.280	(-0.137, 0.701)	0.121	(-0.298, 0.604)
	Vitamin C	-0.169	(-1.196, 0.623)	-0.021	(-1.131, 0.738)
	Maltose	**0.640**	**(0.164, 1.126)**	**0.877**	**(0.260, 1.400)**
	Palmitelaidic trans fatty acid	**-0.530**	**(-1.008, -0.043)**	**-0.716**	**(-1.209, -0.140)**
	Acrylamide	**0.780**	**(0.362, 1.206)**	**0.800**	**(0.382, 1.231)**
	α^+	1.541	(1.104, 2.040)	2.275	(1.489, 3.208)
	p_1	—	—	0.637	(0.420, 0.797)
	p_2	—	—	0.357	(0.197, 0.570)
	p_3	—	—	0.006	(0.000, 0.027)
	$\tilde{w}$	—	—	0.914	(0.791, 0.991)
	WAIC	1686.2		1662.3	

data. Further improvements could be obtained considering an even more flexible distribution for Π, that is the extended flexible Dirichlet [11].

Table 2 Confusion matrix for clustering based on the FDMReg model compared to the PAM algorithm.

		FDMReg	
		1	2
PAM	1	70	16
	2	0	12

References

1. Amato, K.: An introduction to microbiome analysis for human biology applications. Am. J. Hum. Biol. **29** (2017)

2. Arumugam, M. et al.: Enterotypes of the human gut microbiome. Nature. **473**, 174–180 (2011)
3. Ascari, R., Migliorati, S.: A new regression model for overdispersed binomial data accounting for outliers and an excess of zeros. Stat. Med. **40**(17), 3895–3914 (2021)
4. Chen, J., Li, H.: Variable selection for sparse Dirichlet-multinomial regression with an application to microbiome data analysis. Ann. Appl. Stat. **7**(1), 418–442 (2013)
5. Koeth, R. A. et al.: Intestinal microbiota metabolism of L-carnitine, a nutrient in red meat, promotes atherosclerosis. Nat. Med. **19**(5) (2013)
6. McCullagh, P., Nelder, J. A.: Generalized Linear Models. Chapman & Hall (1989)
7. Migliorati, S., Ongaro, A., Monti, G. S.: A structured Dirichlet mixture model for compositional data: inferential and applicative issues. Stat. Comput. 27(4), 963–983, 2017.
8. Morgan, X. C., Huttenhower, C.: Human microbiome analysis. PloS Computational Biology. **8**(12) (2012)
9. Ongaro, A., Migliorati, S.: A generalization of the Dirichlet distribution. J. Multivar. Anal. **114**, 412–426 (2013)
10. Neal, R. M.: An improved acceptance procedure for the hybrid Monte Carlo algorithm. Tech. Rep. (1994)
11. Ongaro, A., Migliorati, S., Ascari, R.: A new mixture model on the simplex. Stat. Comput. **30**(4), 749–770 (2020)
12. Qin, J., Li, Y., Cai, Z., Li, S., Zhu, J., Zhang, F., Liang, S., Zhang, W., Guan, Y., Shen, D., Peng, Y.: A metagenome-wide association study of gut microbiota in type 2 diabetes. Nature. 490 (2012)
13. Stan Development Team: Stan Modeling Language Users Guide and Reference Manual (2017)
14. Turnbaugh, P. J. et al.: A core gut microbiome in obese and lean twins. Nature. 457 (2009)
15. Vehtari, A., Gelman, A., Gabry, J.: Practical Bayesian model evaluation using leave-one-out cross-validation and WAIC. Stat. Comput. **27**(5), 1413–1432 (2017)
16. Wadsworth, W. D., Argiento, R., Guindani, M., Galloway-Pena, J., Shelburne, S. A., Vannucci, M.: An integrative Bayesian Dirichlet-multinomial regression model for the analysis of taxonomic abundances in microbiome data. BMC Bioinformatics. **18**(94) (2017)
17. Watanabe, S.: A widely applicable Bayesian information criterion. J. Mach. Learn. Tech. **14**(1), 867–897 (2013)
18. Wu., G. D. et al.: Linking long-term dietary patterns with gut microbial enterotypes. Science. 334, 105–109 (2011)
19. Xia, F., Chen, J., Fung, W. K., Li, H.: A logistic normal multinomial regression model for microbiome compositional data analysis. Biometrics. **69**(4), 1053–1063 (2013)

Stability of Mixed-type Cluster Partitions for Determination of the Number of Clusters

Rabea Aschenbruck, Gero Szepannek, and Adalbert F. X. Wilhelm

Abstract For partitioning clustering methods, the number of clusters has to be determined in advance. One approach to deal with this issue are stability indices. In this paper several stability-based validation methods are investigated with regard to the *k-prototypes* algorithm for mixed-type data. The stability-based approaches are compared to common validation indices in a comprehensive simulation study in order to analyze preferability as a function of the underlying data generating process.

Keywords: cluster stability, cluster validation, mixed-type data

1 Introduction

In cluster analysis practice, it is common to work with mixed-type data (i.e. numerical and categorical variables), while in theoretical development the research is traditionally often restricted to numerical data. A comprehensive overview on cluster analysis based on mixed-type data is given in [1]. To cluster these mixed-type data, a popular approach is the *k-prototypes* algorithm, an extension of the popular *k-means* algorithm, as proposed in [2] and implemented in [3].

As for all partitioning clustering methods, the number of clusters has to be specified in advance. In the past, several validation methods have been identified for the

Rabea Aschenbruck (✉)
Stralsund University of Applied Sciences, Zur Schwedenschanze 15, 18435 Stralsund, Germany,
e-mail: rabea.aschenbruck@hochschule-stralsund.de

Gero Szepannek
Stralsund University of Applied Sciences, Zur Schwedenschanze 15, 18435 Stralsund, Germany,
e-mail: gero.szepannek@hochschule-stralsund.de

Adalbert F.X. Wilhelm
Jacobs University Bremen, Campus Ring 1, 28759 Bremen, Germany,
e-mail: A.Wilhelm@jacobs-university.de

© The Author(s) 2023
P. Brito et al. (eds.), *Classification and Data Science in the Digital Age*,
Studies in Classification, Data Analysis, and Knowledge Organization,
https://doi.org/10.1007/978-3-031-09034-9_6

k-prototypes algorithm to enable the rating of clusters and to determine the index optimal number of clusters. A brief overview is given in Section 2, followed by an examination of the investigated stability indices to improve clustering mixed-type data[1]. In Section 3, a simulation study has been conducted in order to compare the performance of stability indices as well as a new proposed adjustment, and additionally to rate the performance with respect to internal validation indices. Finally, a summary, which does not state a superiority of the stability-based approaches over internal validation indices in general, and an outlook are given in Section 4.

2 Stability of Cluster Partitions

The assessment of cluster quality can be used for the comparison of clusters resulting from different methods or from the same method but with different input parameters, e.g., with a different number of clusters. Especially the latter has already been an important issue in partitioning clustering many decades ago [5]. Since then, some work has been done on this subject. Hennig [6] points out, that nowadays some literature uses the term *cluster validation* exclusively for methods that decide about the optimal number of clusters, in the following named *internal validation*. An overview of internal validation indices is given, e.g., in [7] or [8]. In [9], a set of internal cluster validation indices for mixed-type data to determine the number of clusters for the *k-prototypes* algorithm was derived and analyzed. In the following, stability indices are presented, before they are compared to each other and additionally to internal validation indices in Section 3. Since cluster stability is a model agnostic method, the indices are applicable to any clustering algorithm and not limited to numerical data [10].

A partition S splits data $Y = \{y_1, \ldots, y_n\}$ into K groups $S_1, \ldots, S_K \subseteq Y$. The focus of this paper is on the evaluation and rating of cluster partitions with so-called stability indices. To calculate these, as discussed by Dolnicar and Leisch [11] or mentioned by Fang and Wang [12], $b \in \{1, \ldots, B\}$ bootstrap samples Y^b (with replacement, see e.g. [13]) from the original data set Y are drawn. For every bootstrap sample Y^b, a cluster partition $S^b = \{S_1^b, \ldots, S_{L_b}^b\}$ is determined. For the validation of the different results of these bootstrap samples, the set of points from the original data set that are also part of the b-th bootstrap sample $X^b = Y \cap Y^b$ is used, where n_b is the size of X^b. Furthermore $C^b = \{S_k \cap X^b | k = 1, \ldots, K\}$ and $D^b = \{S_l^b \cap X^b | l = 1, \ldots, L_b\}$, with $B_C^{\star}$ being the number of bootstrap samples for which $C^b \neq \emptyset$, and n_{S_k}, $n_{C_k^b}$, $n_{S_l^b}$, and $n_{D_l^b}$ with $k \in \{1, \ldots, K\}, l \in \{1, \ldots, L_b\}$ are the numbers of objects in cluster group S_k, C_k^b, S_l^b and D_l^b, respectively.

In 2002, Ben-Hur et al. [14] presented stability-based methods, which can be used to define the optimal number of clusters. In their work, the basis for the calculation of the stability indices is a binary matrix P^{C^b}, which represents the cluster partition C^b in the following way

---

[1] The mentioned and analyzed stability indices will extend the R package clustMixType [4].

$$P_{ij}^{C^b} = \begin{cases} 1, & \text{if objects } x_i^b, x_j^b \in X^b \text{ are in the same cluster and } i \neq j, \\ 0, & \text{otherwise.} \end{cases} \tag{1}$$

With P^{D^b} defined analogously, the dot product of the two cluster partitions C^b and D^b is defined as $D(P^{C^b}, P^{D^b}) = \sum_{i,j} P_{ij}^{C^b} P_{ij}^{D^b}$. This leads to a Jaccard coefficient based index of two cluster partitions C^b and D^b

$$Stab_J(P^{C^b}, P^{D^b}) = \frac{D(P^{C^b}, P^{D^b})}{D(P^{C^b}, P^{C^b}) + D(P^{D^b}, P^{D^b}) - D(P^{C^b}, P^{D^b})}. \tag{2}$$

Hennig proposed a so-called local stability measure for every cluster group in a cluster partition based on the Jaccard coefficient as well [15]. To obtain one stability value $Stab_{J;cw}$ for the whole partition, the weighted mean of the cluster-wise values with respect to the size of the cluster groups is determined. Another stability-based index presented by Ben-Hur et al., based on the simple matching coefficient, is called Rand index [16] and defined as

$$Stab_R(P^{C^b}, P^{D^b}) = 1 - \frac{1}{n^2} \|P^{C^b} - P^{D^b}\|^2. \tag{3}$$

Additionally, they present the stability index based on a similarity measure, which was originally mentioned by Fowlkes and Mallows [17],

$$Stab_{FM}(P^{C^b}, P^{D^b}) = \frac{D(P^{C^b}, P^{D^b})}{\sqrt{D(P^{C^b}, P^{C^b}) D(P^{D^b}, P^{D^b})}}. \tag{4}$$

For determination of the number of clusters, Ben-Hur et al. proposed the analysis of the distribution of index values calculated between pairs of clustered sub-samples, where high pairwise similarities indicate a stable partition. The authors' suggested aim is examining the transition from a stable to an unstable clustering state. In the simulation study, this qualitative criterion was numerically approximated by the differences in the areas under these curves. Furthermore, von Luxburg [18] published an approach to obtain the cluster partition stability based on the minimal matching distance, where the minimum is taken over all permutations of the K labels of clusters. Straightforward, the distances are summarized by their mean to obtain $Instab_L(P^{C^b}, P^{D^b})$ respectively $Stab_L(P^{C^b}, P^{D^b}) = 1 - Instab_L(P^{C^b}, P^{D^b})$.

3 Simulation Study

In order to compare the stability indices of the cluster partition and afterwards with respect to the internal validation indices, a simulation study was conducted. In the following, the setup and execution of this simulation study starting with the data generation is briefly presented, and subsequently the results are evaluated.

3.1 Data Generation and Execution of Simulation Study

The simulation study is based on artificial data, which are generated for different scenarios. In Table 1, the features that define the data scenarios and their corresponding parameter values are listed. Since a full factorial design is used, there are 120 different data settings in the conducted simulation study.[2] The selection of the considered features follow the characteristics of the simulation study in [19] and were extended with respect to the ratio of the variable types as in [20].

Table 1 Features and the associated feature specifications used to generate the data scenarios.

data parameter	feature specification	short
number of clusters	2, 4, 8	nC
clusters of equal size (FALSE: randomly drawn sizes)	TRUE, FALSE	symm
number of variables	2, 4, 8	nV
ratio of factor to numerical variables	0.25, 0.5, 0.75	fac_prop
overlap between cluster groups	0, 0.05, 0.1	overlap

The clusters of the 200 observations are defined by the the feature settings. Each variable can either be *active* or *inactive*. For the numerical variables, *active* means drawing values from the normal distribution $X_1 \sim \mathcal{N}(\mu_1, 1)$, with random $\mu_1 \in \{0, ..., 20\}$, and *inactive* means drawing from $X_0 \sim \mathcal{N}(\mu_0, 1)$ with $\mu_0 = 2 \cdot q_{1-\frac{v}{2}} - \mu_1$, where q_α is the α-quantile of $\mathcal{N}(\mu_1, 1)$ and $v \in \{0.05, 0.1\}$. This results in an overlap of v for the two normal distributions. To achieve an overlap of $v = 0$, the inactive variable is drawn from $\mathcal{N}(\mu_1 - 10, 1)$. Furthermore, each factor variable has two levels, l_0 and l_1. The probability for drawing l_0 for an active variable is v and $(1 - v)$ for level l_1. For an inactive variable, the probability for l_0 is $(1 - v)$ and v for l_1.

Below, the code structure of the simulation study is presented. For each of the 120 data scenarios, a repetition of $N = 10$ runs was performed. This should mitigate the influence of the random initialization of the *k-prototypes* algorithm. For the range of two up to nine cluster groups, the stability indices are determined based on bootstrap samples as suggested in [21]. In order to rank the performance of the stability-based indices, the internal validation indices were also determined on the same data.

Pseudo-Code Simulation Study

```
for(every data situation){
  for(i in 1:N){ # 10 iterations to mitigate/soften random influences
    data <- create.data(data situation)
    for(q in 2:9){
      output <- kproto(data, k = q, nstarts = 20)
      # stability-based indices determined with the usage of 100 bootstrap samples
      stab_val_method <- stab_kproto(output, B = 100, method)
      int_val_method <- validation_kproto(output, method) # internal validation
    }
    # determine optimal cluster size for every method
    cs_method <- max/min(int_val_method or stab_val_method)
  }
}
```

[2] There is no data scenario with two variables and eight cluster groups. Additionally, if there are two variables, obviously only the 0.5 ratio between factor and numerical variables is possible.

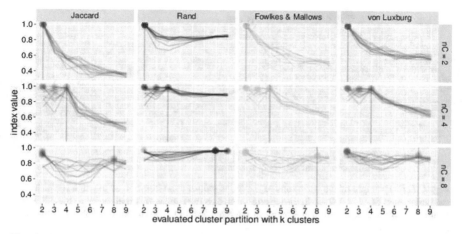

Fig. 1 The evaluations of the four stability-based cluster indices are presented. There are ten repetitions of rating the data situation for k clusters in the range of two to nine and the index-optimal number of clusters is highlighted. The parameters of the underlying data structure are nV = 8, fac_prop = 0.5, overlap = 0.1 and symm = FALSE. The number of clusters nC in the data structure varies row-wise.

3.2 Analysis of the Results

Figure 1 shows exemplary results of the simulation study for three different data scenarios over the 10 repetitions. Each row of the figure shows a different data scenario and each column shows one of the four stability-based indices. The first row is related to a data scenario with two clusters (marked by a vertical green line). Each plot shows the examined number of clusters and the determined index value for the 10 repetitions. The maximum index value for each repetition is highlighted with a larger dot and marks the index-optimal number of clusters of this repetition. It can be seen that all of the four different indices detected the two clusters in the underlying data structure. Rows two and three show the evaluations of data with cluster partitions of four and eight clusters, respectively. It can be seen that the generated number of clusters is not always rated as index optimal (for example, with four clusters, two or three clusters were often also evaluated as optimal). Since the results shown here are representative for all scenarios, the four cluster indices and their interpretation were examined in more detail.

In the left part of Figure 2, different transformations of the index values are presented. Besides the standard index values (green line), the numerical approximation of the approach of Ben-Hur et al. mentioned above is also shown (red line). For the Jaccard-based evaluation, the proposed cluster-wise stability determination by Hennig is presented in orange. Additionally, we propose an adjustment of the index values (hereinafter referred to as *new adjust*), similar to [22], to take into account not only the magnitude of the index but also the local slope: The index value scaled with the geometric mean of the changes to the neighbor values is presented in dark green.

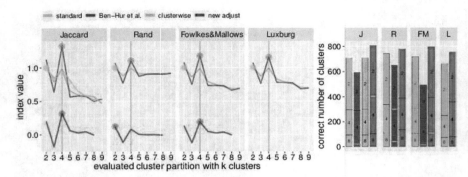

Fig. 2 *Left:* Example of the variations of the index values at an iteration of the data scenario with the parameters nC = 4, nV = 8, fac_prop = 0.5, overlap = 0.1 and symm = FALSE. *Right:* Proportion of correct determinations, partitioned according to the different number of clusters in the underlying data structure.

Again, for each variation of the indices, the index optimal value is highlighted. The numerically determined index values according to the approach of Ben-Hur et al. gain no benefit, thus it can be concluded that the quantification is not appropriate for the purpose and that further research is required. The cluster-wise stability determination of the Jaccard index also does not seem to improve the determination of the number of clusters to a large extent. Obviously, the local slope in the example in Figure 2 is strengthened for four evaluated cluster groups by the new adjustment that leads to a determination of four cluster groups (which is the generated number of clusters). Since only one iteration of one data scenario is shown on the left, the sum of correct determined number of clusters with respect to the generated number of clusters is shown on the right hand side of Figure 2. These sums for two, four and eight clusters in the underlying data structure point out the improvement of the proposed adjustment of the index values. Especially for more than two clusters, the rate of correctly determined numbers of clusters can be increased.

Finally, the internal validation indices were comparatively examined. For analyzing the outcome of the simulation study, the determined index optimal numbers of clusters are shown in Table 2. While the comparison for two clusters in the underlying data shows a slight advantage for the stability-based indices, especially for eight clusters the preference is in favor of the internal validation indices. To gain a better understanding of the mean success rate of determining the correct number of clusters for each data scenario, Figure 3 further shows the results of a linear regression on the various data parameters. It can be seen that in most cases there is not too much difference between the considered methods. The stability-based indices do a better job of determining the number of clusters for data with equally large cluster groups. Obviously, a larger number of variables causes a better determination of the number of clusters. The largest variation in the influence on the proportion of correct determination can be seen for the parameter *number of clusters*. The more cluster groups are available in the underlying data structure, the worse the determination becomes (especially for the stability-based indices and the indices Ptbiserial and Tau).

Table 2 Determined number of clusters for all data scenarios with $nC \in \{2, 4, 8\}$, summarized by the stability-based as well as internal validation indices and the evaluated number of clusters.

clusters	**2**	3	4	5	6	7	8	9	2	3	**4**	5	6	7	8	9	2	3	4	5	6	7	**8**	9
J_newadj	**403**	17	0	0	0	0	0	0	47	74	**298**	1	0	0	0	0	90	70	27	16	16	26	**104**	11
R_newadj	**391**	18	5	0	1	1	1	3	56	99	**258**	3	2	0	0	2	38	68	22	17	16	32	**133**	34
FM_newadj	**402**	17	1	0	0	0	0	0	50	80	**289**	1	0	0	0	0	88	71	26	16	15	26	**106**	12
L_newadj	**394**	21	5	0	0	0	0	0	53	83	**282**	2	0	0	0	0	100	97	31	20	16	16	**76**	4
CIndex	**313**	13	2	2	1	4	18	67	7	27	**344**	13	3	2	5	19	2	0	2	4	22	28	**211**	91
Dunn	**386**	24	4	2	0	1	1	2	39	56	**307**	8	7	3	0	0	19	9	17	7	37	53	**190**	28
Gamma	**343**	9	1	0	1	2	14	50	9	16	**356**	15	3	1	5	15	2	1	4	4	16	16	**198**	119
GPlus	**319**	8	1	0	0	0	9	83	6	10	**319**	12	5	2	15	51	2	1	1	4	14	12	**175**	151
McClain	**71**	3	1	1	5	12	57	270	0	0	**17**	4	4	13	87	295	0	0	0	0	0	9	**34**	317
Ptbiserial	**400**	11	6	0	3	0	0	0	72	120	**225**	3	0	0	0	0	31	62	79	65	55	39	**26**	3
Silhouette	**388**	3	1	4	4	5	8	7	14	37	**348**	7	0	0	8	6	6	0	3	1	12	46	**220**	72
Tau	**391**	16	9	0	4	0	0	0	68	144	**205**	3	0	0	0	0	33	82	119	68	40	14	**3**	1

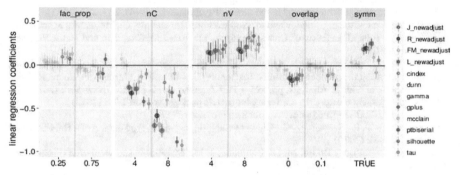

Fig. 3 Linear regression coefficients for the parameters of the five data set features, where coefficients whose confidence intervals contain 0 are displayed in transparent.

4 Conclusion

The aim of this study was to investigate the determination of the optimal number of clusters based on stability indices. Several variations of analysis methods of stability-based index values were presented and comparatively analyzed in a simulation study. The proposed adjustment of the index values with respect not only to their magnitude but also to the local slope was able to improve the standard stability indices, especially for a smaller number of clusters. The simulation study did not show any general superiority of stability-based approaches over internal validation indices.

In the future, the various methods of analyzing the stability-based index values should be examined in more detail, e.g., taking into account the Adjusted Rand Index. For this purpose, further research may address the characteristics of the evaluated curves more precisely, or further extend the approach of Ben-Hur et al. as a quantitative determination method, which has not been done yet.

References

1. Ahmad, A., Khan, S.: Survey of state-of-the-art mixed data clustering algorithms. IEEE Access, 31883–31902 (2019)
2. Huang, Z.: Extension to the k-Means algorithm for clustering large data sets with categorical values. Data Min. Knowl. Discov. **2**(6), 283–304 (1998)
3. Szepannek, G.: clustMixType: User-friendly clustering of mixed-type data in R. The R J. **10**(2), 200–208 (2018)
4. Szepannek, G., Aschenbruck, R.: clustMixType: k-prototypes clustering for mixed variable-type data. R package version 0.2-15 (2021)
   `https://CRAN.R-project.org/package=clusterMixType`
5. Thorndike, R. L.: Who belongs in the family. Psychometrika **18**(4), 267–276 (1953)
6. Hennig, C.: Clustering strategy and method selection. In: Hennig, C., Meila, M. , Murtagh, F., Rocci, R. (eds.) Handbook of Cluster Analysis, pp. 703–730. Chapman and Hall/CRC, New York (2015)
7. Halkidi, M., Vazirgiannia, M., Hennig, C.: Method-independent indices for cluster validation and estimating the number of clusters. In: Hennig, C., Meila, M. , Murtagh, F., Rocci, R. (eds.) Handbook of Cluster Analysis, pp. 595–618. Chapman and Hall/CRC, New York (2015)
8. Desgraupes, B.: clusterCrit: clustering indices. R package version 1.2.8 (2018)
   `https://CRAN.R-project.org/package=clusterCrit`
9. Aschenbruck, R., Szepannek, G.: Cluster validation for mixed-type data. Arch. Data Sci., Ser. A **6**(1), 1–12 (2020)
10. Lange, T., Roth, V., Braun, M. L., Buhmann, J. M.: Stability-based validation of clustering solutions. Neural. Comput. **16**(6), 1299–1323 (2004)
11. Dolnicar, S., Leisch, F.: Evaluation of structure and reproducibility of cluster solutions using bootstrap. Mark. Lett. **21**, 83–101 (2010)
12. Fang, Y., Wang, J.: Selection of the number of clusters via the bootstrap method. Comput. Stat. Data Anal. **56**(3), 468–477 (2012)
13. Mucha, H.-J., Bartel, H.-G.: Validation of k-means clustering: why is bootstrapping better than subsampling. Arch. Data Sci., Ser. A **2**(1), 1–14 (2017)
14. Ben-Hur, A., Elisseeff, A., Guyon, I.: A stability based method for discovering structure in clustered data. In: Pac. Symp. Biocomput. **2002**, 6–17 (2001)
15. Hennig, C.: Cluster-wise assessment of cluster stability. Comput. Stat. Data Anal. **52**(1), 258–271 (2007)
16. Rand, W. M.: Objective criteria for the evaluation of clustering methods. J. Am. Stat. Assoc. **66**(336) 846–850 (1971)
17. Fowlkes, E. B., Mallows, C. L.: A method for comparing two hierarchical clusterings. J. Am. Stat. Assoc. **78**(383) 553–569 (1983)
18. von Luxburg, U.: Clustering stability: an overview. Found. Trends® Mach. Learn. **2**(3), 235–274 (2010)
19. Dangl, R., Leisch, F.: Effects of resampling in determining the number of clusters in a data set. J. Classif. **37**(3), 558–583 (2020)
20. Jimeno, J., Roy, M., Tortora, C.: Clustering mixed-type data: a benchmark study on KAMILA and k-prototypes. In: Chadjipadelis, T., Lausen, B., Markos, A., Lee, T.R., Montanari, A., Nugent, R. (eds.) Data Analysis and Rationality in a Complex World, 83–91, Springer International Publishing, Cham (2021)
21. Leisch, F.: Resampling methods for exploring cluster stability. In: Hennig, C., Meila, M., Murtagh, F., Rocci, R. (eds.) Handbook of Cluster Analysis, pp. 637–652. Chapman and Hall/CRC, New York (2015)
22. Ilies, J., Wilhelm, A. F. X.: Projection-based partitioning for large, high-dimensional datasets. J. Comp. Graph. Stat. **19**(2), 474–492 (2010)

A Review on Official Survey Item Classification for Mixed-Mode Effects Adjustment

Afshin Ashofteh and Pedro Campos

Abstract The COVID-19 pandemic has had a direct impact on the development, production, and dissemination of official statistics. This situation led National Statistics Institutes (NSIs) to make methodological and practical choices for survey collection without the need for the direct contact of interviewing staff (i.e. remote survey data collection). Mixing telephone interviews (CATI) and computer-assisted web interviewing (CAWI) with direct contact of interviewing constitute a new way for data collection at the time COVID-19 crisis. This paper presents a literature review to summarize the role of statistical classification and design weights to control coverage errors and non-response bias in mixed-mode questionnaire design. We identified 289 research articles with a computerized search over two databases, Scopus and Web of Science. It was found that, although employing mixed-mode surveys could be considered as a substitution of traditional face-to-face interviews (CAPI), proper statistical classification of survey items and responders is important to control the nonresponse rates and coverage error risk.

Keywords: mixed-mode official surveys, item classification, weighting methods, clustering, measurement error

Afshin Ashofteh (✉)
Statistics Portugal (Instituto Nacional de Estatística, Departamento de Metodologia e Sistemas de Infomação) and NOVA Information Management School (NOVA IMS) and MagIC, Universidade Nova de Lisboa, Lisboa, Portugal, e-mail: afshin.ashofteh@ine.pt

Pedro Campos
Statistics Portugal (Instituto Nacional de Estatística, Departamento de Metodologia e Sistemas de Infomação) and Faculty of Economics, Universidade do Porto, and LIAAD INESC TEC, Portugal, e-mail: pedro.campos@ine.pt

© The Author(s) 2023
P. Brito et al. (eds.), *Classification and Data Science in the Digital Age*,
Studies in Classification, Data Analysis, and Knowledge Organization,
https://doi.org/10.1007/978-3-031-09034-9_7

1 Introduction

This paper provides a summary of a systematic literature review of the role of classification variables and weighting methods of mixed-mode surveys in minimizing the measurement error, coverage error, and nonresponse bias.

Before the COVID-19 pandemic, the statistical adjustment of mode-specific measurement effects was studied by many scholars. However, after the pandemic, survey methodologists made a strong effort to meet the challenges of new restrictions for collecting data with proper quality [1]. Data collection with mixing different modes by considering their contribution to the overall published statistics was considered as a solution by NSIs. The methodologists have been trying to use technology, data science, and mixed-device surveys to decrease the expected coverage error and nonresponse bias with new target populations at the time of pandemic rather than the traditional interviewer-assisted and paper survey modes [2]. This coverage error is caused by the changes of the target population from the general population to the general population accessible with technological devices. Te Braak et. al. [3] highlighted how the representativeness of self-administered online surveys is expected to be impacted by decreased response rates. Their research demonstrates that a huge group of respondents dropout selectively and that this selectivity varies depending on the dropout moment and demographic categorical information.

According to the studies in Statistics Portugal, using classification methods by categorical variables and applying the repeated weighting techniques seem to be fruitful to estimate and adjust for mode and device effects. Fortunately, many authors discussed the use of weights in statistical analysis [4]. It is important to improve inference in cases where mixed-mode effects are combined with measurement errors caused by primary data collection on categorical variables and socio-demographic information. On one side that the categorical variables are collected with the help of responders (primary data), the survey mode has a strong impact on answering behaviors and answering conditions. Respondents might evaluate some of the new categorical variables as sensitive information or privacy intrusive. They may not be willing to share these personal data by telephone or technological devices, which are necessary for statistical classification. Additionally, for NSIs, also the new data collection channels are costly and redesign of the survey estimation methodology is time consuming. On the other side, the categorical variables should be available in sampling frames (secondary data) and the coverage error is the main concern. For instance, in CATI surveys of Statistics Portugal after COVID-19, the population was considered as belonging to the following categories: (i) households with a listed landline telephone, (ii) households that do not have a telephone but use only a mobile telephone, and (iii) households that do not have a telephone at all (or whose number is unknown). We could expect these households with very different socioeconomic characteristics, and new methods of classification or clustering as helpful methods for measurement error adjustment at the time of the pandemic. However, if they are different in the important categorical variables of our survey, then a weighting solution could amplify a part of the sample, which does not represent the population. As a result, statistical classification would be another source of bias instead of solving the problem. Therefore, we could expect two approaches. First, we could ignore classification, simply because we consider the groups are homogeneous and the weighting could be recommended to adjust for COVID-19 pandemic situation and non-observation errors. Second, the groups or responders are different and we need categorical variables. In this case, the non-observation errors of CATI and CAWI could not be covered by changing only the weights and we have to recommend CAPI to collect categorical information and apply both clustering and weighting together to have a reasonable coverage by mixed modes.

This study undergoes a systematic literature review on this topic guided by the following question. What is the best methodology or modified estimation strategy to mitigate the mode-effects problems based on design weighting and classification? To answer this question, we performed a systematic review analysis limited to the following databases: Web of Science, Scopus, and working papers from NSIs. We only considered papers written in English. This article is organized as follows: Section 2 presents the methodology of research that maps keyword identification search, databases, and bibliometric analysis. In Section 3, we present the results, identifying the PRISMA flow diagram, characteristics of the articles, author co-authorship analysis, as well as the Keywords occurrence over the years. In Section 4, we discuss the content analysis. Section 5 is about the main conclusions and finally, in Section 6, the main research gaps and future works are outlined.

2 Methods

To accomplish the research, the preferred reporting items for systematic reviews and meta-analysis methodology were adopted. The algorithm of the paper selection from databases (Scopus and WOS) was based on screening started by search keywords ((mixed-mode* OR "Mode effect*") AND (weighting OR weight* OR classification) AND ("Measurement error*" OR "Non-response bias" OR "Data quality" OR "response rate*") AND (capi OR "Computer Assisted Personal Interview*" OR cawi OR "Assisted Web Interview*" OR cati OR "Computer Assisted Telephone Interview*" OR "web survey*" OR "mail survey*" OR "telephone survey*")) and then the result was filtered by "official statistics". The results of the two databases were merged, and then duplication was removed. For bibliometric analysis, the Mendeley open-source tool was used to extract metadata and eliminate duplicates. For network analysis, the VOSviewer open-source tool has been applied to visualize the extracted information from the data set and obtain the quantitative and qualitative outcomes. After assessing the eligibility, books and review papers were omitted from results and relevant articles picked up from databases. The final dataset was selected according to the visual abstract in Figure 2, which shows detailed information about this systematic literature review.

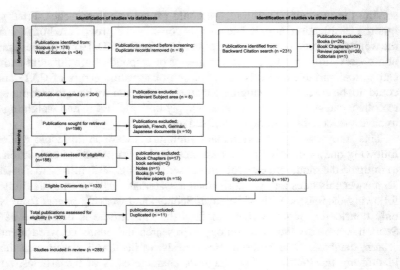

Fig. 1 Literature review flow diagram. (Source: Author's preparation).

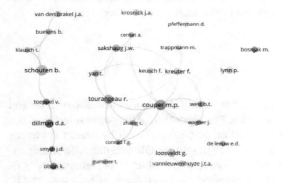

Fig. 2 Density visualization analysis of the 22 leader authors who have at least 3 papers.

3 Results

The 28 leader authors who had at least 4 papers are presented in Figure 2. Author occurrence analysis was performed by applying the VOSviewer research tool for network analysis. The top three leader authors were Mick P. Couper with 14 articles, Barry Schouten with 14 articles, and Roger Tourangeau with 11 articles. With the help of VOSviewer, keywords' analysis was accomplished. We analyzed the co-occurrence of author keywords with the full counting method. In the first step, we select one for the minimum occurrence of a keyword and the result was 711 keywords. We could see the application of keywords over years (Figure 3). Some of the keywords were not exactly the same, but their use and meaning were the same.

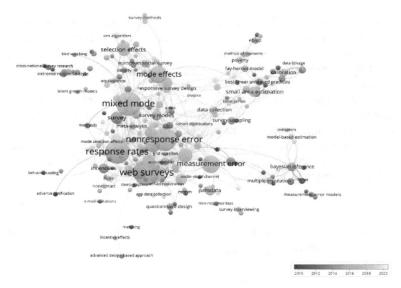

Fig. 3 Application of keywords over years.

We decided to match similar words to make the output clearer. Choosing the full counting method resulted in a total of 592 authors meeting the threshold.

4 Content Analysis

The studies emphasize the dramatic change in mixed-mode strategies in the last decades based on design-based and model-assisted survey sampling, time series methods, small area estimation [6], and high expectation to undergo further changes especially after the magnificent experience of NSIs, trying new modes after COVID-19 pandemic [7].

The problem is about mixed-mode effects and calibration, and briefly, we could follow several approaches such as design weighting to find sampling weights, nonresponse weighting adjustment, and calibration. The design weight of a unit may be interpreted as the number of units from population represented by a specific sample unit. Most surveys, if not all, suffer from nonresponse in item or unit. Auxiliary information could be used to improve the quality of design-weighted estimates. An auxiliary variable must have at least two characteristics to be considered in calibration: (i) It must be available for all sample units; and (ii) Its population total must be known.

The categorical variables from the demographic information of nonrespondents such as education level, age, income, location, language, and marital status could help the survey methodologists to categorize the target population and recognize the best sequence of the modes [8]. Van Berkel et al. [9] considered nine strata in their classification tree by using age, ethnicity, urbanization, and income as explanatory variables. Re-interview design and inverse regression estimator (IREG) are among the best approaches to improve measurement bias by using related auxiliary information [10].

The focus of this approach is on the weights of estimators rather than the bias from the measurements. For an estimator, we could consider $y_{i,m}$ the measurement obtained from unit i through mode m. The $y_{i,m}$ consists of u_i as the observed value for respondent i, an additive mode-dependent measurement bias b_m, and a mode-dependent measurement variance $\varepsilon_{i,m}$ with an expected value equal to zero. Equation (1) shows the measurement error model.

$$y_{i,m} = u_i + b_m + \varepsilon_{i,m} \tag{1}$$

If we consider two different modes m and $\dot{m}$, then the differential measurement error between these two modes is given by

$$y_{i,m} - y_{i,\dot{m}} = (b_m - b_{\dot{m}}) + (\varepsilon_{i,m} - \varepsilon_{i,\dot{m}}) \tag{2}$$

The expected value of $(b_m - b_{\dot{m}})$ is the differential measurement bias. If we consider $\hat{t}_y$ as an estimation of the total of variable y according to its observations in different modes $y_{i,m}$, then

$$\hat{t}_y = \sum_{i=1}^{n} \omega_i y_{i,m} \tag{3}$$

where ω_i is a survey weight assigned to unit i with n the number of respondents. From a combination of equations (2) and (3), and taking the expectation over the measurement error model (1), we would have

$$E\left(\hat{t}_y\right) = E\left(\sum_{i=1}^{n} \omega_i y_{i,m}\right) = \sum_{i=1}^{n} \omega_i u_{i,m} + \sum_{i=1}^{n} b_m \omega_i \partial_{i,m} + \sum_{i=1}^{n} \omega_i \partial_{i,m} E\left(\varepsilon_{i,m}\right) \tag{4}$$

with $\partial_{i,m} = 1$ if unit i responded through mode m, and zero otherwise. Since $E\left(\varepsilon_{i,m}\right) = 0$

$$E\left(\hat{t}_y\right) = E\left(\sum_{i=1}^{n} \omega_i y_{i,m}\right) = \sum_{i=1}^{n} \omega_i u_{i,m} + \sum_{i=1}^{n} \omega_i \partial_{i,m} b_m \tag{5}$$

stating that the expected total of the survey estimate for Y consists of the estimated true total of U, plus true total of b_m from data collected through mode m. Since b_m is an unobserved mode-dependent measurement bias, $\sum_{i=1}^{n} \omega_i \partial_{i,m} b_m$ in equation (5) indicates the existence of an unknown mode-dependent bias for estimation of t_y. According to Equation (5), there is an unknown measurement bias in sequential mixed-mode designs that might be adjusted by different estimators. Data obtained via a re-interview design or a sub-set of respondents to the first stage of a sequential mixed-mode survey provides necessary auxiliary information to adjust measurement bias in sequential mixed-mode surveys. Klausch et al [10] propose six different estimators and show that an inverse version of regression estimator (IREG) performs well under all considered scenarios. The idea of IREG is to use re-interview data to estimate the inverse slope of ordinary or generalized least squares linear regression of benchmark measurements y^{mb} on y^{mj} as follows [11]

$$y_i^{mj} = \hat{\beta}_0 + \hat{\beta}_1 y_i^{mb} \tag{6}$$

and estimate the measurement of target variable by applying the inverse of $\hat{\beta}_1$ in the following estimator, so-called inverse regression estimator

$$\hat{y}_{r_{mm}}^{ireg} = \frac{1}{(\hat{N}_{m_1} + \hat{N}_{m_2})} \left(\sum_{i=1}^{n_{mb}} d_i y_i^{mb} + \sum_{i=1}^{m_j} d_i \left(\hat{y}_{re}^{mb} - \frac{1}{\hat{\beta}_1} \left(\hat{y}_{re}^{mj} - y_i^{mj} \right) \right) \right) b, j = 1, 2; b \neq j \tag{7}$$

where $\hat{y}_{re}^{mj}$ and $\hat{y}_{re}^{mb}$ are the respondents means of focal and benchmark mode outcome in the re-interview and d_i denotes the design weight of the sample design. For a detailed presentation and discussion of the methods see Chapter 8.5 in [12]. However, for longitudinal studies with different modes at different time points, the effect of time on the respondents would make it difficult to estimate the pure mixed-mode effect especially for volatile classification variables such as the address for immigrants. The solution could be conducting the survey on parallel or separate samples to evaluate the time effect and mode effect separately.

In practice, Statistics Portugal has been using the available information of a sampling frame as a part of FNA (the dwellings national register database) at the time of COVID-19. The situation was considered as telephone numbers are linked to a sample drawn from a population register in FNA for the samples for CATI rotation-scheme surveys such as Labor Force Survey. In 2020, the Labour Force Survey (LFS) in Portugal as a mandatory survey for the member states within the EU was adjusted for undercover of the percentage of households with a listed landline telephone. As a result, the comparison of these surveys after and before COVID-19 shows the usefulness of the discussed methodologies. In 2021, the successful CAWI mode census by Statistics Portugal shows respondents tend to favor the web-based questionnaire to avoid the risk of COVID-19 infection with a face-to-face interview. It shows the potential change in the mode tendency by responders.

5 Conclusions

COVID-19 crisis led to new solutions on item classification for mixed-mode effects adjustment, such as applying mode calibration to population subgroups by categorical variables such as gender, regions, age groups, etc. Studies offer sequential mixed-mode design started with CAWI as the cheapest mode supported by an initial postal mail or telephone contact and possible cash incentive. With a lag, follow up the non-respondents with giving them a choice between CAPI and CATI according to their specific classification group and demographic information, such as education level, age, income, location, language, and marital status. It is fruitful to reduce the cost and increase the accuracy simultaneously.

This study showed that sample frames might need updates for necessary categorical information, which are based on choices made several years ago. Additionally, more research studies seem necessary for ethics concerns, privacy regulations, and standards for using categorical variables and classification information in social mixed-mode surveys and official statistics.

References

1. Ashofteh, A., Bravo, J. M.: A study on the quality of novel coronavirus (COVID-19) official datasets. Stat. J. IAOS, **36**(2), 291–301, (2020) doi: 10.3233/SJI-200674
2. Ashofteh, A., Bravo, J. M.: Data science training for official statistics: A new scientific paradigm of information and knowledge development in national statistical systems. Stat. J. IAOS, **37**(3), 771–789, (2021) doi: 10.3233/SJI-200674
3. Te Braak, P., Minnen, J., Glorieux, I.: The representativeness of online time use surveys. Effects of individual time use patterns and survey design on the timing of survey dropout. J. Off. Stat., **36**(4), 887–906, (2020)
4. Szymkowiak, M., Wilak, K.: Repeated weighting in mixed-mode censuses. Econ. Bus. Rev., **7**(1), 26–46, (2021)
5. Zax, M., Takahashi, S.: Cultural influences on response style: comparisons of Japanese and American college students. J. Soc. Psychol., **71**(1), 3–10, (1967)
6. Pfeffermann, D.: New important developments in small area estimation. Stat. Sci., 28(1), 40–68, (2013)
7. Toepoel, V., de Leeuw, E., Hox, J.: Single- and Mixed-Mode Survey Data Collection. SAGE Res. Methods Found, (2020) doi: 10.4135/9781526421036876933
8. Kim, S., Couper, M. P.: Feasibility and quality of a national RDD smartphone web survey: comparison with a cell phone CATI survey. Soc. Sci. Comput. Rev., **39**(6), 1218–1236, (2021)
9. Van Berkel, K., Van Der Doef, S., Schouten, B.: Implementing adaptive survey design with an application to the Dutch health survey. J. Off. Stat., **36**(3), 609–629, (2020) doi: 10.2478/jos-2020-0031
10. Klausch, T., Schouten, B., Buelens, B., van den Brakel, J.: Adjusting measurement bias in sequential mixed-mode surveys using re-interview data. J. Surv. Stat. Methodol., **5**(4), 409–432, (2017) doi: 10.1093/jssam/smx022
11. Särndal, C. E., Lundström, S.: Estimation in surveys with nonresponse. Estimation in surveys with nonresponse. John Wiley (2005) doi: 10.1002/0470011351
12. Schouten, B., Brakel, J. van den, Buelens, B., Giesen, D., Luiten, A., Meertens, V.: Mixed-Mode Official Surveys. Chapman and Hall/CRC (2021) doi: 10.1201/9780429461156

Clustering and Blockmodeling Temporal Networks – Two Indirect Approaches

Vladimir Batagelj

Abstract Two approaches to clustering and blockmodeling of temporal networks are presented: the first is based on an adaptation of the clustering of symbolic data described by modal values and the second is based on clustering with relational constraints. Different options for describing a temporal block model are discussed.

Keywords: social networks, network analysis, blockmodeling, symbolic data analysis, clustering with relational constraints

1 Temporal Networks

Temporal networks described by *temporal quantities* (TQs) were introduced in the paper [2]. We get a *temporal network* $N_T = (V, L, T, P, W)$ by attaching the *time* T to an ordinary network, where V is the set of nodes, L is the set of links, P is the set of node properties, W is the set of link weights, and $T = [T_{min}, T_{max})$ is a linearly ordered set of time points $t \in T$ which are usually integers or reals.

In a temporal network nodes/links activity/presence, nodes properties, and links weights can change through time. These changes are described with TQs. A TQ is described by a sequence $a = [(s_r, f_r, v_r) : r = 1, 2, \ldots, k]$ where $[s_r, f_r)$ determines a time interval and v_r is the value of the TQ a on this interval. The set $T_a = \bigcup_r [s_r, f_r)$ is called the *activity set* of a. For $t \notin T_a$ its value is *undefined*, $a(t) = ⌘$.

Assuming that for every $x \in \mathbb{R} \cup \{⌘\} : x + ⌘ = ⌘ + x = x$ and $x \cdot ⌘ = ⌘ \cdot x = ⌘$ we can extend the addition and multiplication to TQs

Vladimir Batagelj (✉)
IMFM, Jadranska 19, 1000 Ljubljana, Slovenia & IAM UP, Muzejski trg 2, 6000 Koper, Slovenia & HSE, 11 Pokrovsky Bulvar, 101000 Moscow, Russian Federation,
e-mail: `vladimir.batagelj@fmf.uni-lj.si`

© The Author(s) 2023
P. Brito et al. (eds.), *Classification and Data Science in the Digital Age*,
Studies in Classification, Data Analysis, and Knowledge Organization,
https://doi.org/10.1007/978-3-031-09034-9_8

63

$$(a + b)(t) = a(t) + b(t) \quad \text{and} \quad T_{a+b} = T_a \cup T_b$$
$$(a \cdot b)(t) = a(t) \cdot b(t) \quad \text{and} \quad T_{a \cdot b} = T_a \cap T_b$$

Let $T_V(v) \subseteq \mathcal{T}$, $T_V \in \mathcal{P}$, be the activity set for a node $v \in \mathcal{V}$ and $T_L(\ell) \subseteq \mathcal{T}$, $T_L \in \mathcal{W}$, the activity set for a link $\ell \in \mathcal{L}$. The following *consistency condition* must be fulfilled for activity sets: If a link $\ell(u, v)$ is active at the time point t then its end-nodes u and v should be active at the time point t : $T_L(\ell(u, v)) \subseteq T_V(u) \cap T_V(v)$.

In the following we will need

1. *Total*: $\text{total}(a) = \sum_i (f_i - s_i) \cdot v_i$
2. *Average*: $\text{average}(a) = \dfrac{\text{total}(a)}{|T_a|}$ where $|T_a| = \sum_i (f_i - s_i)$
3. *Maximum*: $\max(a) = \max_i v_i$

To support the computations with TQs we developed in Python the libraries TQ and Nets, see `https://github.com/bavla/TQ` .

2 Traditional (Generalized) Blockmodeling Scheme

A *blockmodel* (BM) [11] consists of structures obtained by identifying all units from the same cluster of the clustering / *partition* $\mathbf{C} = \{C_i\}$, $\pi(v) = i \Leftrightarrow v \in C_i$. Each pair of clusters (C_i, C_j) determines a block consisting of links linking C_i to C_j. For an exact definition of a blockmodel we have to be precise also about which blocks produce an arc in the *reduced graph* on classes and which do not, what is the *weight* of this arc, and in the case of generalized BM, of what *type*. The reduced graph can be represented by relational matrix, called also *image matrix*.

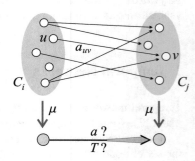

Fig. 1 Blockmodel.

To develop a BM method we specify a criterion function $P(\mu)$ measuring the "error" of the BM μ. We can introduce additional knowledge by constraining the partitions to a set Φ of feasible partitions. We are searching for a partition $\pi^* \in \Phi$ such that the corresponding BM μ^* minimizes the criterion function $P(\mu)$.

3 BM of Temporal Networks

For an early attempt of temporal network BM see [2, 5]. To the traditional BM scheme we add the time dimension. We assume that the network is described using temporal quantities [2] for nodes/links activity/presence, and some nodes properties and links weights. Then also the BM partition π is described for each node v with a temporal quantity $\pi(v, t)$: $\pi(v, t) = i$ means that in time t node v belongs to cluster i. The structure and activity of clusters $C_i(t) = \{v : \pi(v, t) = i\}$ can change through time, but they preserve their identity.

For the BM μ the clusters are mapped into BM nodes $\mu : C_i \to [i]$. To determine the BM we still have to specify how the links from C_i to C_j are represented in the BM – in general, for the model arc $([i], [j])$, we have to specify two TQs: its *weight* a_{ij} and, in the case of generalized BM, its *type* τ_{ij}. The weight can be an object of a different type than the weights of the block links in the original temporal network.

We assume that in a temporal network $\mathcal{N} = (\mathcal{V}, \mathcal{L}, \mathcal{T}, \mathcal{P}, \mathcal{W})$ the links weight is described by a TQ $w \in \mathcal{W}$. In the following years we intend to develop BM methods case by case.

1. constant partition – nodes stay in the same cluster all the time:

    a. indirect approach based on clustering of TQs: $p(v) = \sum_{u \in N(v)} w(v, u)$, hierarchical clustering and leaders;
    b. indirect approach by conversion to the *clustering with relation constraint* (CRC);
    c. direct approach by (local) optimization of the criterion function P over Φ

2. dynamic partition – nodes can move between clusters through time. The details are still to be elaborated.

In this paper, we present approaches for cases 1.a and 1.b.

In the literature there exist other approaches to BM of temporal networks. A recent overview is available in the book [12].

3.1 Adapted Symbolic Clustering Methods

In [8] we adapted traditional leaders [13, 10] and agglomerative hierarchical [14, 1] clustering methods for clustering of modal-valued symbolic data. They can be almost directly applied for clustering units described by variables that have for their values temporal quantities.

For a unit X_i, each variable V_j is described with a size h_{ij} and a temporal quantity $\mathbf{x}_{ij}$, $X_{ij} = (h_{ij}, \mathbf{x}_{ij})$. In our algorithms we use *normalized* values of temporal variables $V' = (h, \mathbf{p})$ where

$$\mathbf{p} = [(s_r, f_r, p_r) : r = 1, 2, \ldots, k] \qquad \text{and} \qquad p_r = \frac{v_r}{h}$$

In the case, when $h = \text{total}(\mathbf{x})$, the normalized TQ $\mathbf{p}$ is essentially a probability distribution.

Both methods create cluster representatives that are represented in the same way.

3.2 Clustering of Temporal Network and CRC

To use the CRC in the construction of a nodes partition we have to define a dissimilarity measure $d(u,v)$ (or a similarity $s(u,v)$) between nodes. An obvious solution is $s(u,v) = f(w(u,v))$, for example

1. *Total activity*: $s_1(u,v) = \text{total}(w(u,v))$
2. *Average activity*: $s_2(u,v) = \text{average}(w(u,v))$
3. *Maximal activity*: $s_3(u,v) = \max(w(u,v))$

We can transform a similarity $s(u,v)$ into a dissimilarity by $d(u,v) = \frac{1}{s(u,v)}$ or $d(u,v) = S - s(u,v)$ where $S > \max_{u,v} s(u,v)$. In this way, we transformed the temporal network partitioning problem into a clustering with relational constraints problem [6, 360–369]. It can be efficiently solved also for large sparse networks.

3.3 Block Model

Having the partition π, to produce a BM we have to specify the values on its links. There are different options for model links weights $a(([i],[j]))$.

1. *Temporal quantities*: $a(([i],[j])) = \text{activity}(C_i, C_j) = \sum_{u \in C_i, v \in C_j} w(u,v)$, for $i \neq j$, and $a(([i],[i])) = \frac{1}{2}\text{activity}(C_i, C_i)$.
2. *Total intensities*: $a_t(([i],[j])) = \text{total}(a(([i],[j])))$.
3. *Geometric average intensities*: $a_g(([i],[j])) = \dfrac{a_t(([i],[j]))}{\sqrt{|C_i| \cdot |C_j|}}$.

4 Example: September 11th Reuters Terror News

The *Reuters Terror News* network was obtained from the CRA (Centering Resonance Analysis) networks produced by Steve Corman and Kevin Dooley at Arizona State University. The network is based on all the stories released during 66 consecutive days by the news agency Reuters concerning the September 11 attack on the U.S., beginning at 9:00 AM EST 9/11/01.

The nodes, $n = 13332$, of this network are important words (terms). For a given day, there is an edge between two words iff they appear in the same utterance (for details see the paper [9]). The network has $m = 243447$ edges. The weight of an edge is its daily frequency. There are no loops in the network. The network Terror News is undirected – so will be also its BM.

The Reuters Terror News network was used as a case network for the Viszards visualization session on the Sunbelt XXII International Sunbelt Social Network Conference, New Orleans, USA, 13-17. February 2002. It is available at `http://vlado.fmf.uni-lj.si/pub/networks/data/CRA/terror.htm` .

We transformed the Pajek version of the network into NetsJSON format used in libraries TQ and Nets. For a temporal description of each node/word for clustering we took its activity (sum of all TQs on edges adjacent to a given node v)

$$\text{act}(v) = \sum_{u \in N(v)} w(v : u).$$

Our leaders' and hierarchical clustering methods are compatible – they are based on the same clustering error criterion function. Usually, the leaders' method is used to reduce a large clustering problem to up to some hundred units. With hierarchical clustering of the leaders of the obtained clusters, we afterward determine the "right" number of clusters and their representatives.

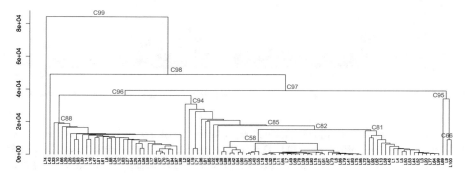

Fig. 2 Hierarchical clustering of 100 leaders in Terror News.

To cluster all 13332 words (nodes) in Terror News we used the adapted leaders' method searching for 100 clusters. We continued with the hierarchical clustering of the obtained 100 leaders. The result is presented in the dendrogram in Figure 2.

Fig. 3 Word clouds for clusters $C58$ and $C81$.

To get an insight into the content of a selected cluster we draw the corresponding word cloud based on the cluster's leader. In Figure 3 the word clouds for clusters $C58$ and $C81$ ($|C58| = 1396, |C81| = 2226$) are presented.

We can also compare the activities of pairs of clusters by considering the overlap of p-components (probability distributions) of their leaders. In Figure 4, we compare cluster $C58$ with cluster $C81$, and cluster $L96$ with cluster $C66$. In the right diagram some values are outside the display area: $L96[15] = 0.3524$, $C66[4] = 0.1961$, $C66[5] = 0.2917$.

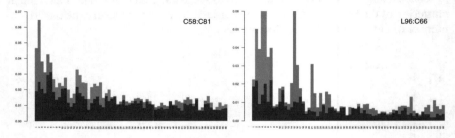

Fig. 4 Comparing activities of clusters (blue – first cluster, red – second cluster, violet – overlap).

We decided to consider in the BM the clustering of Terror News into 5 clusters $\mathbf{C} = \{C94, C88, C95, L43, L74\}$. The split of cluster C95 gives clusters of sizes 325 and 629 (for sizes, see the right side of Figure 5). Both clusters C94 and C88 have a chaining pattern at their top levels.

Because of large differences in the cluster sizes, it is difficult to interpret the total intensities image matrix. An overall insight into the BM structure we get from the geometric average intensities image matrix (right side) and the corresponding BM network (cut level 0.3), left side of Figure 5.

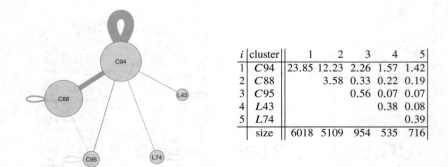

i	cluster	1	2	3	4	5
1	$C94$	23.85	12.23	2.26	1.57	1.42
2	$C88$		3.58	0.33	0.22	0.19
3	$C95$			0.56	0.07	0.07
4	$L43$				0.38	0.08
5	$L74$					0.39
	size	6018	5109	954	535	716

Fig. 5 Block model and image matrix.

A more detailed BM is presented by the activities (*p*-components) image matrix in Figure 6.

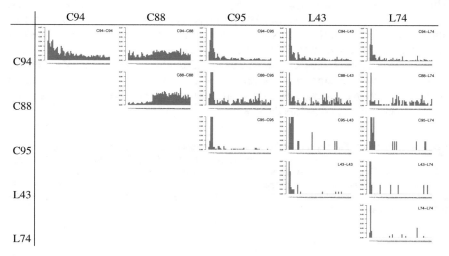

Fig. 6 BM represented as *p*-components of temporal activities of links between pairs of clusters.

A more compact representation of a temporal BM is a heatmap display of this matrix in Figure 7. Because of some relatively very large values, it turns out that the display of the matrix with logarithmic values provides much more information.

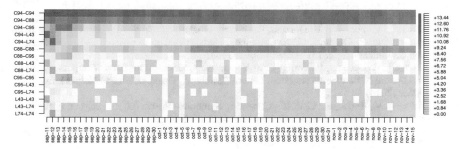

Fig. 7 BM heatmap with log2 values.

To the Terror News network, we applied also the clustering with relational constraints approach. Because of the limited space available for each paper, we can not present it here. A description of the analysis with the corresponding code is available at `https://github.com/bavla/TQ/wiki/BMRC` .

5 Conclusions

The presented research is a work in progress. It only deals with the two simplest cases of temporal blockmodeling. We provided some answers to the problem of normalization of model weights TQs when comparing them and some ways to present/display the temporal BMs.

We used different tools (R, Python, and Pajek) to obtain the results. We intend to provide the software support in a single tool – probably in Julia. We also intend to create a collection of interesting and well-documented temporal networks for testing and demonstrating the developed software.

Acknowledgements The paper contains an elaborated version of ideas presented in my talks at the XXXX Sunbelt Social Networks Conference (on Zoom), July 13-17, 2020 and at the EUSN 2021 – 5th European Conference on Social Networks, Naples (on Zoom), September 6-10, 2021.

This work is supported in part by the Slovenian Research Agency (research program P1-0294 and research projects J1-9187, J1-2481, and J5-2557), and prepared within the framework of the HSE University Basic Research Program.

References

1. Anderberg, M.R.: Cluster Analysis for Applications. Academic Press, New York (1973)
2. Batagelj, V., Praprotnik, S.: An algebraic approach to temporal network analysis based on temporal quantities. Soc. Netw. Anal. Min. **6**(1), 1–22 (2016)
3. Batagelj, V., Ferligoj, A.: Clustering relational data. In: Gaul, W., Opitz, O., Schader, M. (Eds.) Data Analysis / Scientific Modeling and Practical Application, pp. 3–15. Springer (2000)
4. Batagelj, V.: Generalized Ward and related clustering problems. In: Bock H.-H. (ed) Classification and Related Methods of Data Analysis, pp. 67–74. North-Holland, Amsterdam (1988)
5. Batagelj, V., Ferligoj, A., Doreian, P.: Indirect blockmodeling of 3-way networks. In: Brito, P., Bertrand, P., Cucumel, G., de Carvalho, F. (eds.) Selected Contributions in Data Analysis and Classification, pp. 151–159. Springer (2007)
6. Batagelj, V., Doreian, P., Ferligoj, A., Kejžar, N.: Understanding Large Temporal Networks and Spatial Networks: Exploration, Pattern Searching, Visualization and Network Evolution. Wiley (2014)
7. Batagelj, V., Kejžar, N.: Clamix – Clustering Symbolic Objects (2010) Program in R https://r-forge.r-project.org/projects/clamix/
8. Kejžar, N., Korenjak-Černe, S., Batagelj, V.: Clustering of modal-valued symbolic data. Adv. Data Anal. Classif. **15**, pp. 513—541 (2021)
9. Corman, S. R., Kuhn, T., McPhee, R. D., Dooley, K. J.: Studying complex discursive systems: Centering resonance analysis of communication. Hum. Commun. Res. **28**(2), 157–206 (2002)
10. Diday, E.: Optimisation en Classification Automatique. Tome 1.,2. INRIA, Rocquencourt (in French) (1979)
11. Doreian, P., Batagelj, V., Ferligoj, A.: Generalized Blockmodeling. Structural Analysis in the Social Sciences. Cambridge University Press (2005)
12. Doreian, P., Batagelj, V., Ferligoj, A. (Eds.) Advances in Network Clustering and Blockmodeling. Wiley (2020)
13. Hartigan, J. A.: Clustering Algorithms. Wiley-Interscience, New York (1975)
14. Ward, J. H.: Hierarchical grouping to optimize an objective function. J. Am. Stat. Asso.c **58**, 236–244 (1963)

Latent Block Regression Model

Rafika Boutalbi, Lazhar Labiod, and Mohamed Nadif

Abstract When dealing with high dimensional sparse data, such as in recommender systems, co-clustering turns out to be more beneficial than one-sided clustering, even if one is interested in clustering along one dimension only. Thereby, co-clusterwise is a natural extension of clusterwise. Unfortunately, all of the existing approaches do not consider covariates on both dimensions of a data matrix. In this paper, we propose a *Latent Block Regression Model* (LBRM) overcoming this limit. For inference, we propose an algorithm performing simultaneously co-clustering and regression where a linear regression model characterizes each block. Placing the estimate of the model parameters under the maximum likelihood approach, we derive a Variational Expectation-Maximization (VEM) algorithm for estimating the model's parameters. The finality of the proposed VEM-LBRM is illustrated through simulated datasets.

Keywords: co-clustering, clusterwise, tensor, data mining

1 Introduction

The *cluster-wise* linear regression algorithm CLR (or Latent Regression Model) is a finite mixture of regressions and one of the most commonly used methods for simultaneous learning and clustering [14, 5]. It aims to find clusters of entities to minimize the overall sum of squared errors from regressions performed over these clusters. Specifically, $\mathbf{X} = [x_{ij}] \in \mathbb{R}^{n \times v}$ is the covariate matrix and $\mathbf{Y} \in \mathbb{R}^{n \times 1}$ the response vector. The *cluster-wise* method aims to find g clusters $C_1, \ldots, C_g$ and regression coefficients $\boldsymbol{\beta}^{(k)} \in \mathbb{R}^{d \times 1}$ by minimizing the following objective function $\sum_{k=1}^{g} \sum_{i \in C_k} (y_i - \sum_{j=1}^{v} \beta_j^{(k)} x_{ij} + b_k)^2$ where:

- y_i is the value of the dependent variable for subject/observation i defined by $\mathbf{x}_i = (x_{i1}, \ldots, x_{id})$,
- x_{ij} is the value of the j-th independent variable for subject/observation i,
- $\beta_j^{(k)}$ is the j-th multiple regression coefficient and b_k is the *intercept*.

Rafika Boutalbi (✉)
Institute for Parallel and Distributed Systems, Analytic Computing, University of Stuttgart, Germany, e-mail: rafika.boutalbi@ipvs.uni-stuttgart.de

Lazhar Labiod · Mohamed Nadif
Centre Borelli UMR 9010, Université Paris Cité, France,
e-mail: lazhar.labiod@u-paris.fr;mohamed.nadif@u-paris.fr

© The Author(s) 2023
P. Brito et al. (eds.), *Classification and Data Science in the Digital Age*,
Studies in Classification, Data Analysis, and Knowledge Organization,
https://doi.org/10.1007/978-3-031-09034-9_9

Various adjustments have been made to this model to improve its performance in terms of clustering and prediction. In our contribution, we propose to embed the co-clustering in the model.

Co-clustering is a simultaneous clustering of both dimensions of a data matrix that has proven to be more beneficial than traditional one-sided clustering, especially when dealing with sparse data. When dealing with high dimensional data sparse or not, co-clustering turns out to be more valuable than one-sided clustering [1, 13], even if one is interested in clustering along one dimension only. In [4] the authors proposed the SCOAL approach (Simultaneous Co-clustering and Learning model), leading to co-clustering and prediction for binary data; they generalized the model to continuous data. However, this model does not take into account the sparsity of data in the sense that it does not lead to homogeneous blocks. The obtained results in terms of *Mean Square Error* (MSE) are good, but in terms of co-clustering (homogeneity of co-clusters), no analysis has been presented. This model is also related to the soft PDLF (Predictive Discrete Latent Factor) model [2], where the value of response y_{ij}'s in each co-cluster is modeled as a sum $\beta^T x_{ij} + \delta_{k\ell}$ where β is a global regression model. In contrast, $\delta_{k\ell}$ is a co-cluster specific offset. More recently, in [17] the authors proposed an algorithm taking into account only row covariates information to realize co-clustering and regression simultaneously. To this end, the authors are based on the latent block models [8]. In our contribution, we propose to rely also on this model but considering row and column covariates.

The proposed Latent Block Regression Model (LBRM) is an extension of finite mixtures of regression models where the co-clustering is embedded. It allows us to deal with co-clustering and regression simultaneously while taking into account covariates. To estimate the parameters we rely on a *Variational* Expectation-Maximization algorithm [7] referred to as VEM–LBRM.

2 From Clusterwise Regression to Co-clusterwise Regression

2.1 Latent Block Model (LBM)

Given an $n \times d$ data matrix $\mathbf{X} = (x_{ij}, i \in I = \{1, \ldots, n\}; j \in J = \{1, \ldots, d\})$. It is assumed that there exists a partition on I and a partition on J. A partition of $I \times J$ into $g \times m$ blocks will be represented by a pair of partitions $(\mathbf{z}, \mathbf{w})$. The k-th row cluster corresponds to the set of rows i such that $z_{ik} = 1$ and $z_{ik'} = 0 \; \forall k' \neq k$. Thereby, the partition represented by $\mathbf{z}$ can be also represented by a matrix of elements in $\{0, 1\}^g$ satisfying $\sum_{k=1}^{g} z_{ik} = 1$. Similarly, the ℓ-th column cluster corresponds to the set of columns j and the partition $\mathbf{w}$ can be represented by a matrix of elements in $\{0, 1\}^m$ satisfying $\sum_{\ell=1}^{m} w_{j\ell} = 1$.

Considering the Latent Block Model (LBM) [6], it is assumed that each element x_{ij} of the $k\ell$th block is generated according to a parameterized probability density function (pdf) $f(x_{ij}; \alpha_{k\ell})$. Furthermore, in the LBM the univariate random variables x_{ij} are assumed to be conditionally independent given $(\mathbf{z}, \mathbf{w})$. Thereby, the conditional pdf of $\mathbf{X}$ can be expressed as $P(z_{ik} = 1, w_{j\ell} = 1|\mathbf{X}) =$

$P(z_{ik} = 1|\mathbf{X})P(w_{j\ell} = 1|\mathbf{X})$. From this hypothesis, we then consider the latent block model where the two sets I and J are considered as random samples and the row, and column labels become latent variables. Therefore, the parameter of the latent block model is $\boldsymbol{\Theta} = (\boldsymbol{\pi}, \boldsymbol{\rho}, \boldsymbol{\alpha})$, with $\boldsymbol{\pi} = (\pi_1, \ldots, \pi_g)$ and $\boldsymbol{\rho} = (\rho_1, \ldots, \rho_m)$ where $(\pi_k = P(z_{ik} = 1), k = 1, \ldots, g)$, $(\rho_\ell = P(w_{j\ell} = 1), \ell = 1, \ldots, m)$ are the mixing proportions and $\boldsymbol{\alpha} = (\alpha_{k\ell}; k = 1, \ldots g, \ell = 1, \ldots, m)$ where $\alpha_{k\ell}$ is the parameter of the distribution of block $k\ell$. Considering that the complete data are the vector $(\mathbf{X}, \mathbf{z}, \mathbf{w})$, i.e, we assume that the latent variable $\mathbf{z}$ and $\mathbf{w}$ are known, the resulting complete data log-likelihood of the latent block model $L_C(\mathbf{X}, \mathbf{z}, \mathbf{w}, \boldsymbol{\Theta}) = \log f(\mathbf{X}, \mathbf{z}, \mathbf{w}; \boldsymbol{\Theta})$ can be written as follows

$$\sum_{k=1}^{g} z_k \log \pi_k + \sum_{\ell=1}^{m} w_\ell \log \rho_\ell + \sum_{i=1}^{n} \sum_{j=1}^{d} \sum_{k=1}^{g} \sum_{\ell=1}^{m} z_{ik} w_{j\ell} \log \phi_{k\ell}(x_{ij}; \alpha_{k\ell}).$$

where the π_k's and ρ_ℓ's denote the proportions of row and columns clusters respectively; see for instance [8]. Note that the complete-data log-likelihood breaks into three terms: the first one depends on proportions of row clusters, the second on proportions of column clusters and the third on the pdf of each block or co-cluster. The objective is then to maximize the function $L_C(\mathbf{z}, \mathbf{w}, \boldsymbol{\Theta})$.

2.2 Latent Block Regression Model (LBRM)

For co-clustering of continuous data, the Gaussian latent block model can be used. For instance, note that it is easy to show that the minimization of the well-known criterion of $||\mathbf{X} - \mathbf{z}\boldsymbol{\mu}\mathbf{w}^T||^2 = \sum_{k=1}^{g} \sum_{\ell=1}^{m} \sum_{i|z_{ik}=1} \sum_{j|w_{j\ell}=1} (x_{ij} - \mu_{k\ell})^2$ where $\mathbf{z} \in \{0, 1\}^{n \times g}$, $\mathbf{w} \in \{0, 1\}^{d \times m}$ and $\boldsymbol{\mu} \in \mathbb{R}^{g \times m}$ is associated to Latent block Gaussian model whith $\alpha_{k\ell} = (\mu_{k\ell}, \sigma_{k\ell}^2)$, the proportions of row clusters and column clusters are equal and in addition the variances of blocks are identical [9]. Note that 1) the characteristic of the latent block model is that the rows and the columns are treated symmetrically 2) the estimation of the parameters requires a variational approximation [7, 17]. In the sequel, we see how can we integrate a regression model. Hereafter, we propose a novel Latent Block Regression model for co-clustering and learning simultaneously. The model considers the response matrix $\mathbf{Y} = [y_{ij}] \in \mathbb{R}^{n \times d}$ and the covariate tensor $\mathcal{X} = [1, \mathbf{x}_{ij}] \in \mathbb{R}^{n \times d \times v}$ where n is the number of rows, d the number of columns, and v the number of covariates. Figure 1 presents data structure for the proposed model LBRM.

In the following we propose the integration of mixture of regression [5] per block in the Latent Block model (LBM) considering the distribution $\Phi(y_{ij}|\mathbf{x}_{ij}; \lambda_{k\ell})$. We assume in the following the normality of Φ,

$$\Phi(y_{ij}|\mathbf{x}_{ij}; \lambda_{k\ell}) = p(y_{i,j}|\mathbf{x}_{ij}, \boldsymbol{\beta}_{k\ell}, \sigma_{k\ell}) = (2\pi\sigma_{k\ell}^2)^{-0.5} \exp\left\{-\frac{1}{2\sigma_{k\ell}^2}(y_{ij} - \boldsymbol{\beta}_{k\ell}^\top \mathbf{x}_{ij})^2\right\}$$

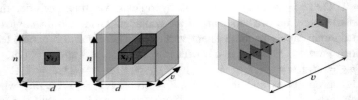

Fig. 1 Data representation for proposed model.

With the LBRM model, the parameter Ω is composed of row and column proportions π, ρ respectively, $\beta = \{\beta_{11}, \ldots, \beta_{gm}\}$ with $\beta_{k\ell}^\top = (\beta_{k\ell}^0, \beta_{k\ell}^1, \ldots, \beta_{k\ell}^v)$ where $\beta_{k\ell}^0$ represents the intercept of regression and $\sigma = \{\sigma_{11}, \ldots, \sigma_{gm}\}$. The classification log-likelihood can be written:

$$\sum_{i,k} z_{ik} \log \pi_k + \sum_{j,\ell} w_{j\ell} \log \rho_\ell - \frac{1}{2} \sum_{k,\ell} z_{.k} w_{.\ell} \log(\sigma_{k\ell}^2) - \frac{1}{2\sigma_{k\ell}^2} \sum_{i,j,k,\ell} z_{ik} w_{j\ell} (y_{ij} - \beta_{k\ell}^\top \mathbf{x}_{ij})^2$$

with $z_{.k} = \sum_i z_{ik}$ et $w_{.\ell} = \sum_j w_{j\ell}$.

3 Variational EM Algorithm

To estimate Ω, the EM algorithm [3] is a candidate for this task. It maximizes the log-likelihood $f(\mathcal{X}, \Omega)$ w.r. to Ω iteratively by maximizing the conditional expectation of the complete data log-likelihood $L_C(\mathbf{z}, \mathbf{w}; \Omega)$ w.r. to Ω, given a previous current estimate $\Omega^{(c)}$ and the observed data $\mathbf{x}$. Unfortunately, difficulties arise owing to the dependence structure among the variables x_{ij} of the model. To solve this problem an approximation using the [12] interpretation of the EM algorithm can be proposed; see, e.g., [7, 8]. Hence, the aim is to maximize the following lower bound of the log-likelihood criterion: $F_C(\tilde{\mathbf{z}}, \tilde{\mathbf{w}}; \Omega) = L_C(\tilde{\mathbf{z}}, \tilde{\mathbf{w}}, \Omega) + H(\tilde{\mathbf{z}}) + H(\tilde{\mathbf{w}})$ where $H(\tilde{\mathbf{z}}) = -\sum_{i,k} \tilde{z}_{ik} \log \tilde{z}_{ik}$ with $\tilde{z}_{ik} = P(z_{ik} = 1|\mathcal{X})$, $H(\tilde{\mathbf{w}}) = -\sum_{j,\ell} \tilde{w}_{j\ell} \log \tilde{w}_{j\ell}$ with $\tilde{w}_{j\ell} = P(w_{j\ell} = 1|\mathcal{X})$, and $L_C(\tilde{\mathbf{z}}, \tilde{\mathbf{w}}; \tilde{\Omega})$ is the fuzzy complete data log-likelihood (up to a constant). $L_C(\tilde{\mathbf{z}}, \tilde{\mathbf{w}}; \Omega)$ is given by

$$L_C(\mathbf{z}, \mathbf{w}, \Omega) = \sum_{i,k} \tilde{z}_{ik} \log \pi_k + \sum_{j,\ell} \tilde{w}_{j\ell} \log \rho_\ell - \frac{1}{2} \sum_{k,\ell} \tilde{z}_{.k} \tilde{w}_{.\ell} \log(\sigma_{k\ell}^2)$$
$$- \frac{1}{2\sigma_{k\ell}^2} \sum_{i,j,k,\ell} \tilde{z}_{ik} \tilde{w}_{j\ell} (y_{ij} - \beta_{k\ell}^\top \mathbf{x}_{ij})^2$$

The maximization of $F_C(\tilde{\mathbf{z}}, \tilde{\mathbf{w}}, \Omega)$ can be reached by realizing the three following optimization: update $\tilde{\mathbf{z}}$ by $\mathrm{argmax}_{\tilde{\mathbf{z}}} F_C(\tilde{\mathbf{z}}, \tilde{\mathbf{w}}, \Omega)$, update $\tilde{\mathbf{w}}$ by $\mathrm{argmax}_{\tilde{\mathbf{w}}} F_C(\tilde{\mathbf{z}}, \tilde{\mathbf{w}}, \Omega)$, and update Ω by $\mathrm{argmax}_{\Omega} F_C(\tilde{\mathbf{z}}, \tilde{\mathbf{w}}, \Omega)$. In what follows, we detail the Expectation (E) and Maximization (M) step of the Variational EM algorithm for tensor data.

E-step. It consists in computing, for all i, k, j, ℓ the posterior probabilities $\tilde{z}_{ik}$ and $\tilde{w}_{j\ell}$ maximizing $F_C(\tilde{z}, \tilde{w}, \Omega)$ given the estimated parameters $\Omega_{k\ell}$. It is easy to show that, the posterior probability $\tilde{z}_{ik}$ maximizing $F_C(\tilde{z}, \tilde{w}, \Omega)$ is given by: $\tilde{z}_{ik} \propto \pi_k \exp\left(\sum_{j,\ell} \tilde{w}_{j\ell} \log\left(p(y_{ij}|\mathbf{x}_{ij}, \beta_{k\ell}, \sigma_{k\ell})\right)\right)$. In the same manner, the posterior probability $\tilde{w}_{j\ell}$ is given by: $\tilde{w}_{j\ell} \propto \rho_\ell \exp\left(\sum_{i,k} \tilde{z}_{ik} \log\left(p(y_{ij}|\mathbf{x}_{ij}, \beta_{k\ell}, \sigma_{k\ell})\right)\right)$

M-step. Given the previously computed posterior probabilities $\tilde{z}$ and $\tilde{w}$, the M-step consists in updating , $\forall k, \ell$, the parameters of the model $\pi_k, \rho_\ell, \mu_{k\ell}$ and $\lambda_{k\ell}$ maximizing $F_C(\tilde{z}, \tilde{w}, \Omega)$. Using the computed quantities from step E, the maximization step (M-step) involves the following closed-form updates.

- Taking into account the constraints $\sum_k \pi_k = 1$ and $\sum_\ell \rho_\ell = 1$, it is easy to show that $\pi_k = \frac{\sum_i \tilde{z}_{ik}}{n} = \frac{\tilde{z}_{.k}}{n}$ and $\rho_\ell = \frac{\sum_j \tilde{w}_{j\ell}}{d} = \frac{\tilde{w}_{.\ell}}{d}$.
- The update of $\lambda_{k\ell}$ which is formed by $(\beta_{k\ell}, \sigma_{k\ell})$ where can be given by simple derivates of $F_C(\tilde{z}, \tilde{w}, \Omega)$ with respect to $\beta_{k\ell}$ and $\sigma_{k\ell}$ respectively. This leads to

$$\beta_{k\ell} = \left(\sum_{i,j} \tilde{z}_{ik} \tilde{w}_{j\ell} y_{ij} \mathbf{x}_{ij}\right)\left(\sum_{i,j} \tilde{z}_{ik} \tilde{w}_{j\ell} \mathbf{x}_{ij} \mathbf{x}_{ij}^\top\right)^{-1} , \sigma_{k\ell}^2 = \frac{\sum_{i,j} \tilde{z}_{ik} \tilde{w}_{j\ell} (y_{ij} - \beta_{k\ell}^\top \mathbf{x}_{ij})^2}{\sum_{i,j} \tilde{z}_{ik} \tilde{w}_{j\ell}}.$$

The proposed algorithm for tensor data referred to as `VEM-LBRM` alternates the two previously described steps Expectation-Maximization. At the convergence, a hard co-clustering is deduced from the posterior probabilities.

4 Experimental Results

First, we evaluate the proposed `VEM-LBRM` on three synthetic datasets in terms of co-clustering and regression. We compare `VEM-LBRM` with some clustering and regression methods namely `Global model` which is a single multiple linear regression model performed on all observations, `K-means`, `Clusterwise`, `Co-clustering` and `SCOAL`. We retain two widely used measures to assess the quality of clustering, namely the Normalized Mutual Information (NMI) [16] and the Adjusted Rand Index (ARI) [15]. Intuitively, NMI quantifies how much the estimated clustering is informative about the true clustering. The ARI metric is related to the clustering accuracy and measures the degree of agreement between an estimated clustering and a reference clustering. Both NMI and ARI are equal to 1 if the resulting clustering is identical to the true one. On the other hand, we use RMSE (Root MSE) and MAE (Mean Absolute Error) metrics to evaluate the precision of prediction while RMSE is a loss function which is suitable for Gaussian noises when MAE uses the absolute value which is less sensitive to extreme values.

We generated tensor data $\mathbf{X}$ with size $200 \times 200 \times 2$ according to Gaussian model per block. In the simulation study, we considered three scenarios by varying the regression parameters — the examples have blocks with different regression collinearity and different co-clusters structure complexity. The parameters for each example are reported in Tables 1. In Figures 2 and 3 are depicted the true regression planes and the true simulated response matrix $\mathbf{Y}$.

Table 1 Parameters generation for examples.

Dataset	Example 1		Example 2		Example 3	
	$\pi = [0.35, 0.35, 0.3]$, $\rho = [0.55, 0.45]$					
σ	$\sigma = 5$		$\sigma = 7$		$\sigma = 7$	
Σ	$\Sigma = \begin{bmatrix} 1 & 0 \\ 0 & 1 \end{bmatrix}$		$\Sigma = \begin{bmatrix} 2 & 0.3 \\ 0.3 & 2 \end{bmatrix}$		$\Sigma = \begin{bmatrix} 1 & 2 \\ 2 & 1 \end{bmatrix}$	
Co-clusters	$\beta_{k\ell}$	$\mu_{k\ell}$	$\beta_{k\ell}$	$\mu_{k\ell}$	$\beta_{k\ell}$	$\mu_{k\ell}$
Cluster (1,1)	[1, -10, 1]	[5,20]	[1, -10, 1]	[5,20]	[1, -10, 1]	[5,20]
Cluster (1,2)	[10, 4, 13]	[5,10]	[1, -10, 1]	[5,10]	[1, -10, 1]	[5,10]
Cluster (2,1)	[3, 20, -2]	[10,20]	[1, -10, 1]	[10,20]	[1, -10, 1]	[5,30]
Cluster (2,2)	[-5, -2, -6]	[10,10]	[7, 5, -10]	[10,10]	[7, 5, -10]	[20,10]
Cluster (3,1)	[-10, 20, 10]	[20,20]	[7, 5, -10]	[20,20]	[7, 5, -10]	[20,20]
Cluster (3,2)	[7, 5, -10]	[20,10]	[7, 5, -10]	[20,10]	[7, 5, -10]	[20,30]

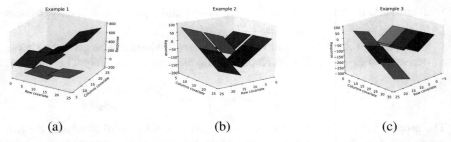

(a) (b) (c)

Fig. 2 Synthetic data: True regression plans according to the chosen parameters.

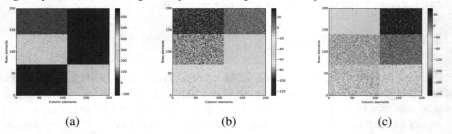

(a) (b) (c)

Fig. 3 Synthetic data: True co-clustering according to the chosen parameters.

In our illustrations, we consider co-clustering and regression challenges. All metrics concerning rows and columns are computed by averaging on ten random training, and testing data split using an 80% vs. 20% of training and validation data. Thereby, we compare VEM-LBRM with Global model (which is a multiple linear regression), K-means, Clusterwise by reshaping the tensor to matrix with size $N \times v$ where $N = n \times d$. On the other hand, the VEM algorithm for co-clustering is applied on response matrix $\mathbf{Y}$. Furthermore, for clustering algorithms, the RMSE, MAE, and R-squared are computed by applying linear regression on each obtained co-cluster. In Table, 2 are reported the performances for all algorithms. The missing values represent measures that cannot be computed by the corresponding models. From these comparisons, we observe that whether the block structure is easy to identify or not, the ability of VEM-LBRM to outperform other algorithms.

To go further, note that in [11], the authors reformulated the clusterwise and introduced the linear cluster-weighted model (CWM) in a statistical setting and showed that it is a general and flexible family of mixture models. They included in

Table 2 (co)-clustering and prediction: mean and sd in parentheses.

Examples	Algorithms	Regression					Clustering			
		RMSE		MAE		Rsquare	ARI		NMI	
		Training	Test	Training	Test	Avg.	Row	Col	Row	Col
Example1	Global model	164.38	164.05	145.29	145.05	0.46	-	-	-	-
		(0.03)	(0.49)	(0.08)	(0.71)	(0.0)	-	-	-	-
	K-means	49.62	49.51	34.86	34.91	0.8	0.61	-	0.49	-
		(60.2)	(67.48)	(33.56)	(35.79)	(0.02)	0.02	-	0.03	-
	Clusterwise	154.57	154.47	127.77	127.93	0.52	0.07	-	0.01	-
	($g = 3$)	(0.01)	(0.36)	(0.03)	(0.45)	(0.0)	0.0	-	0.0	-
	Co-clustering	10.86	10.83	7.29	7.29	0.88	0.84	1.0	0.71	1.0
	($g = 3$)	(14.76)	(14.36)	(4.67)	(4.59)	(0.0)	0.01	0.0	0.04	0.0
	SCOAL	14.99	14.92	10.45	10.41	0.99	0.91	1.0	0.84	1.0
	($g = 3$, $m= 2$)	(207.56)	(208.91)	(89.48)	(90.55)	(0.0)	0.01	0.0	0.04	0.0
	VEM-LBRM	**7.1**	**7.06**	**5.29**	**5.26**	**0.99**	**0.95**	**1.0**	**0.92**	**1.0**
	($g = 3$, $m= 2$)	(**17.71**)	(**16.86**)	(**6.8**)	(**6.32**)	(**0.0**)	**(0.01)**	**(0.0)**	**(0.03)**	**(0.0)**
Example2	Global model	29.15	29.21	24.64	24.68	0.34	-	-	-	-
		(0.04)	(0.15)	(0.04)	(0.12)	(0.0)	-	-	-	-
	K-means	10.43	10.49	7.73	7.77	0.71	0.56	-	0.45	-
		(0.25)	(0.24)	(0.17)	(0.16)	(0.01)	0.0	-	0.0	-
	Clusterwise	18.54	18.62	11.33	11.38	0.73	0.15	-	0.16	-
	($g = 3$)	(0.09)	(0.27)	(0.06)	(0.14)	(0.0)	0.0	-	0.0	-
	Co-clustering	7.5	7.49	5.89	5.9	0.8	0.95	1.0	0.94	1.0
	($g = 3$)	(1.35)	(1.38)	(0.82)	(0.86)	(0.07)	0.14	0.0	0.17	0.0
	SCOAL	12.63	12.69	8.75	8.81	0.81	0.97	1.0	0.94	1.0
	($g = 3$, $m= 2$)	(12.57)	(12.81)	(7.38)	(7.58)	(0.35)	0.1	0.0	0.17	0.0
	VEM-LBRM	**6.99**	**6.99**	**5.57**	**5.57**	**0.96**	**1.0**	**1.0**	**1.0**	**1.0**
	($g = 3$, $m= 2$)	(**0.01**)	(**0.04**)	(**0.01**)	(**0.02**)	(**0.0**)	**(0.0)**	**(0.0)**	**(0.0)**	**(0.0)**
Example3	Global model	45.38	45.24	38.33	38.21	0.49	-	-	-	-
		(0.06)	(0.24)	(0.07)	(0.26)	(0.0)	-	-	-	-
	K-means	10.47	10.41	7.44	7.42	0.83	0.54	-	0.45	-
		(1.73)	(1.74)	(1.08)	(1.08)	(0.08)	0.01	-	0.01	-
	Clusterwise	23.09	23.18	12.09	12.15	0.87	0.09	-	0.09	-
	($g = 3$)	(1.84)	(2.02)	(1.23)	(1.29)	(0.02)	0.0	-	0.0	-
	Co-clustering	9.48	9.39	6.98	6.93	0.73	0.74	1.0	0.7	1.0
	($g = 3$)	(0.16)	(0.22)	(0.01)	(0.02)	(0.02)	0.04	0.0	0.08	0.0
	SCOAL	27.32	27.14	16.82	16.73	0.57	0.98	1.0	0.96	1.0
	($g = 3$, $m= 2$)	(41.97)	(41.83)	(24.13)	(24.16)	(0.93)	0.07	0.0	0.12	0.0
	VEM-LBRM	**7.21**	**7.21**	**5.71**	**5.71**	**0.99**	**0.98**	**1.0**	**0.96**	**1.0**
	($g = 3$, $m= 2$)	(**0.68**)	(**0.7**)	(**0.42**)	(**0.42**)	(**0.0**)	**(0.07)**	**(0.0)**	**(0.12)**	**(0.0)**

the classical model of clusterwise the probability $\Phi'(\mathbf{x}_i|\mathbf{\Omega}_k)$ to model the covariates, whereas the classical cluster-wise model the output only using $\Phi(y_i|\mathbf{x}_i; \lambda_k)$. They prove that sufficient conditions for model identifiability are provided under a suitable assumption of Gaussian covariates [10]. We can include in LBRM a joint probability $\Phi'(\mathbf{x}_{ij}|\mathbf{\Omega}_{k\ell})$ where $\mathbf{\Omega}_{k\ell} = [\boldsymbol{\mu}_{k\ell}, \boldsymbol{\Sigma}_{k\ell}]$ to evaluate its impact in terms of clustering and regression. Figure 4 presents the graphical model of LBRM and its extension. The first experiments on real datasets give encouraging results.

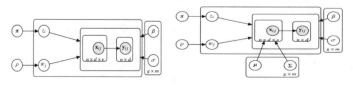

Fig. 4 Graphical model of LBRM (left) and its extension (right).

5 Conclusion

Inspired by the flexibility of the latent block model (LBM), we proposed extending it to tensor data aiming at both tasks: co-clustering and prediction. This model (LBRM) gives rise to a variational EM algorithm for co-clustering and prediction referred to as VEM-LBRM. This algorithm which can be viewed as the co-clusterwise algorithm can easily deal with sparse data. Empirical results on synthetic data showed that VEM-LBRM does give more encouraging results for clustering and regression than some algorithms that are devoted to one or both tasks simultaneously. For future work, we plan to develop the extension of LBRM and apply the proposed models for the recommender system task.

Acknowledgements Our work is funded by the German Federal Ministry of Education and Research under Grant Agreement Number 01IS19084F (XAPS).

References

1. Affeldt, S., Labiod, L., Nadif, M.: Regularized bi-directional co-clustering. Statistics and Computing, **31**(3), 1-17 (2021)
2. Agarwal, D., and Merugu, S.: Predictive discrete latent factor models for large scale dyadic data. In: SIGKDD, pp. 26–35 (2007)
3. Dempster, A. P., Laird, N. M., Rubin, D. B.: Maximum likelihood from incomplete data via the EM algorithm. Journal of the Royal Statistical Society Series B (Methodological), **39**(1), 1–22 (1977)
4. Deodhar, M., Ghosh, J.: A framework for simultaneous co-clustering and learning from complex data. In: SIGKDD, pp. 250–259 (2007)
5. DeSarbo, W. S., and Cron, W. L.: A maximum likelihood methodology for clusterwise linear regression. Journal of Classification, **5**(2), 249–282 (1988)
6. Govaert, G., Nadif, M.: Clustering with block mixture models. Pattern Recognition, **36**, 463-473, (2003)
7. Govaert, G., Nadif, M.: An EM algorithm for the block mixture model. IEEE Transactions on Pattern Analysis and Machine Intelligence, **27**(4), 643–647 (2005)
8. Govaert, G., Nadif, M.: Block clustering with Bernoulli mixture models: Comparison of different approaches. Computational Statistics & Data Analysis, 3233–3245 (2008)
9. Govaert, G., Nadif, M.: Co-clustering: Models, Algorithms and Applications. John Wiley & Sons (2013)
10. Ingrassia, S., Minotti, S. C., Punzo, A.: Model-based clustering via linear cluster-weighted models. Computational Statistics & Data Analysis, **71**, 159–182 (2014)
11. Ingrassia, S., Minotti, S. C., Vittadini, G.: Local statistical modeling via a cluster-weighted approach with elliptical distributions. In: Journal of Classification, **29**(3), 363–401 (2012)
12. Neal, R. M., Hinton, G. E.: A view of the EM algorithm that justifies incremental, sparse, and other variants. In Learning in Graphical Models, pp. 355–368. Springer (1998)
13. Salah, A., Nadif, M.: Directional co-clustering. Advances in Data Analysis and Classification, **13**(3), 591-620 (2019)
14. Späth, H.: Algorithm 39 clusterwise linear regression. Computing, **22**(4), 367–373 (1979)
15. Steinley, D.: Properties of the Hubert–Arable Adjusted Rand Index. Psychological Methods, **9**(3), 386 (2004)
16. Strehl, A., Ghosh, J.: Cluster ensembles—a knowledge reuse framework for combining multiple partitions. Journal of Machine Learning Research, **3**, 583–617 (2002)
17. Vu, D., Aitkin, M.: Variational algorithms for biclustering models. Computational Statistics & Data Analysis, **89**, 12–24 (2015)

Using Clustering and Machine Learning Methods to Provide Intelligent Grocery Shopping Recommendations

Nail Chabane, Mohamed Achraf Bouaoune, Reda Amir Sofiane Tighilt, Bogdan Mazoure, Nadia Tahiri, and Vladimir Makarenkov

Abstract Nowadays, grocery lists make part of shopping habits of many customers. With the popularity of e-commerce and plethora of products and promotions available on online stores, it can become increasingly difficult for customers to identify products that both satisfy their needs and represent the best deals overall. In this paper, we present a grocery recommender system based on the use of traditional machine learning methods aiming at assisting customers with creation of their grocery lists on the MyGroceryTour platform which displays weekly grocery deals in Canada. Our recommender system relies on the individual user purchase histories, as well as the available products' and stores' features, to constitute intelligent weekly grocery lists. The use of clustering prior to supervised machine learning methods allowed us to identify customers profiles and reduce the choice of potential products of interest for each customer, thus improving the prediction results. The highest average F-score of 0.499 for the considered dataset of 826 Canadian customers was obtained using the Random Forest prediction model which was compared to the Decision Tree, Gradient Boosting Tree, XGBoost, Logistic Regression, Catboost, Support Vector Machine and Naive Bayes models in our study.

Keywords: clustering, dimensionality reduction, grocery shopping recommendation, intelligent shopping list, machine learning, recommender systems

Nail Chabane · Mohamed Achraf Bouaoune · Reda Amir Sofiane Tighilt ·
Vladimir Makarenkov (✉)
Université du Québec à Montreal, 405 Rue Sainte-Catherine Est, Montreal, Canada,
e-mail: chabane.nail_amine@courrier.uqam.ca
e-mail: bouaoune.mohamed_achraf@courrier.uqam.ca
e-mail: tighilt.reda@courrier.uqam.ca;makarenkov.vladimir@uqam.ca

Bogdan Mazoure
McGill University and MILA - Quebec AI Institute, 845 Rue Sherbrooke O, Montreal, Canada,
e-mail: bogdan.mazoure@mail.mcgill.ca

Nadia Tahiri
University of Sherbrooke, 2500 Bd de l'Université, Sherbrooke, Canada,
e-mail: Nadia.Tahiri@USherbrooke.ca

© The Author(s) 2023
P. Brito et al. (eds.), *Classification and Data Science in the Digital Age*,
Studies in Classification, Data Analysis, and Knowledge Organization,
https://doi.org/10.1007/978-3-031-09034-9_10

1 Introduction

Grocery shopping is a common activity that involves different factors such as budget and impulse purchasing pressure [1]. Customers typically rely on a mental or digital list to facilitate their grocery trips. Many of them show a favorable interest towards tools and applications that help them manage their grocery lists, while keeping them updated with special offers, coupons and promotions [2, 3]. Major retailers throughout the world typically offer discounts on different products every week in order to improve sales and attract new customers. This very common practice leads to the fact that thousands of items go on special simultaneously across different retailers at a given week. The resulting information overload often makes it difficult for customers to quickly identify the deals that best suit their needs, which can become a source of frustration [4]. To address this problem, many grocery stores have taken advantage of the popularity of e-commerce to set up their own websites featuring various functionalities, including Recommender Systems, to assist customers during the shopping process.

Recommender Systems (RSs) [5] are tools and techniques that offer personalized suggestions to users based on several parameters (e.g. their past behavior). RSs have recently become a field of interest for researchers and retailers as many e-commerces, online book stores and streaming platforms have started to offer this service on their websites (e.g. Amazon, Netflix and Spotify). Here, we recall some recent works in this field. Faggioli et al. [6] used the popular Collaborative Filtering (CF) approach to predict the customer's next basket in a context of grocery shopping, taking into account the recency parameter. When comparing their model with the CF baseline models, Faggioli et al. observed a consistent improvement of their prediction results. Che et al. [7] used attention-based recurrent neural networks to capture both inter- and intra-basket relationships, thus modelling users' long-term preferences dynamic short-term decisions.

Content-based recommendation has also proven efficient in the literature, as demonstrated by Xia et al. [8] who proposed a tree-based model for coupons recommendation. By processing their data with undersampling methods, the authors were able to increase the estimated click rate from 1.20% to 7.80% as well as to improve significantly the F-score results using Random Forest Classifier and the recall results using XGBoost. Dou [9] presented a statistical model to predict whether a user will buy or not buy an item using Yandex's CatBoost method [10]. Dou relied on contextual and temporal features as well as on some session features, such as the time of visit of specific web pages, to demonstrate the efficiency of CatBoost in this context. Finally, Tahiri et al. [11] used recurrent and feedforward neural networks (RNNs and FFNs) in combination with non-negative matrix factorization and gradient boosting trees to create intelligent weekly grocery baskets to be recommended to the users of MyGroceryTour. Tahiri et al. considered different (from our study) features characterizing the users of MyGroceryTour to provide their predictions, with the best F-score results of 0.37 obtained from the augmented dataset.

2 Materials and Methods

2.1 Data Considered

In this section we describe the dataset obtained from MyGroceryTour website used in our research. MyGroceryTour [11] is a Canadian grocery shopping website and database available in both English and French languages. The main purpose of the website is to present weekly specials offered by the major grocery retailers in Canada. It allows users to display grocery products available in their living area, compare their products over different stores as well as to build their grocery shopping baskets based on the provided insights. MyGroceryTour users can easily archive and manage their grocery lists and access them at any given time.

In this study, we considered 826 MyGroceryTour users with varying numbers of grocery baskets (between 3 and 100 baskets were available per user). The grocery baskets contained different products added by users when they were creating their weekly shopping lists. In our recommender system (i.e current basket prediction experiment), we have considered the following features:

- *user_id* : unique user identifier (numerical)
- *basket_id* : unique basket identifier (numerical)
- *product_id* : unique product identifier (numerical)
- *category* : category of the product (categorical)
- *price* : price of the product (numerical)
- *special* : discount on the product (in %) compared to regular price (numerical)
- *distance_min* : minimal distance between user's home and the closest store where the product was available (numerical)
- *distance_mean* : mean distance between user's home and all stores where the product was available (numerical)
- *availability* : availability of the product at different stores (binary)

In addition, we engineered the *total_bought* feature which represents, for each product, the total number of times it has been bought over all users.

2.2 Data Normalization

Data normalization is an important data preprocessing step in both unsupervised and supervised machine learning [12] as well as in data mining [13]. Prior to feeding the data to our models we rescaled the available features using z-score standardization. Thus, all rescaled features had the mean of 0 and the standard deviation of 1:

$$z(x_f) = \frac{x_f - \mu_f}{\sigma_f}, \tag{1}$$

where x_f is the original value of the observation at feature f, μ_f is the mean and σ_f is the standard deviation of f.

2.3 Further Data Preprocessing Steps

In order to determine which weekly products could be recommended to a given user we propose to classify them using both clustering (unsupervised learning) and traditional supervised machine learning methods. The final recommendation is obtained based on the availability of the products, the data on the products' regular prices and available discounts, as well as on the user's shopping history. In our context, the baskets contain only the products bought by the users. The information about the other available products (not selected by the user at the moment he/she organized his/her shopping basket) is also available on MyGroceryTour. It has been used to create a large class of available items that were not bought by the user.

While we considered the items bought by a given user as positive feedback, we regarded the items that were available to this user at the time of the order, but not acquired by him/her, as a negative feedback. For an order of size P, if T is the total amount of items available to the user at the time of the order, the negative feedback N for that order is $N = T - P$. In this context, N usually represents thousands of products, while P is typically inferior to 50. This difference in size between positive and negative feedback can lead to a situation of imbalanced training data and could result in an important loss in performance. Similarly to Xia et al. [8], we applied an undersampling method to balance our data instead of considering all of the available disregarded items as the negative feedback.

To identify customer profiles and perform a preselection of products that are susceptible to be of interest to a given user, we first carried out the clustering of the normalized original dataset (the K-means [14] and DBSCAN [15] data partitioning algorithms were used). Then, we limited the choice of the items offered to a given user to the products purchased by the members of his/her cluster. By doing so, we managed to reduce the amount of products which could be recommended to the user and thus minimize eventual classifications mistakes. The clustering phase is detailed in the Subsection 2.4. Then traditional machine learning methods were used to provide the final weekly recommendation. The size S of the weekly basket recommended to a given user was equal to the mean size of his/her previous shopping baskets. As the number of items to be recommended by the machine learning methods was often greater than S, we retained as final recommendation the top S items, ranked according to the confidence score (i.e. the probability estimate for a given observation, computed using the *predict_proba* function from the *scikit-learn* [16] library).

2.4 Data Clustering

In this section, we present the steps we carried out to obtain the clusters of users. As explanatory features used to generate clusters, we considered the mean prices and mean specials of the products purchased by the user as well as a new feature, called here the fidelity ratio FR_u, which is meant to give insight on whether a given user u has a favorite store where he/she makes most of his/her grocery purchases. FR_u is defined as follows:

$$FR_u = \frac{X_{max,u} - \frac{1}{(n-1)} \sum_{i=2}^{n} X_{i,u}}{X_{total,u}}, \qquad (2)$$

where $X_{max,u}$ is the total number of products bought by user u at the store where he/she made most of his/her purchases, n ($n>1$) is the total number of stores visited by user u, and $X_{total,u}$ ($X_{total,u} = X_{max,u} + \sum_{i=2}^{n} X_{i,u}$) is the total number of products purchased by user u over all stores he/she visited. A high fidelity ratio means that user u buys most of his/her products at the same store, whereas a low fidelity ratio indicates that user u buys his/her products at different stores. When user u purchases all of his/her products at the same store ($X_{max,u} = X_{total,u}$ and $n = 1$), the fidelity ratio equals 1. It equals 0 when he/she purchases the same number of products at different stores.

The K-means [14] and DBSCAN [15] algorithms were used to perform clustering. Here we present the results of DBSCAN, as the clusters provided by DBSACAN had less entity overlap than those provided by K-means. The main advantage of DBSCAN is that this density-based algorithm is able to capture clusters of any shape.

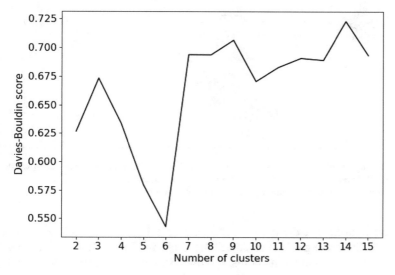

Fig. 1 Davies-Bouldin cluster validity index variation with respect to the number of clusters.

We used the Davies-Bouldin (DB) [17] cluster validity index to determine the number of clusters in our dataset. The Davies-Bouldin index is the average similarity between each cluster C_i for $i = 1, ..., k$ and its most similar counterpart C_j. It is calculated as follows:

$$DB = \frac{1}{k} \sum_{i=1}^{k} \max_{i \neq j} R_{ij}, \qquad (3)$$

where R_{ij} is the similarity measure between clusters calculated as $(d_i + d_j)/\delta_{ij}$, where d_i (d_j) is the the mean distance between objects of cluster C_i (C_j) and the cluster centroid and δ_{ij} is the distance between the centroids of clusters C_i and C_j.

Figure 1 illustrates the variation of the Davies-Bouldin cluster validity index whose lowest (i.e. best) value was reached for our dataset with 6 clusters. The resulting clusters are represented in Figure 2. After performing the data clustering, we applied the t-SNE [18] dimensionality reduction method for data visualisation purposes. Since t-SNE is known to preserve the local structure of the data but not the global one, we used the PCA initialization parameter to mitigate this issue.

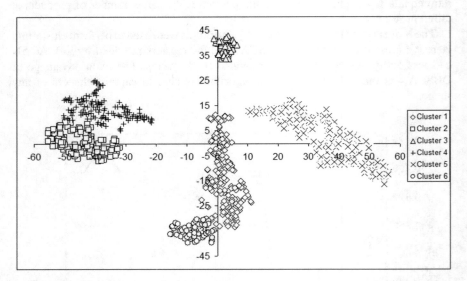

Fig. 2 Clustering results : Clustering obtained with DBSCAN with the best number of clusters according to the Davies-Bouldin index. Data reduction was performed using t-SNE. The 6 clusters of customers found by DBSCAN are represented by different symbols.

We have noticed that the users in Cluster 1 (see Fig. 2) are fairly sensitive to specials and have a high fidelity score, the users in Cluster 2 mostly purchase products on special in different stores, the users in Cluster 3 seem to be sensitive to the total price of their shopping baskets, Cluster 4 includes the users who are sensitive to specials but have a low fidelity score, Cluster 5 includes the users who are not very attracted by specials but are rather loyal to their favorite store(s), and the users in Cluster 6 tend to buy products on special and have high fidelity scores.

3 Application of Supervised Machine Learning Methods

To predict the products to be recommended for the current weekly basket, we used the following supervised machine learning methods: Decision Tree, Random Forest,

Gradient Boosting Tree, XGBoost, Logistic Regression, Catboost, Support Vector Machine and Naive Bayes. These methods were used through their *scikit-learn* implementations [16]. Due to the lack of large datasets we did not use deep learning models in our study. We decided to use these classical machine learning methods because they are usually recommended to work with smaller datasets contrary to their deep leaning counterparts. Also, deep leaning algorithms usually don't handle properly mixed types of features present in our data. Most of the methods we used are the ensemble methods, i.e. they use multiple replicates to reduce the variance. The F-score results provided by each method without (using all products available) and with clustering (using only the products purchased by the cluster members) are presented in Table 1.

As shown in Table 1, Random Forest outperformed the other competing methods without and with data clustering, providing the average F-scores of 0.494 and 0.499 (obtained over all users), respectively. Tree-based models relying on gradient boosting performed relatively well and could possibly give better results with a different data processing. We can also notice that all the methods, except CatBoost, benefited from the data clustering process.

Table 1 F-scores provided by ML methods without and with clustering of MyGroceryTour users.

Machine learning methods	Results without clustering	Results with clustering
CatBoost	0.438	0.438
Decision Tree	0.463	0.468
Gradient Boosting Tree	0.488	0.495
Logistic Regression	0.474	0.478
Naive Bayes	0.433	0. 436
Random Forest	**0.494**	**0.499**
SVM-RBF	0.392	0.397
XGBoost	0.476	0.481

4 Conclusion

In this paper, we presented a novel recommender system that is intended to predict the content of the customer's weekly basket depending on his/her purchase history. Our system is also able to predict the store(s) where the purchase(s) will take place. The clustering step allowed us to identify customer profiles and to improve the F-score result for every tested machine learning model, except CatBoost. Using our methodology and the new data available on MyGroceryTour, we were able to improve the F-score performance by the margin of 0.129, compared to the results obtained by Tahiri et al. [11]. Our model is able to predict products that will be purchased again or acquired for the first time by a given user, but it is not yet able to predict the optimal quantity for each product to be bought. Another important issue is how to provide plausible recommendations for customers without shopping history (i.e. the cold start problem). We will tackle these important issues in our future work.

References

1. Vincent-Wayne, M., Aylott, R.: An exploratory study of grocery shopping stressors. Int. J. Retail. Distrib. Manag. **26**, 362–373 (1998)
2. Newcomb, E., Pashley, T., Stasko, J.: Mobile computing in the retail arena. In: Proceedings of the SIGCHI conference on Human factors in computing systems, pp. 337-344. Association for Computing Machinery, New York (2003)
3. Sourav, B., Floréen, P., Forsblom, A., Hemminki, S., Myllymäki, P., Nurmi, P., Pulkkinen, T., Salovaara, A.: An Intelligent Mobile Grocery Assistant. In: 2012 Eighth International Conference on Intelligent Environments, pp. 165-172. IEEE, Guanajuato (2012)
4. Park, Y., Chang, K.: Individual and group behavior-based customer profile model for personalized product recommendation. Expert Systems with Applications **36**(2), 1932-1939 (2009)
5. Ricci, F., Rokach, L., Shapira, B.: Recommender Systems: Introduction and Challenges. In: Recommender Systems Handbook, pp. 1-34. Springer, Boston (2015)
6. Faggioli, G., Mirko P., Fabio A.: Recency aware collaborative filtering for next basket recommendation. In: Proceedings of the 28th ACM Conference on User Modeling, Adaptation and Personalization, pp. 80-87. Association for Computing Machinery, New York (2020)
7. Che et al.: Inter-basket and intra-basket adaptive attention network for next basket recommendation. IEEE Access **7**, 80644-80650 (2019)
8. Xia, Y. Giuseppe, D. F., Shikhar, V., Ankur, D.: A content-based recommender system for e-commerce offers and coupons. In: Proceedings of the SIGIR 2017 eCom workshop. eCOM@ SIGIR, Tokyo (2017)
9. Dou, X.: Online purchase behavior prediction and analysis using ensemble learning. In : 2020 IEEE 5th International Conference on Cloud Computing and Big Data Analytics (ICCCBDA), pp. 532-536. IEEE (2020)
10. Prokhorenkova, L., Gleb G., Aleksandr V., Anna V. D., Andrey G.: CatBoost: unbiased boosting with categorical features (2017) Available via arXiv.
11. Tahiri, N., Mazoure, B. and Makarenkov, V.: An intelligent shopping list based on the application of partitioning and machine learning algorithms. In: Proceedings of the 18th Python in Science Conference (SCIPY 2019), pp. 85-92. Austin, Texas (2019)
12. Kotsiantis, S. B., Kanellopoulos, D., Pintelas, P. E.: Data preprocessing for supervised leaning. Int. J. Comput. Sci. **2**, 111-117 (2006)
13. García, S., Luengo, J., Herrera, F.: Data Preprocessing in Data Mining. Springer, Cham, Switzerland (2015)
14. MacQueen, J.: Some methods for classification and analysis of multivariate observations. In: Proceedings of the fifth Berkeley symposium on mathematical statistics and probability, vol. 1, no. 14, pp. 281-297 (1967)
15. Ester, M., Kriegel, H. P., Sander, J., Xu, X.: A Density-Based Algorithm for Discovering Clusters in Large Spatial Databases with Noise. In: Proceedings of the 2nd International Conference on Knowledge Discovery and Data Mining. AAAI Press, pp. 226–231. AAAI Press, Portland, Oregon (1996)
16. Pedregosa et al.: Scikit-learn: Machine Learning in Python. JMLR **12**, 2825-2830 (2011)
17. Davies, D. L., Bouldin, D. W.: A Cluster Separation Measure. In: IEEE Transactions on Pattern Analysis and Machine Intelligence, vol. PAMI-1, no. 2, pp. 224-227 (1979)
18. Van der Maaten, L. J. P., Hinton, G.: Visualizing High-Dimensional Data Using t-SNE. J. Mach. Learn. Res. **9**, 2579-2605 (2008)

COVID-19 Pandemic: a Methodological Model for the Analysis of Government's Preventing Measures and Health Data Records

Theodore Chadjipadelis and Sofia Magopoulou

Abstract The study aims to investigate the associations between the government's response measures during the COVID-19 pandemic and weekly incidence data (positivity rate, mortality rate and testing rate) in Greece. The study focuses on the period from the detection of the first case in the country (26th February 2020) to the first week of 2022 (08th January 2022). Data analysis was based on Correspondence Analysis on a fuzzy-coded contingency table, followed by Hierarchical Cluster Analysis (HCA) on the factor scores. Results revealed distinct time periods during which interesting interactions took place between control measures and incidence data.

Keywords: hierarchical cluster analysis, correspondence analysis, COVID-19, evidence-based policy making

1 Introduction

The present study focuses on the period of the COVID-19 pandemic in Greece, from the detection of the first case of COVID-19 to the first week of 2022. This period can be divided into five distinct phases. The first phase extends from the beginning of 2020 until the first lockdown, i.e., from the first case reported in Greece until the end of the first quarantine period in May 2020. The second phase concerns the interim period from June to October 2020, when the pandemic indices improved, and policies were loosened for the opening of tourism. The third phase concerns the second lockdown and the evolution of the pandemic in the country from November 2020 to April 2021, when the first vaccination period of the adult population took place. The fourth phase includes the interim period from May 2021 to October

Theodore Chadjipadelis (✉)
Aristotle University of Thessaloniki, Greece, e-mail: `chadji@polsci.auth.gr`

Sofia Magopoulou
Aristotle University of Thessaloniki, Greece, e-mail: `sofimago@polsci.auth.gr`

© The Author(s) 2023
P. Brito et al. (eds.), *Classification and Data Science in the Digital Age*,
Studies in Classification, Data Analysis, and Knowledge Organization,
https://doi.org/10.1007/978-3-031-09034-9_11

2021, where a general stabilization of the number of cases occurred, while the last period refers to a significant increase in the number of cases from November 2021 to January 2022.

Overall, from March 2020 to January 2022, a total of 1,79 million cases of COVID-19 were recorded in Greece (Figure 1) and a total of 22,635 deaths. Vaccination coverage is as of January 2022 over 65% of country's population, i.e., 7,241,468 fully vaccinated citizens.

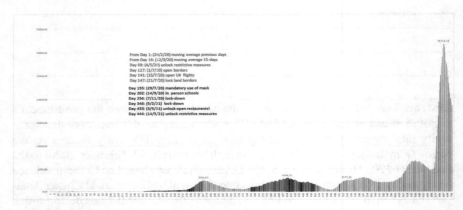

Fig. 1 Record of cases of COVID-19 in Greece (March 2020-January 2022).

In this study, a combination of multivariate data analysis methods was employed to analyze COVID-19-related data so as to assess the quality of decision-making outputs during the crisis and improve evidence-based decision-making processes. Section 2 presents the methodology and describes the data sources and the data analysis workflow. Section 3 presents the study results and Section 4 discusses the results and proposes methodological tools and presents the paper conclusions.

2 Methodology

2.1 Data

For the study purposes, data were obtained from the Oxford Covid-19 Government Response Tracker (OxCGRT) and were combined with self-collected Covid-19 data for Greece [3] daily updated in Greek. The Oxford Covid-19 Government Response Tracker (OxCGRT) collects publicly available information reflecting government response from 180 countries since 1 January 2020 [4]. The tracker is based on data for 23 indicators. In this study, two groups of indicators were considered: Containment & Closure and Health Systems in the case of Greece. The first group of indicators refers to "collective" level policies and measures, such as school closures and restriction in mobility, while the second refers to "individual" level policies and measures, such as testing and vaccination. Specifically, the collective level indicators refer to policies taken by the governments' and reflect on a collective level on the society: school closing, workplace closing, cancelation of public events, restrictions on gathering, closure of public transport, stay at home requirements, internal movement restrictions and international travel controls. The health system policies primarily touch upon the individual level and specifically refer to: public information campaigns, testing, contact tracing, healthcare facilities, vaccines' investments, facial coverings, vaccination and protection of the elderly people. All collective-level indicators (C1 to C8) were summed to yield a total score (ranging from 0 to 16). Similarly, individual-level indicators (H1 to H3 and H6 to H8) were summed to compute a total score (ranging from to 12).

The self-collected data refer to positive cases, number of Covid-19-related deaths, number of tests and total number of vaccinations administered. These data have been recorded daily since March 2020 from public announcements by official and verified sources. A total of 94 time points were considered in the present study, corresponding to weekly data (Monday was used as a reference). Three quantitative indicators were derived, a positivity index (#cases / #tests), a mortality index (#deaths / #cases) and a testing index (#tests / #people). The number of vaccinations is not used in the present study because the vaccination process began in January 2021 and the administration of the booster dose began in September 2021. The final data set consisted of five indicators: two ordinal total scores, and three quantitative indices.

2.2 Data Analysis

A four-step data analysis strategy was adopted. In the first step, the three quantitative variables (positivity rate, mortality rate and testing rate) were transformed into ordinal variables, via a method used in [7] (see Step 1) transformation of continuous variables into ordinal categorical variables, with minimum information loss. Three ordinal variables were derived. In the second step, the five ordinal variables (i.e., the three recoded variables and the two ordinal total scores), were fuzzy-coded into three categories each, using the barycentric coding scheme proposed in [7]. This scheme has been recently evaluated in the context of hierarchical clustering in [7] and was applied with the DIAS Excel add-in [6]. Barycentric coding allows us to convert an m-point ordinal variable into an n-point fuzzy-coded variable [6, 7]. In other words, the transformation of the three quantitative variables into ordinal variables resulted in a generalized 0-1 matrix (fuzzy-coded matrix), where for each variable we obtain the estimated probability for each category. A drawback of the proposed approach is that the ordinal information in the 5 ordinal variables is lost.

The third step involved the application of Correspondence Analysis (CA) on the fuzzy-coded table with the 94 weeks as rows and the fifteen fuzzy categories as columns (see [1] for a similar approach). The number of significant axes was determined based on percentage of inertia explained and the significant points on each axis were determined based on the values of two statistics that accompany standard CA output; quality of representation (COR) greater than 200 and contribution (CTR) greater than 1000/(n + 1), where n is the total number of categories (i.e., 15 in our case). In the final step, Hierarchical Cluster Analysis (HCA) using Benzecri's chi-squared distance and Ward's linkage criterion [2, 8] was employed to cluster the 94 points (weeks) on the CA axes obtained from the previous step. The number of clusters was determined upon the empirical criterion of the change in the ratio of between-cluster inertia to total inertia, when moving from a partition with r clusters to a partition with r − 1 cluster [8]. Lastly, we interpret the clusters after determining the contribution of each indicator to each cluster. All analyses were conducted with the M.A.D. [Méthodes de l'Analyse des Données] software [5].

3 Results

Correspondance Analysis resulted in four significant axes, which explain 74.91% of the total inertia (Figure 2). For each axis, we describe the main contrast between groups of categories based on their coordinates, COR and CTR values (Figure 3). "Low and moderate mortality rates" and "high factor testing rates" define a pole on the 1st axis, which is opposed to "average and high levels of "individual" measures". On the second axis, "low positivity rate" and "average levels of collective measures" define a pole, while "average and high positivity rate" and "high levels of collective measures" define the opposite pole. The third axis is characterized by "moderate and high mortality rate", "high levels of collective measures" and "average levels of individual measures" that are opposed to "average levels of collective measures". On the fourth axis, "average levels of collective measures" are opposed to "average testing rate" and "high levels of collective measures".

Total Inertia 0,62704				
Axis	Inertia	%Inertia	Cumulative	Histogram
1	0,1739028	27,73	27,73	|**
2	0,1136495	18,12	45,86	|*******************************
3	0,1066425	17,01	62,87	|*****************************
4	0,0755233	12,04	74,91	|*******************
5	0,0526040	8,39	83,30	|**************
6	0,0401367	6,40	89,70	|**********
7	0,0307749	4,91	94,61	|********
8	0,0163050	2,60	97,21	|****
9	0,0113139	1,80	99,01	|***
10	0,0061758	0,98	100,00	|**
11	0,0000054	0,00	100,00	|*
12	0,0000036	0,00	100,00	|*

Fig. 2 Explained inertia by axis.

IND	Variables	Level	#G1	COR	CTR	#G2	COR	CTR	#G3	COR	CTR	#G4	COR	CTR
VAR01	cases/tests	low	-62	14	3	463	827	240	156	93	29	11	0	1
VAR02	cases/tests	average	-31	0	1	-583	324	139	-373	132	61	-398	150	98
VAR03	cases/tests	high	352	46	19	-1203	540	334	-91	3	3	649	157	146
VAR11	deaths/cases	low	113	258	13	9	1	1	-170	579	48	28	16	2
VAR12	deaths/cases	average	-694	188	46	-111	4	2	971	388	146	-183	13	8
VAR13	deaths/cases	high	-902	148	43	21	0	1	1468	394	185	-211	8	6
VAR21	tests/people	low	-395	839	137	-82	36	10	-101	54	15	70	26	10
VAR22	tests/people	average	585	117	39	388	51	26	-211	15	9	-1219	510	388
VAR23	tests/people	high	1841	774	519	184	7	9	737	124	135	494	55	86
VAR31	containment & closure	low	-97	18	1	130	33	2	-105	21	2	161	52	5
VAR32	containment & closure	average	-291	109	29	489	311	123	-426	236	100	436	248	148
VAR33	containment & closure	high	143	111	16	-240	306	64	207	230	51	-219	255	80
VAR41	health system	low	-602	111	27	-619	118	45	347	37	15	314	30	17
VAR42	health system	average	-736	360	81	-29	0	0	815	442	162	83	4	3
VAR43	health system	high	169	374	26	56	41	4	-162	339	39	-40	20	4

Fig. 3 Category coordinates on the four CA axes (#G), quality of representation (COR) and contribution (CTR). COR values greater than 200 and CTR values greater than 1000 / 16 = 62.5 are shown in yellow. Positive coordinates are shown in green and negative in pink.

Hierarchical Cluster Analysis on the factor scores resulted in seven clusters using the empirical criterion for cluster determination (see Section 2.2). The corresponding dendrogram is shown in Figure 4. The seven nodes in the figure that correspond to the seven clusters are 182, 181, 175, 177, 171, 181, 133 and 179. Cluster content reflects the different periods (phases) presented in the introductory section.

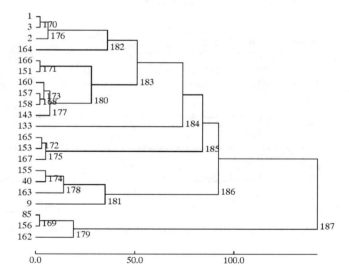

Fig. 4 Dendrogram of the HCA.

The first cluster (182) combines data points from March 2020, the onset of the pandemic with data points from a period following the summer of 2020 (October and November). This cluster is characterized by high positivity rate, low testing rate, high levels of "collective" measures (containment & closure) and low levels of "individual" measures (health system). The second cluster (181) contains data points from April and May 2020 and is characterized by low positivity rate, average to high mortality rate, low testing rate, high levels of "collective" measures (containment & closure) and average levels of "individual" measures (health system). The third cluster (175) combines summer months of 2020 and 2021. This cluster is characterized by low positivity rate, low testing rate and average levels of "collective" measures (containment & closure). The fourth cluster (177) marks the period of December 2020 and the period of spring of 2021, with average positivity rate and high levels of "collective" measures (containment & closure). The fifth cluster (171) refers to the period from December 2020 to February 2021, but also includes August 2021, with high levels of "collective" measures (containment & closure). The sixth cluster (133) refers to the period following the summer of 2021 (September and October 2021). In this cluster, average positivity rates were observed but also strict containment and closure measures.

Lastly, the seventh cluster (179) refers to November and December 2021, including also January 2022, with high positivity and high testing rates, while high levels of containment and closure and health system measures were observed. Figure 5 shows the contributions of each indicator in each cluster.

INT	Variable	Level	Cluster						
			133	171	175	177	179	181	182
VAR01	cases/tests	low				2,6888		3,6136	
VAR02	cases/tests	average	8,884				8,884		
VAR03	cases/tests	high					1,9892		13,556
VAR11	deaths/cases	low							
VAR12	deaths/cases	average						10,5652	
VAR13	deaths/cases	high						9,2727	
VAR21	tests/people	low				1,9732		2,6465	2,3519
VAR22	tests/people	average							
VAR23	tests/people	high					21,3527		
VAR31	containment & closure	low							
VAR32	containment & closure	average				6,8689			
VAR33	containment & closure	high	3,8875	2,2049		1,7047	2,5037	2,3608	1,971
VAR41	health system	low							10,3651
VAR42	health system	average						12,7955	
VAR43	health system	high					1,8392		

Fig. 5 Cluster description (contribution values of the indicators in each cluster - node).

4 Discussion

Based on the study results, we can argue that, when it comes to measures and real time data following a situation such as the pandemic, "the chicken and egg" dilemma arises. The question is whether "collective" and "individual" measures affect daily incidence data or the inverse (i.e., that the daily data lead to measures). We conclude that in fact the two should be perceived as working in conjunction and not independently from one another. The analysis showed that lower positivity rate is accompanied by average levels of measures from the government at both the "individual" and the "collective" level. Furthermore, higher positivity rate is accompanied by higher levels of measures, as a response. With regard to mortality rate, we observed that higher mortality invokes higher levels of "collective" measures and average levels of "individual" measures, whereas average levels of "collective" measures are associated with higher mortality rate.

It is therefore evident that when it comes to decision making in crisis situations, a systematic collection, analysis and use of data is linked to more effective government response overall. Therefore, evidence-based policy making should be linked to crisis management. This paper presents a first attempt to capture an ongoing phenomenon and therefore it is crucial that the collection and analysis of data will be complemented until the end of the phenomenon.

References

1. Aşan, Z., Greenacre, M.: Biplots of fuzzy coded data. Fuzzy Sets and Systems, **183**(1), 57–71 (2011)
2. Benzècri, J. P.: L'Analyse des Données. 2. L'Analyse des Correspondances. Dunod, Paris (1973)
3. Chadjipadelis, T.: Facebook profile (2022).
   https://www.facebook.com/theodore.chadjipadelis
4. Hale, T., Petherick, A., Phillips, T., Webster, S.: Variation in government responses to COVID-19. Blavatnik School of Government Working Paper, 31, 2020-11 (2020)
5. Karapistolis, D.: Software Method of Data Analysis MAD. (2010)
   http://www.pylimad.gr/
6. Markos, A., Moschidis, O., Chadjipadelis, T.: Hierarchical clustering of mixed-type data based on barycentric coding (2022) https://arxiv.org/submit/4142768
7. Moschidis, O., Chadjipadelis, T.: A method for transforming ordinal variables. In: Palumbo, F., Montanari, A., Vichi, M. (eds) Data Science, pp. 285-294. Studies in Classification, Data Analysis, and Knowledge Organization. Springer, Cham. (2017)
   https://doi.org/10.1007/978-3-319-55723-6_22
8. Papadimitriou, G., Florou, G.: Contribution of the Euclidean and chi-square metrics to determining the most ideal clustering in ascending hierarchy (in Greek). In Annals in Honor of Professor I. Liakis, 546-581. University of Macedonia, Thessaloniki (1996)

pcTVI: Parallel MDP Solver Using a Decomposition into Independent Chains

Jaël Champagne Gareau, Éric Beaudry, and Vladimir Makarenkov

Abstract Markov Decision Processes (MDPs) are useful to solve real-world probabilistic planning problems. However, finding an optimal solution in an MDP can take an unreasonable amount of time when the number of states in the MDP is large. In this paper, we present a way to decompose an MDP into Strongly Connected Components (SCCs) and to find dependency chains for these SCCs. We then propose a variant of the Topological Value Iteration (TVI) algorithm, called *parallel chained TVI* (pcTVI), which is able to solve independent chains of SCCs in parallel leveraging modern multicore computer architectures. The performance of our algorithm was measured by comparing it to the baseline TVI algorithm on a new probabilistic planning domain introduced in this study. Our pcTVI algorithm led to a speedup factor of 20, compared to traditional TVI (on a computer having 32 cores).

Keywords: Markov decision process, automated planning, strongly connected components, dependancy chains, parallel computing

1 Introduction

Automated planning is a branch of Artificial Intelligence (AI) aiming at finding optimal plans to achieve goals. One example of problems studied in automated planning is the electric vehicle path-planning problem [1]. Planning problems with non-deterministic actions are known to be much harder to solve. Markov Decision

Jaël Champagne Gareau (✉)
Université du Québec à Montréal, Canada, e-mail: champagne_gareau.jael@uqam.ca

Éric Beaudry
Université du Québec à Montréal, Canada, e-mail: beaudry.eric@uqam.ca

Vladimir Makarenkov
Université du Québec à Montréal, Canada, e-mail: makarenkov.vladimir@uqam.ca

© The Author(s) 2023
P. Brito et al. (eds.), *Classification and Data Science in the Digital Age*,
Studies in Classification, Data Analysis, and Knowledge Organization,
https://doi.org/10.1007/978-3-031-09034-9_12

Processes (MDPs) are generally used to solve such problems leading to probabilistic models of applicable actions [2].

In probabilistic planning, a solution is generally a policy, i.e., a mapping specifying which action should be executed in each observed state to achieve an objective. Usually, dynamic programming algorithms such as Value Iteration (VI) are used to find an optimal policy [3]. Since VI is time-expensive, many improvements have been proposed to find an optimal policy faster, using for example the Topological Value Iteration (TVI) algorithm [4]. However, very large domains often remain out of reach. One unexplored way to reduce the computation time of TVI is by taking advantage of the parallel architecture of modern computers and by decomposing an MDP into independent parts which could be solved concurrently.

In this paper, we show that state-of-the-art MDP planners such as TVI can run an order of magnitude faster when considering task-level parallelism of modern computers. Our main contributions are as follows:

- An improved version of the TVI algorithm, *parallel-chained TVI* (pcTVI), which decomposes MDPs into independent chains of strongly connected components and solves them concurrently.
- A new parametric planning domain, *chained-MDP*, and an evaluation of pcTVI's performance on many instances of this domain compared to the VI, LRTDP [5] and TVI algorithms.

2 Related Work

Many MDP solvers are based on the Value Iteration (VI) algorithm [3], or more precisely on asynchronous variants of VI. In asynchronous VI, MDP states can be backed up in any order and do not need to be considered the same number of times. One way to take advantage of this is by assigning a priority to every state and by considering them in priority order.

Several state-of-the-art MDP algorithms have been proposed to increase the speed of computation. Many of them are able to focus on the most promising parts of MDP through heuristic search algorithms such as LRTDP [5] or LAO* [6]. Some other MDP algorithms use partitioning methods to decompose the state-space in smaller parts. For example, the P3VI (Partitioned, Prioritized, Parallel Value Iteration) algorithm partitions the state-space, uses a priority metric to order the partitions in an approximate best solving order, and solves them in parallel [7]. The biggest disadvantage of P3VI is that the partitioning is done on a case-by-case basis depending on the planning domain, i.e., P3VI does not include a general state-space decomposition method. The inter-process communication between the solving threads also incurs an overhead on the computation time. The more recent TVI (Topological Value Iteration) algorithm [4] also decomposes the state-space, but does it by considering the topological structure of the underlying graph of the MDP, making it more general than P3VI. Unfortunately, to the best of our knowledge, no parallel version of TVI has been proposed in the literature.

3 Problem Definition

There exist different types of MDP, including Finite-Horizon MDP, Infinite-Horizon MDP and Stochastic Shortest Path MDP (SSP-MDP) [2]. The first two of them can be viewed as special cases of SSP-MDP [8]. In this work, we focus on SSP-MDPs, which we describe formally in Definition 1 below.

Definition 1 A *Stochastic Shortest Path MDP* (SSP-MDP) is given by a tuple (S, A, T, C, G), where:

- S is a finite set of states;
- A is a finite set of actions;
- $T: S \times A \times S \rightarrow [0, 1]$ is a transition function, where $T(s, a, s')$ is the probability of reaching state s' when applying action a while in state s;
- $C: S \times A \rightarrow \mathbb{R}^+$ is a cost function, where $C(s, a)$ gives the cost of applying the action a while in state s;
- $G \subseteq S$ is the set of goal states (which can be assumed to be sink states).

We generally search for a policy $\pi: S \rightarrow A$ that tells us which action should be executed at each state, such that an execution following the actions given by π until a goal is reached has a minimal expected cost. This expected cost is given by a value function $V^\pi: S \rightarrow \mathbb{R}$. The Bellman Optimality Equations are a system of equations satisfied by any optimal policy.

Definition 2 The Bellman Optimality Equations are the following:

$$
V(s) = \begin{cases} 0, & \text{if } s \in G, \\ \min_{a \in A} \left[C(s, a) + \sum_{s' \in S} T(s, a, s')V(s) \right], & \text{otherwise.} \end{cases}
$$

The expression between square brackets is called the *Q-value* of a state-action pair:

$$
Q(s, a) = C(s, a) + \sum_{s' \in S} T(s, a, s')V(s).
$$

When an optimal value function $V^\star$ has been computed, an optimal policy $\pi^\star$ can be found greedily:

$$
\pi^\star(s) = \operatorname{argmin}_{a \in A} Q^\star(s, a).
$$

Most MDP solvers are based on dynamic programming algorithms like Value Iteration (VI), which update iteratively an arbitrarily initialized value function until convergence with a given precision ϵ. In the worst case, VI needs to do $|S|$ sweeps of the state space, where one sweep consists in updating the value estimate of every state using the Bellman Optimality Equations. Hence, the number of state updates (called a *backup*) is $O(|S|^2)$. When the MDP is acyclic, most of these backups are wasteful, since the MDP can in this situation be solved using only $|S|$ backups (ordered in reverse topological order), thus allowing one to find an optimal policy in $O(|S|)$ [8].

4 Parallel-chained TVI

In this section, we describe an improvement to the TVI algorithm, named pcTVI (Parallel-Chained Topological Value Iteration), which is able to solve an MDP in parallel (as P3VI). pcTVI uses the decomposition proposed by TVI, known to give good performance on many planning domains. We start by summarizing how the original TVI algorithm works.

First, TVI uses Kosaraju's graph algorithm on a given MDP to find the strongly connected components (SCCs) of its graphical structure (the graph corresponding to its all-outcomes determinization).The SCCs are found by Kosaraju's algorithm in reverse topological order, which means that for every $i < j$, there is no path from a state in the ith SCC to a state in the jth SCC. This property ensures that every SCC can be solved separately by VI sweeps if previous SCCs (according to the reverse topological order) have already been solved. The second step of TVI is thus to solve every SCC one by one in that order. Since TVI divides the MDP in multiple subparts, it maximizes the usefulness of every state backup by ensuring that only useful information (i.e., converged state values) is propagated through the state-space.

Unfortunately, TVI can only solve one SCC at a time. Since modern computers have many computing units (cores) which can work in parallel, we could theoretically solve many SCCs in parallel to greatly reduce computation time. Instead of choosing SCCs to solve in parallel arbitrarily or using a priority metric (as in P3VI), which incur a computational overhead to propagate the values between the threads, we want to consider their topological order (as in TVI) to minimize redundant or useless computations. One way to share the work between the processes is to find independent chains of SCCs which can be solved in parallel. The advantage of independent chains is that no coordination and communication is needed between the SCCs, which both removes some running-time overhead and simplifies the implementation.

The Parallel-Chained TVI algorithm we propose (Algorithm 1) works as follows. First, we find the graph G corresponding to the graphical structure of the MDP, decompose it into SCCs, and find the reverse topological order of the SCCs (as in TVI, but we use Tarjan's algorithm instead of Kosaraju's algorithm since it is about twice as fast). We then build the condensation of the graph G, i.e., the graph G_c whose vertices are SCCs of G, where an edge is present between two vertices scc_1 and scc_2 if there exists an edge in G between a state $s_1 \in scc_1$ and a state $s_2 \in scc_2$. We also store the reversed edges in G_c and a counter c_{scc} on every vertex scc which indicates how many incoming neighbors have not yet been computed. We use this (usually small) graph G_c to detect which SCCs are ready to be considered (the SCCs whose incoming neighbors have all been determined with precision ϵ, i.e., the SCCs whose associated counter c_{scc} is 0). When a new SCC is ready, it is inserted into a work queue from which the waiting threads acquire their next task.

Algorithm 1 Parallel-Chained Topological Value Iteration

1: **procedure** PCTVI(M: MDP, t: Number of threads)
2:     ▷ Find the SCCs of M
3:     $G \leftarrow$ GRAPH(M)   ▷ G implicitly shares the same data structures as M
4:     $SCCs \leftarrow$ TARJAN(G)   ▷ SCCs are found in reverse topological order
5:
6:     ▷ Build the graph of SCCs of G
7:     $G_c \leftarrow$ GRAPHCONDENSATION($G, SCCs$)
8:
9:     ▷ Solve in parallel independent SCCs
10:    $Pool \leftarrow$ CREATETHREADPOOL(t)   ▷ Create t threads
11:    $V \leftarrow$ NEWVALUEFUNCTION()   ▷ Arbitrarily initialized; Shared by all threads
12:    $Q \leftarrow$ CREATEQUEUE()   ▷ Shared by all threads
13:    INSERT(Q, HEAD($SCCs$))   ▷ The goal SCC is inserted in the queue
14:    **while** NOTEMPTY(Q) **do**   ▷ Only one thread runs this loop
15:        $scc \leftarrow$ EXTRACTNEXTITEM(Q)
16:        **for all** $neighbor \in$ NEIGHBORS(scc) **do**
17:            Decrement NUMINCOMINGNEIGHBORS($neighbor$)
18:            **if** NUMINCOMINGNEIGHBORS($neighbor$) = 0 **then**
19:                ASSIGNTASKTOAVAILABLETHREAD($Pool$, PARTIALVI(M, V, scc))
20:                PUSH(Q, scc)   ▷ Neighbors of scc are ready to be considered next
21:            **end if**
22:        **end for**
23:    **end while**
24:
25:    ▷ Compute and return an optimal policy using the computed value function
26:    $\Pi \leftarrow$ GREEDYPOLICY(V)
27:    **return** Π
28: **end procedure**

---

5 Empirical Evaluation

In this section, we evaluate empirically the performance of pcTVI, comparing it to the three following algorithms: (1) VI – the standard dynamic programming algorithm (here we use its asynchronous round-robin variant), (2) LRTDP – a well-known heuristic search algorithm, and (3) TVI – the Topological Value Iteration algorithm described in Section 4. In the case of LRTDP, we carried out the admissible and domain-independent $h_{\min}$ heuristic, first described in the original paper introducing LRTDP [5]:

$$h_{\min}(s) = \begin{cases} 0, & \text{if } s \in G. \\ \min_{a \in A_s} \left[C(s, a) + \min_{s' \in succ_a(s)} V(s') \right], & \text{otherwise,} \end{cases}$$

where A_s denotes the set of applicable actions in state s and $succ_a(s)$ is the set of successors when applying action a at state s. The four competing algorithms (VI, TVI, LRTDP and pcTVI) were implemented in C++ by the authors of this paper and compiled using the GNU g++ compiler (version 11.2). All tests were performed on a computer equipped with four Intel Xeon E5-2620V4 processors (each of them having 8 cores at 2.1 GHz, for a total of 32 cores). For every test domain, we measured the running time of the four compared algorithms carried out until convergence to an ϵ-optimal value function (we used $\epsilon = 10^{-6}$). Every domain was tested 15 times with randomly generated MDP instances. To minimize random factors, we report the median values obtained over these 15 MDP instances.

Since there is no standard MDP domain in the scientific literature suitable to benchmark a parallel MDP solver, we propose a new general parametric MDP domain that we use to evaluate the algorithms. This domain, which we call chained-MDP, uses 5 parameters: (1) k, the number of independent chains $\{c_1, c_2, \ldots, c_k\}$ in the MDP; (2) n_{scc}, the number of SCCs $\{scc_{i,1}, scc_{i,2}, \ldots, scc_{i,n_{scc}}\}$ in every chain c_i; (3) n_{sps}, the number of states per SCC; (4) n_a the number of applicable actions per state, and (5) n_e the number of probabilistic effects per action. The possible successors $succ(s)$ of a state s in $scc_{i,j}$ are states in $scc_{i,j}$ and either the states in $scc_{i,j+1}$ if it exists, or the goal state otherwise. When generating the transition function of a state-action pair (s, a), we sampled n_e states uniformly from $succ(s)$ with random probabilities. In each of our tests, we used $n_{scc} = 2$, $n_a = 5$ and $n_e = 5$. A representation of a Chained-MDP instance is shown in Figure 1.

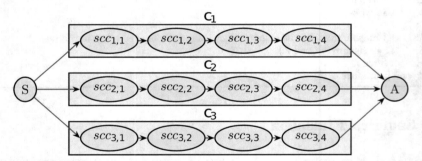

Fig. 1 A chained-MDP instance where $n_c = 3$ and $n_{scc} = 4$. Each ellipse represents a strongly connected component.

Figure 2 presents the obtained results for the Chained-MDP domain when varying the number of states and fixing the number of chains (32). We can observe that when the number of states is small, pcTVI does not provide an important advantage over the existing algorithms since the overhead of creating and managing the threads is taking most of the possible gains. However, as the number of states increases, the gap in the running time between pcTVI and the three other algorithms increases. This indicates that pcTVI is particularly useful on very large MDPs, which are usually needed when considering real-world domains.

Figure 3 presents the obtained results for the same Chained-MDP domain when varying the number of chains and fixing the number of states (1M). When the number of chains increases, the total number of SCCs implicitly increases (which also implies the number of states per SCC decreases). This explains why each tested

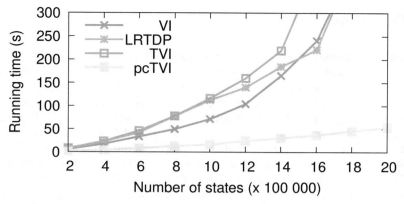

Fig. 2 Average running times (in s) for the Chained-MDP domain with varying number of states and fixed number of chains (32).

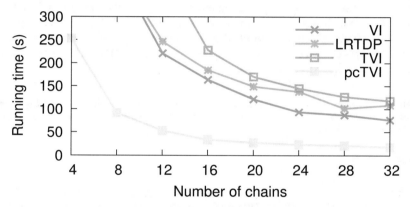

Fig. 3 Average running times (in s) for the Chained-MDP domain with varying number of chains and fixed number of states (1M).

algorithms becomes faster (TVI becomes faster by design, since it solves SCCs one-by-one without doing useless state backups, and VI and LRTDP become faster due to an increased locality of the considered states in memory, which improves cache performance). The performance of pcTVI increases as the number of chains increases (for the same reason as the others algorithms, but also due to increased parallelization opportunities). We can also observe that for domains with 4 chains only, pcTVI still clearly outperforms the other methods. This means that pcTVI does not need a highly parallel server CPU and can be used on standard 4-core computer.

6 Conclusion

The main contributions of this paper are two-fold. First, we presented a new algorithm, pcTVI, which is, to the best of our knowledge, the first MDP solver that takes into account both the topological structure of the MDP (as in TVI) and the parallel capacities of modern computers (as in P3VI). Second, we introduced a new parametric planning domain, Chained-MDP, which models any situation where different strategies (corresponding to a chain) can reach a goal, but where, once committed to a strategy, it is not possible to switch to a different one. This domain is ideal to evaluate the parallel performance of an MDP solver. Our experiments indicate that pcTVI outperforms the other competing methods (VI, LRTDP, and TVI) on every tested instance of the Chained-MDP domain. Moreover, pcTVI is particularly effective when the considered MDP has many SCC chains (for increased parallelization opportunities) of large size (for decreased overhead of assigning small tasks to the threads). As future work, we plan to investigate ways of pruning provably suboptimal actions, which would allow more SCCs to be found. While this paper focuses on the automated planning side of MDPs, the proposed optimization and parallel computing approaches could also be applied when using MDPs with Reinforcement Learning and other ML algorithms.

Acknowledgements This research has been supported by the *Natural Sciences and Engineering Research Council of Canada* (NSERC) and the *Fonds de Recherche du Québec — Nature et Technologies* (FRQNT).

References

1. Champagne Gareau, J., Beaudry E., Makarenkov, V.: A fast electric vehicle planner using clustering. In: Stud. in Classif., Data Anal., and Knowl. Organ., **5**, 17-25. Springer (2021)
2. Mausam, Kolobov, A.: Planning with Markov Decision Processes: An AI Perspective. Morgan & Claypool (2012)
3. Bellman, R.: Dynamic Programming. Prentice Hall (1957)
4. Dai, P., Mausam, Weld, D. S., Goldsmith, J.: Topological value iteration algorithms. J. Artif. Intell. Res., **42**, 181-209 (2011)
5. Bonet, B., Geffner, H.: Labeled RTDP: Improving the convergence of real-time dynamic programming. In: Proc. of ICAPS, pp. 12-21 (2013)
6. Hansen, E., Zilberstein, S.: LAO*: A heuristic search algorithm that finds solutions with loops. Artif. Intell., **129**(1-2), 35-62 (2001)
7. Wingate, D., Seppi, K.: P3VI: A partitioned, prioritized, parallel value iterator. In: Proc. of the Int. Conf. on Mach. Learn. (ICML), 863-870 (2004)
8. Bertsekas, D.: Dynamic Programming and Optimal Control, vol. 2. Athena scientific Belmont, MA (2001)

Three-way Spectral Clustering

Cinzia Di Nuzzo and Salvatore Ingrassia

Abstract In this paper, we present a spectral clustering approach for clustering *three-way data*. Three-way data concern data characterized by three modes: n units, p variables, and t different occasions. In other words, three-way data contain a $t \times p$ observed matrix for each statistical observation. The units generated by simultaneous observation of variables in different contexts are usually structured as three-way data, so each unit is basically represented as a matrix. In order to cluster the n units in K groups, the spectral clustering application to three-way data can be a powerful tool for unsupervised classification. Here, one example on real three-way data have been presented showing that spectral clustering method is a competitive method to cluster this type of data.

Keywords: spectral clustering, kernel function, three-way data

1 Introduction

Spectral clustering methods are based on the graph theory, where the units are represented by the vertices of an undirected graph and the edges are weighted by the pairwise similarities coming from a suitable kernel function, so the clustering problem is reformulated as a graph partition problem, see e.g. [16, 6]. The spectral clustering algorithm is a very powerful method for finding non-convex clusters of data, moreover, it is a handy approach for handling high-dimensional data since it works on a transformation of the raw data having a smaller dimension than the space of the original data.

Cinzia Di Nuzzo (✉)
Department of Statistics, University of Roma La Sapienza, Piazzale Aldo Moro, 5, 00185 Roma, Italy, e-mail: cinzia.dinuzzo@uniroma1.it

Salvatore Ingrassia
Department of Economics and Business, University of Catania, Piazza Università, 2, 95131 Catania, Italy, e-mail: s.ingrassia@unict.it

© The Author(s) 2023
P. Brito et al. (eds.), *Classification and Data Science in the Digital Age*,
Studies in Classification, Data Analysis, and Knowledge Organization,
https://doi.org/10.1007/978-3-031-09034-9_13

Three-way data derives from the observation of various attributes measured on a set of units in different situations; some examples are longitudinal data on multiple response variables and multivariate spatial data. Three-way data can also derive from temporal measurements of a feature vector, thus having the dataset composed of three modes: n units (matrices), p variables (columns), and t times (rows). Clustering of three-way data has attracted a growing interest in literature, see e.g. [14], [1]; model-based clustering of three-way data has been introduced by [15] in the framework of matrix-variate normal mixtures; recent papers include [9] handle on parsimonious models for modeling matrix data; [11] introduce two matrix-variate distributions, both the elliptical heavy-tailed generalization of the matrix-variate normal distribution; [12] deal with three-way data clustering using matrix-variate cluster-weighted models (MV-CWM); and, [13] consider an application to educational data via mixtures of parsimonious matrix-normal distribution.

In this paper, we present a spectral clustering approach for clustering *three-way data* and a suitable kernel function between matrices is introduced. As a matter of fact, the data matrices represent the vertices of the graph, consequently, the edges must be weighted by a single value.

The rest of the paper is organized as follows: in Section 2 the spectral clustering method is summarized; in Section 3 a method to select the parameters in the spectral clustering algorithm is described; in Section 4 the three-way spectral clustering with a new kernel function are introduced; in Section 5 an application based on real three-way data is presented. Finally, in Section 5 we provide concluding remarks.

2 Spectral Clustering

Spectral clustering algorithm for two-way data has been described in [8, 16, 6]. Here, we summarize the main step of this algorithm.

Let $V = \{x_1, x_2, \ldots, x_n\}$ be a set of points in $X \subseteq \mathbb{R}^p$. In order to group the data V in K cluster, the first step concerns the definition of a symmetric and continuous function $\kappa : X \times X \rightarrow [0, \infty)$ called the *kernel function*. Afterwards, a *similarity matrix* $W = (w_{ij})$ can be assigned by setting $w_{ij} = \kappa(x_i, x_j) \geq 0$, for $x_i, x_j \in X$. and finally the *normalized graph Laplacian* matrix $L_{\text{sym}} \in \mathbb{R}^{n \times n}$ is introduced

$$L_{\text{sym}} = I - D^{-1/2} W D^{-1/2}, \tag{1}$$

where $D = \text{diag}(d_1, d_2, \ldots, d_n)$ is the *degree matrix* and d_i is the *degree* of the vertex x_i defined as $d_i = \sum_{j \neq i} w_{ij}$ and I denotes the $n \times n$ identity matrix. The Laplacian matrix L_{sym} is positive semi-definite with n non-negative eigenvalues. For a fixed $K \ll n$, let $\{\gamma_1, \ldots, \gamma_K\}$ be the eigenvectors corresponding to the smallest K eigenvalues of L_{sym}. Then, the *normalized Laplacian embedding in the K principal subspace* is defined as the map $\Phi_\Gamma : \{x_1, \ldots, x_n\} \rightarrow \mathbb{R}^K$ given by

$$\Phi_\Gamma(x_i) = (\gamma_{1i}, \ldots, \gamma_{Ki}), \quad i = 1, \ldots, n,$$

where $\gamma_{1i}, \ldots, \gamma_{Ki}$ are the i-th components of $\gamma_1, \ldots, \gamma_K$, respectively. In other words, the function $\Phi_\Gamma(\cdot)$ maps the data from the input space X to a feature space defined by the K principal subspace of L_{sym}. Afterwards, let $Y = (y_1', \ldots, y_n')$ be the $n \times K$ matrix given by the embedded data in the feature space, where $y_i = \Phi_\Gamma(x_i)$ for $i = 1, \ldots, n$. Finally, the embedded data Y are clustered according to some clustering procedure; usually, the k-means algorithm is taken into account in literature. However, to this end Gaussian mixtures have been proposed because they yield elliptical cluster shapes, i.e. more flexible cluster shapes with respect to the k-means, see [2]. Finally, we point out that the performances of other mixture models based on non-Gaussian component densities have been analyzed, but Gaussian mixture models can be considered as a good trade-off between model simplicity and effectiveness, see [3] for details.

3 A Graphical Approach for Parameter Selection

According to spectral clustering algorithm introduced in Section 2, the spectral approach requires to set: i) the number of clusters K, ii) the kernel function κ (with the corresponding parameter). In order to select these quantities, in the following we summarize the method proposed in [4].

To begin with, we point out that the choice of the kernel function affects the entire data structure in the graph, and consequently, the structure of the Laplacian matrix and its eigenvectors. An optimal kernel function should lead to a similarity matrix W having (as much as possible) diagonal blocks: in this case, we get well-separated groups and we are also able to understand the number of groups in that data set by counting the number of blocks. For the sake of simplicity, we consider here the self-tuning kernel introduced by [17]

$$\kappa(x_i, x_j) = \exp\left(-\frac{\|x_i - x_j\|^2}{\epsilon_i \epsilon_j}\right) \tag{2}$$

with $\epsilon_i = \|x_i - x_h\|$, where x_h is the h-th neighbor of point x_i (similarly for ϵ_j). This function allow to get a similarity matrix that does not depend on any parameter so that the algorithm of spectral clustering will be based on the pairwise proximity between units. On the contrary, we need to select the h-th neighbor of the unit in (2).

The main novelty of the joint-graphical approach concerns the analysis of some graphic features of the Laplacian matrix including the shape of the embedded space. Indeed, the embedded data provide useful information for the clustering, in particular the main results in [10] and [5] allow to deduce that if the embedded data assume a cones structure, then the number of clusters is equal to the number of the cones/spikes in the feature space; furthermore, a clearer clustering structure emerges when the spikes are narrower and well separated.

The idea behind the graphical approach is to select the number K of groups and the parameter h in the kernel function from a joint analysis of three main characteristics: the plot of the Laplacian matrix; the maxima values of the eigengaps between two consecutive eigenvalues; the scatter plot of the mapped data in the feature space and in particular the number of spikes counted in the embedded data space.

We remark that we cannot analyze all possible values of $h \in \{1, 2, \ldots, n-1\}$ and hence we choose a suitable subset $\mathcal{H} \subset \{1, 2, \ldots, n-1\}$, in particular we choose $\mathcal{H} = \{1\%, 2\%, 5\%, 10\%, 15\%, 20\%\} \times n \subset \{1, 2, \ldots, n-1\}$, and select $h \in \mathcal{H}$, see the following procedure for details.

Parameter selection (K and h)

Input: data set V, kernel function κ, $\mathcal{H}$.

1. For each h in $\mathcal{H}$, compute the matrix M_s and analyze the block structure in the greyscale plot of M_s.
2. For each h in $\mathcal{H}$, plot the embedded data in the feature space and analyze the shape of the cone structure.
3. If the number of blocks in Step 1 is equal to the number of spikes in Step 2, then set K equal to the number of blocks. Go to Step 5.
4. Otherwise, analyze the eigengap plot.

   a. If this plot shows a unique maximum eigengap for each $h \in \mathcal{H}$, then set K according to this maximum. Go to Step 5.
   b. If this plot shows multiple maxima for different $h \in \mathcal{H}$, select the number of clusters K not to be smaller than the number of tight spikes in the corresponding plot of the embedded data.

5. Select $h \in \mathcal{H}$ such that the clearest orthogonal data structure emerges from the plot of the embedded data.
6. Stop.

Output: K, h.

---

4 Three-way Spectral Clustering

In this section, we propose a spectral approach for clustering three-way data. Three-way data consists of a data set referring to the same sets of units and variables, observed in different situations, i.e., a set of multivariate matrices, that can be organized in three modes: n units, p variables, and t situations. Therefore, given n matrices that represent the vertices of the graph, each matrix is composed by p columns that represent our variables and t rows that represent the time or another feature. So we have a tensor of dimension $n \times t \times p$, thus the dataset is a tensor $\{X\}_{isk}$ for $i = 1, \ldots, n, s = 1, \ldots, t, k = 1, \ldots, p$.

We define a distance function δ_M between two matrices $A, B \in \mathbb{R}^{p \times t}$ such that $\delta_M : R^{t \times p} \times R^{t \times p} \rightarrow [0, +\infty)$ is defined as

$$\delta_M(A, B) := \|A - B\|_F = \sqrt{\sum_{s=1}^{t} \sum_{k=1}^{p} |a_{sk} - b_{sk}|^2} \tag{3}$$

where $\| \cdot \|_F$ is Frobenius norm[1]. Thus the distance between two units in the matrix data X is equal to

$$\delta_M(X_{i_1sk}, X_{i_2sk}) = \sqrt{\sum_{s=1}^{t} \sum_{k=1}^{p} |X_{i_1sk} - X_{i_2sk}|^2}, \qquad \text{for } i_1, i_2 = 1, \ldots, n. \tag{4}$$

For simplicity, in the following, we denote $\delta_M(X_{i_1sk}, X_{i_2sk})$ by $\delta_M(i_1, i_2)$. Moreover, we define the three-way self-tuning kernel function as

$$\kappa_S : X \times X \rightarrow [0, +\infty), \qquad \kappa_S(i_1, i_2) = \exp\left(-\frac{\delta_M(i_1, i_2)}{\epsilon_{i_1} \epsilon_{i_2}}\right) \tag{5}$$

where ϵ_{i_1} and ϵ_{i_2} need to be selected like in the kernel defined in (2).

Afterwards, we compute the similarity matrix W given by $w_{i_1i_2} = \kappa(i_1, i_2)$, so that we can apply the spectral clustering algorithm.

Finally, we point out that, differently from approaches based on mixtures of matrix-variate data, the number of variables of the data set is not a critical issue because the spectral clustering algorithm is based on distance measures.

5 A Real Data Application

We apply the three-way spectral clustering to the analysis of the Insurance data set, available in the splm R package. This dataset was initially introduced by [7] and has recently been analyzed by [12]. The goal is to study the consumption of non-life insurance during the years 1998-2002 in the 103 Italian provinces, so $t = 5$ and $n = 103$. As regards the number of variables, we consider all the variables contained in the data set, so $p = 11$. Thus, we have 103 matrices of dimensions 5×11.

The 103 Italian provinces are divided into north-west (24 provinces), north-east (22 provinces), center (21 provinces), south (23 provinces), and islands (13 provinces).

As regard the choice of K and h, we consider the graphical approach introduced in Section 3. In Figure 1 the geometric features of spectral clustering are plotted as h varies. From the number of blocks of the Laplacian matrix (Figure 1-a)), the first maximum eigengap (Figure 1-b)) and the number of spikes in the feature space (Figure 1-c)), we deduce that the number of clusters is $K = 2$. For the selection of

---

[1] In general, given a matrix $A \in \mathbb{R}^{n \times m}$, with $A = (a_{ij})$ for $i = 1, \ldots, n$ and $j = 1, \ldots, m$. The Frobenius norm is defined by

$$\|A\|_F := \sqrt{\sum_{j=1}^{m} \sum_{i=1}^{n} |a_{ij}|^2}.$$

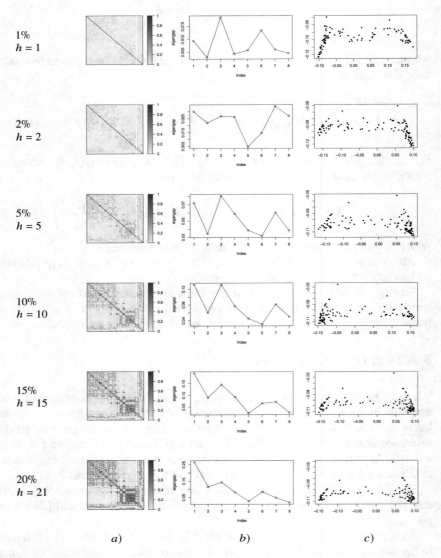

Fig. 1 *Insurance data.* Spectral clustering features: a) plot of Laplacian matrix in greyscale; b) plot of the first eight eigengap values; c) scatterplot of the embedded data along with directions (γ_1, γ_2).

Table 1 *Insurance data.* Table of spectral clustering result.

	NORTHWEST (24 provinces)
Cluster 1	NORTH EAST (22 provinces)
	CENTRE (15 provinces)
	CENTRE (6 provinces)
Cluster 2	SOUTH (23 provinces)
	ISLANDS (13 provinces)

h we choose indifferently $h = 15$ and $h = 21$ because in these cases the maximum eigengap highlights the maximum values corresponding to $K = 2$. In Table 1 the clustering results are presented. This table shows that only 6 center provinces are classified together with the southern provinces. But to be sure that these provinces are neighboring the south provinces, let us analyze spectral clustering results on the map of Italy. Figure 2-a) illustrates the partition deriving from spectral clustering in the political map of Italy, where Italian regions are described by the yellow lines, while the provinces are by the black lines. The result shows a clear separation between center-north Italy and south-insular Italy, in fact, the center-north has a level of insurance penetration close to the European averages, while the South is less developed economically. However, the Massa-Carrara province should belong to the centre-north group. Moreover, we remark that the Rome province, being the capital of Italy, has one socio-economic development comparable to that of north Italy justifying belonging to the centre-north group.

Furthermore, in Figure 2-b) we also represented the partition produced by MN-CWM proposed in [12], we note that the two clustering results are very similar to each other and differ only for one province of central Italy (precisely for the province of Terni). It should also be emphasized that the dataset analyzed by [12] is different from the one analyzed here, since, to avoid excessive parameterization of the models, the authors select only $p = 5$ variables in the data set.

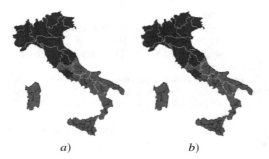

a) b)

Fig. 2 *Insurance data.* a) Three-way spectral clustering; b) Method proposed by [12].

6 Conclusion

In this paper, a spectral approach to cluster three-way data has been proposed. So the data are organized in a tensor and the vertices in the graph are represented by the matrices of dimension $t \times p$. In order to weigh the matrices in the graph, a kernel function based on the Frobenius norm between the matrix difference has been introduced. The performance of the spectral clustering algorithm has been shown in one real three-way data set. Our method is competitive with respect to other clustering methods proposed in the literature to perform matrix-data clustering. Finally, in order to provide suggestions for future research, other kernel functions can be introduced considering different distances with respect to the Frobenius norm.

Acknowledgements This work was supported by the University of Catania grant PIACERI/CRASI (2020).

References

1. Bocci, L., Vicari, D.: ROOTCLUS: Searching for "ROOT CLUSters" in Three-Way Proximity Data. Psychometrika. **84**, 941–985 (2019)
2. Di Nuzzo, C., Ingrassia, S.: A mixture model approach to spectral clustering and application to textual data. Stat. Meth. Appl. Forthcoming (2022)
3. Di Nuzzo, C.: Model selection and mixture approaches in the spectral clustering algorithm. Ph.D. thesis, Economics, Management and Statistics, University of Messina (2021)
4. Di Nuzzo, C., Ingrassia, S.: A joint graphical approach for model selection in the spectral clustering algorithm. Tech. Rep. (2022)
5. Garcia Trillos, N., Hoffman, F., Hosseini, B.: Geometric structure of graph Laplacian embeddings. arXiv preprint arXiv:1901.10651. (2019)
6. Meila, M.: Spectral clustering. In Hennig, C., Meila, M., Murtagh, F., Rocci, R. (eds.). Handbook of Cluster Analysis. Chapman and Hall/CRC (2015)
7. Millo, G., Carmeci, G.: Non-life insurance consumption in Italy: A sub-regional panel data analysis. J. Geogr. Syst. **12**, 1–26 (2011)
8. Ng, A., Jordan, M., Weiss, Y.: On spectral clustering: Analysis and an algorithm. Adv. Neural Inf. Process. Syst. **14** (2002)
9. Sarkar, S., Zhu, X., Melnykov, V., Ingrassia, S.: On parsimonious models for modeling matrix data. Comput. Stat. Data Anal. **142**, 106822 (2020)
10. Schiebinger, G., Wainwright, M. J., Yu, B.: The geometry of kernelized spectral clustering. Ann. Stat. **43**(2), 819–846 (2015a)
11. Tomarchio, S. D., Punzo, A., Bagnato, L.: Two new matrix-variate distributions with application in model-based clustering. Comput. Stat. Data Anal. **152**, 107050 (2020)
12. Tomarchio, S. D., McNicholas, P., Punzo, A.: Matrix normal cluster-weighted models. J. Classif. **38**, 556-575 (2021)
13. Tomarchio, S. D., Ingrassia, S., Melnykov, V.: Modeling students' career indicators via mixtures of parsimonious matrix-normal distributions. Aust. New Zeal. J. Stat. Forthcoming (2022)
14. Vichi, M., Rocci, R., Kiers, H. A. L.: Simultaneous component and clustering models for three-way data: Within and between approaches. J. Classif. **24**, 71–98 (2007)
15. Viroli, C.: Finite mixtures of matrix normal distributions for classifying three-way data. Stat. Comput. **21**, 511–522 (2011)
16. von Luxburg, U.: A tutorial on spectral clustering. Stat. Comput. **17**(4), 395–416 (2007)
17. Zelnik-Manor, L., Perona, P.: Self-tuning spectral clustering. Adv. Neural Inf. Process. Syst. **17** (2004)

Improving Classification of Documents by Semi-supervised Clustering in a Semantic Space

Jasminka Dobša and Henk A. L. Kiers

Abstract In the paper we propose a method for representation of documents in a semantic lower-dimensional space based on the modified Reduced k-means method which penalizes clusterings that are distant from classification of training documents given by experts. Reduced k-means (RKM) enables simultaneously clustering of documents and extraction of factors. By projection of documents represented in the vector space model on extracted factors, documents are clustered in the semantic space in a semi-supervised way (using penalization) because clustering is guided by classification given by experts, which enables improvement of classification performance of test documents.

Classification performance is tested for classification by logistic regression and support vector machines (SVMs) for classes of Reuters-21578 data set. It is shown that representation of documents by the RKM method with penalization improves the average precision of classification by SVMs for the 25 largest classes of Reuters collection for about 5,5% with the same level of average recall in comparison to the basic representation in the vector space model. In the case of classification by logistic regression, representation by the RKM with penalization improves average recall for about 1% in comparison to the basic representation.

Keywords: classification of textual documents, LSA, reduced k-means

Jasminka Dobša (✉)
Faculty of Organization and Informatics, University of Zagreb, Pavlinska 2, 40000 Varaždin, Croatia, e-mail: jasminka.dobsa@foi.hr

Henk A. L. Kiers
Department of Psychology, University of Groningan, Grote Kruisstraat 2/1, 9712 TS Groningen, The Netherlands, e-mail: h.a.l.kiers@rug.nl

121

P. Brito et al. (eds.), *Classification and Data Science in the Digital Age*,
Studies in Classification, Data Analysis, and Knowledge Organization,
https://doi.org/10.1007/978-3-031-09034-9_14

1 Introduction

There are two main families of methods that deal with representation of documents and words that index them: global matrix factorization methods such as Latent Semantic Analysis (LSA) [2] and local context window methods such as the continuous bag of words (CBOW) model and the continuous skip-gram model [8]. The latter use neural networks for learning of representations of words and are intensively explored lately in the scientific community since the development of fast processors has enabled processing of huge amounts of data which resulted in improvements in performance of wide spectra of text mining and natural language tasks. However, representation of words solely by context window methods has a drawback due to the neglect of information about global corpus statistics [9].

In this paper we propose a method for representation of documents by application of a penalized version of the RKM method [4] on a term-document matrix. The corpus of textual documents is represented by a sparse term-document matrix in which entry (i, j) is equal to the weight of the i-th index term for the j-th document. Weights of terms are given by the TfIdf weighting which utilizes local information about the frequency of the i-th term in the j-th document and global information about usage of the i-th term in the entire collection. A benchmark method that utilizes global matrix factorization on term-document matrices is LSA [2] which uses truncated singular value decomposition (SVD) for representation of terms and documents in lower-dimensional semantic space. SVD does not capture the clustering structure of data which motivates application of the RKM.

The rest of the paper is organized as follows: the second section describes related work on representation of documents and words and methods of dimensionality reduction related to RKM. The third section describes the modified RKM method with penalization, while the fourth section describes an experiment on Reuters-21578 data set. In the last section conclusions and directions for further work are given.

2 Related Work

2.1 Representation by Matrix Factorization Methods

A benchmark method among methods that utilize matrix factorization for representation of textual documents is the method of LSA introduced in 1994 [2]. By LSA a sparse term-document matrix is transformed via SVD into a dense matrix of the same term-document type with representations of words (index terms) and documents in a lower-dimensional space. The idea is to map similar documents, or those that describe the same topics, closer to each other regardless of the terms that are used in them. A very efficient application of LSA is in cross-lingual information retrieval where relevant documents for a query in one language are retrieved from a set of documents in another language [7]. According to our knowledge application of methods that simultaneously cluster objects and extract factors in the field of text mining is very limited. In [6] a method is proposed for cross-lingual information retrieval based on the RKM method.

2.2 Neural Network Word Embeddings

Another approach is to learn representations of words, or so called embeddings, by using local context windows. In 2003 Bengio and coauthors [1] proposed a neural probabilistic language model that uses simple neural network architecture to learn distributed representations for each word as well as probability functions for word sequences, expressed in terms of these representations. Mikolov and coautors [8] proposed in 2013 two models based on single-layer neural network architectures: the skip gram-model that predicts context words given the current word and the continuous bag of words model which predicts current words based on the context. In 2014 the GloVe model [9] was proposed, based on the critique that neural network models suffer from the disadvantage that they do not utilize co-occurrence statistics of the entire corpus, but scan only context windows of words ignoring vast amounts of repetition in the data. That model exploits the advantages of global matrix factorization methods by utilization of term-term co-occurrence matrices and local context window methods.

Word embedding can be classified as static such as word2vec [8] and GloVe [9], and contextual, such as ELMo [10] and BERT [5]. Contextual representation is introduced in [10] in order to model characteristics of word use (syntax and semantics) on one side and variation in word representation due to the context in which words are appearing.

2.3 Methods for Simultaneous Clustering and Factor Extraction

A standard procedure for clustering of objects in a lower-dimensional space is tandem analysis which includes projection of data by principal components and clustering of data in a lower-dimensional space. Such an approach was criticized in [3] and [4] since principal components may extract dimensions which do not necessarily significantly contribute to the identification of a clustering structure in the data. As a response, De Soete and Carroll proposed the method of RKM [4] which simultaneously clusters data and extracts the factors of variables by reconstructing the original data with only centroids of clusters in a lower-dimensional space. The algorithm of Factorial k-means (FKM) proposed by Vichi and Kiers [13] has the same aim of simultaneous reduction of objects and variables and it reconstructs the data in a lower-dimensional space by its centroids in the same space. The application of the latter method is limited in text mining since the method is limited to cases in which the number of variables is less than the number of cases. In [11] the RKM and FKM methods are compared using simulations and theoretically in order to identify cases for their application. Timmerman and associates also propose method of Subspace k-means [12] which gives an insight into cluster characteristics in terms of relative positions of clusters given by centroids and the shape of the clusters given by within cluster residuals.

3 Reduced k-Means with Penalization

Let $\mathbf{X}$ be $m \times n$ term-document matrix. We use the following notation:

- $\mathbf{A}$ is an $m \times k$ columnwise orthonormal matrix of extracted factors;
- $\mathbf{M}$ is an $n \times c$ membership matrix, where c is a predefined number of clusters; $m_{ic} = 1$ if object (document) i belongs to cluster c and 0 otherwise;
- $\mathbf{Y}$ is a $c \times k$ matrix which gives centroids of clusters in the lower-dimensional space.

By definition, we suppose that every document in the collection belongs to exactly one cluster. The RKM method minimizes the loss function

$$F(\mathbf{M},\mathbf{A}) = \|\mathbf{X} - \mathbf{A}\mathbf{Y}^T \mathbf{M}^T\|^2 \tag{1}$$

in the least squares sense. The dimension of the lower-dimensional space must be less or equal to the number of clusters. Modified RKM with penalization minimizes the loss function

$$F(\mathbf{M},\mathbf{A}) = \|\mathbf{X} - \mathbf{A}\mathbf{Y}^T \mathbf{M}^T\|^2 + \lambda\|\mathbf{M} - \mathbf{G}\|^2 \tag{2}$$

where $\mathbf{G}$ is $n \times c$ membership matrix based on expert judgements. If c is number of classes then $g_{ic} = 0$ if object (document) i belongs to class c, and 0 otherwise. By the second summand in the loss function we penalize clusterings that are distant from the classes by expert judgements using parameter λ that regularizes the importance of that penalization. We use the alternating least squares (ALS) algorithm analogous to the one in [4] which alternates between corrections of the loading matrix $\mathbf{A}$ in one step and of the membership matrix $\mathbf{M}$ in another. As each of the steps in the ALS algorithm improves the loss function, the algorithm converges to at least a local minimum. By starting the procedure from a large number of random initial estimates and choosing the best solution, the chances of obtaining the global minimum are increased.

4 Experiment

4.1 Design of Experiment

Experiments are conducted for classification on the Reuters-21578 data set, specifically using the ModApte Split which assigns Reuters reports from April 7, 1987 and before to the training set, and after, until end of 1987, to the test set. It consists of 9603 training and 3299 test documents. The collection has 90 classes which contain at least one training and test document. Documents are represented by a bag of words representation. A list of index terms is formed based on terms that appear in at least four documents of the collection, which resulted in a list of 9867 index terms.

Classification is conducted by logistic regression (LR) and SVM algorithm. The basic model is the bag of words representation (full representation), while representations in the lower-dimensional space are obtained by SVD (Latent Sematic Analysis), RKM and RKM with penalization ($\lambda = 0.1, 0.2, 0.4, 0.6$). For RKM and RKM with penalization representations are obtained by applying matrix factorization on the term-document matrix of the training documents, and by projection of test documents on factors given by matrix A in the factorization. RKM is computed for 90 clusters (which corresponds to the number of classes in the collection) using as dimension of the lower-dimensional space $k = 85$, and truncated SVD is computed for $k = 85$ as well. The RKM and RKM with penalization algorithms are run 10 times (with different starting estimates), and the representation and factorization with the minimal loss function is chosen. The optimal cost parameter for LR and SVM is chosen by grid search technique from the set of values 0.1, 0.5, 1, 10, 100 and 1000. For the classification methods, the LiblineaR library in R is used, while RKM and RKM with penalization algorithm are implemented in Matlab.

4.2 Results

Results are given in terms of precision, recall, and F_1 measure of the classification. Recall is proportion of correctly classified samples among all positive samples (i.e., samples actually belonging to the class, according to the expert), while precision is proportion of correctly classified samples among all samples classified as positive by the model. In the Figures 1 and 2, are shown results of average F_1 measures of classification for 5 classes sorted in descending order by their size, i.e. number of train documents (which is 2877 to 389 for classes 1-5, 369 to 181 for classes 6-10, 140 to 111 for classes 11-15, 101 to 75 for classes 16-20, 75 to 55 for classes 21-25, 50 to 41 for classes 26-30, 40 to 37 for classes 31-35, 35 to 24 for classes 36-40, 23 to 19 for classes 41-45, 18 to 16 for classes 46-50, 16 to 13 for classes 51-55, and 13-10 for classes 56-60). Figure 1 shows the results for classification by LR, while Figure 2 for classification by SVM. Only the 60 largest classes are observed since smaller classes (less than 10 training documents) are not interesting for the

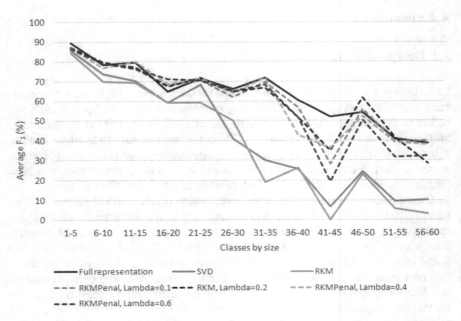

Fig. 1 Average F_1 measure of classification by LR for 5 classes sorted by their size.

research, because for those classes recall is low and it can be expected that full bag of words representation will result in better recognition since classes can possibly be recognized by key words, but not by transformed representations. It can be seen that F_1 measures are comparable for the full representation and various representations by RKM with penalization for both classification algorithms for the biggest 25 classes. For smaller classes results for representation by RKM with penalization are unstable, although for some classes they were better than the basic representation (in the case of LR). Classification for representations obtained by SVM and RKM without penalization resulted in lower F_1 measures for all class sizes.

In Table 1 are shown average precision, recall and F_1 measures for the 25 largest classes for both classification algorithms and all observed representations. In the case of classification by LR the average recall is improved for representation by RKM with penalization (for $\lambda = 0.4$) approximately 1% compared to basic full represen- tation. For classification by SVM average precision is improved for representation by RKM with penalization (for $\lambda = 0.6$) for almost 6% and F_1 measure is improved for representation by RKM with penalization ($\lambda = 0.4$) for 2% in comparison to the basic full representation. The best results are obtained for classification by the SVM algorithm and representation with RKM with penalization with $\lambda = 0.2$ for which precision is improved for 5% with the similar level of recall as in the basic representation.

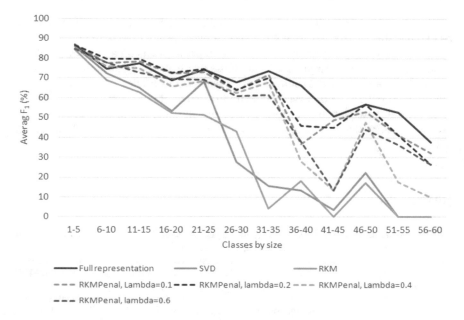

Fig. 2 Average F_1 measure of classification by SVM for 5 classes sorted by their size.

Table 1 Average precision, recall, and F_1 measure of classification for the 25 largest classes.

Class. algorithm	Logistic regression			SVM		
Representation	Precision	Recall	F_1	Precision	Recall	F_1
Full	**86.31**	70.24	76.84	82.76	71.72	76.47
SVD	82.80	64.84	71.42	85.24	61.61	68.99
RKM	80.80	61.10	68.44	82.93	55.66	63.83
RKMPenal, $\lambda = 0.1$	84.24	70.71	76.27	87.24	71.01	77.62
RKMPenal, $\lambda = 0.2$	84.68	71.23	76.72	87.78	**72.16**	**78.57**
RKMPenal, $\lambda = 0.4$	84.72	**71.38**	**76.88**	87.86	64.93	73.87
RKMPenal, $\lambda = 0.6$	85.89	70.40	76.80	**88.40**	66.11	74.75

5 Conclusions and Further Work

In this paper we propose a modification of the RKM method that simultaneously clusters documents and extracts factors on one side, and penalizes clusterings that are distant from the classification of the training documents given by experts on the other side. We show that such a modification enables representation of textual documents in a semantic lower-dimensional space that improves performance of classification. The method is tested for classes of Reuters-21758 data set and compared to the full bag of words representation and the method of LSA. It is also shown that the original RKM method without proposed modification does not have the same effect on classification performance; it has a similar effect as the LSA method.

The proposed representation method can improve precision and recall of classification for sufficiently large classes, i.e. those that have enough training documents to enable capturing of semantic relations and characteristics of classes. A more important effect can be observed in the improvement of precision.

In the future we plan to investigate hybrid models using representation of words by neural language models and application in different domains, such as classification of images.

References

1. Bengio, J., Ducharme, R., Vincet, P., Jauvin, C.: A Neural probabilistic language model. Journal of Machine Learning Research **3**, 1137-1155 (1997)
2. Deerwester, S., Dumas, S. T., Furnas, G.W., Landauer, T. K., Harshman, R. A.: Indexing by latent semantic analysis. Journal of the American Society for Information Science **41(6)**, 381-407 (1990)
3. De Sarbo, W. S., Jedidi, K., Cool, K., Schendel, D.: Simultaneous multidimensional unfolding and cluster analysis: an investigation of strategic groups. Marketing Letters, **2**, 129-146 (1990)
4. De Soete, G., Carroll, J. D.: K-means clustering in a low-dimensional Euclidean space. In: Diday, E., Lechevallier, Y., Schader, M., Bertrand, P., Burtschy, B. (eds.) New Approaches in Classification and Data Analysis. Studies in Classification, Data Analysis, and Knowledge Organization, pp. 212-219. Springer, Heidelberg (1994)
5. Devlin, J., Chang, M., Lee, K., Toutanova, K.: BERT: Pre-training of deep bidirectional transformers for language understanding. In: Proceedings of Annual Conference of the North American Chapter of the Association for Computation Linguistic, pp. 4171-4186, Association for Computational Linguistic (2019)
6. Dobša, J., Mladenić, D., Rupnik, J., Radošević, D., Magdalenić, I.: Cross-language information retrieval by Reduced k-means, International Journal of Computer Information Systems and Industrial Management Applications, **10**, 314-322 (2018)
7. Dumas, S., Letche, T., Littman, M., Landauer, T.: Automatic cross-language retrieval using latent semantic indexing. In: Proceedings of the AAAI spring symposium on cross-language text and speech retrieval, pp. 15-21. American Association for Artificial Intelligence (1997)
8. Mikolov, T., Chen, K., Corrado, G.S., Dean, J.: Efficient estimation of word representations in vector space (2013) Available via arXiv.org
   `https://arxiv.org/abs/1301.3781.Cited21Jan2022`
9. Pennington, J., Socher, R., Manning, C. D.: GloVe: Global vectors for word representation. In: Proceedings of the 2014 Conference on Empirical Methods in Natural Language Processing (EMNLP), pp. 1532-1543, Association for Computational Linguistics, (2014)
10. Peters, M., Neumann, M., Iyyer, M., Gardner, M., Clark, C., Lee, K., Tettlemoyer, L.: Deep contextualized word representations. In Proceedings of the Conference of the North American Chapter of the Association for Computational Linguistics: Human Language Technologies, 1:2227-2237 (2018)
11. Timmerman, M. E. Ceulemans, E., Kiers, H. A. L., Vichi, M: Factorial and Reduced k-means reconsidered. Computational Statistics & Data Analyisis, **54**, 1856-1871 (2010)
12. Timmerman, M. E., Ceulemans, E., De Rover, K., Van Leeuwen, K.: Subspace$k$-means clustering, Behavioural Research, **45**, 1011-1023 (2013)
13. Vichi, M., Kiers, H. A. L.: Factorial k-means analysis for two-way data, Computational Statistics & Data Analysis, **37**, 49-64 (2001)

Trends in Data Stream Mining

João Gama

Abstract Learning from data streams is a hot topic in machine learning and data mining. This article presents our recent work on the topic of learning from data streams. We focus on emerging topics, including fraud detection and hyper-parameter tuning for streaming data. The first study is a case study on interconnected by-pass fraud. This is a real-world problem from high-speed telecommunications data that clearly illustrates the need for online data stream processing. In the second study, we present an optimization algorithm for online hyper-parameter tuning from non-stationary data streams.

Keywords: fraud detection, hyperparameter tuning, learning from data streams

1 Introduction

The developments of information and communication technologies dramatically change the data collection and processing methods. What distinguishes current data sets from earlier ones are automatic data feeds. We do not just have people entering information into a computer. We have computers entering data into each other. In most challenging applications, data are modeled best not as persistent tables, but rather as transient data streams.

This article presents our recent work on the topic of learning from data streams. It is organized into main sections. The first one is a real-world application of data stream techniques to a telecommunications fraud detection problem. It is based on the work presented in [5]. The second topic discusses the problem of hyperparameter tuning in the context of data stream mining. It is based on the work presented in [4].

João Gama (✉)
FEP-University of Porto and INESC TEC
R. Dr. Roberto Frias, Porto, Portugal, e-mail: `jgama@fep.up.pt`

© The Author(s) 2023
P. Brito et al. (eds.), *Classification and Data Science in the Digital Age*,
Studies in Classification, Data Analysis, and Knowledge Organization,
https://doi.org/10.1007/978-3-031-09034-9_15

2 Fraud Detection: a Case Study

The high asymmetry of international termination rates with regard to domestic ones, where international calls have higher charges applied by the operator where the call terminates, is fertile ground for the appearance of fraud in Telecommunications. There are several types of fraud that exploit this type of differential, being the Interconnect Bypass Fraud one of the most expressive [1, 3].

In this type of fraud, one of several intermediaries responsible for delivering the calls forwards the traffic over a low-cost IP connection, reintroducing the call in the destination network already as a local call, using VOIP Gateways. This way, the entity that sent the traffic is charged the amount corresponding to the delivery of international traffic. However, once it is illegally delivered as national traffic, it will not have to pay the international termination fee, appropriating this amount.

Traditionally, the telecom operators analyze the calls of these Gateways to detect the fraud patterns and, once identified, have their SIM cards blocked. The constant evolution in terms of technology adopted on these gateways allows them to work like real SIM farms capable of manipulating identifiers, simulating standard call patterns similar to the ones of regular users, and even being mounted on vehicles to complicate the detection using location information.

The interconnect bypass fraud detection algorithms typically consume a stream S of events, where S contains information about the origin number $A - Number$, the destination number $B - Number$, the associated timestamp, and the status of the call (accomplished or not). The expected output of this type of algorithm is a set of potential fraudulent $A - Numbers$ that require validation by the telecom operator. This process is not fully automated to avoid blocking legit $A - Numbers$ and getting penalties. In the interconnect bypass fraud, we can observe three different types of abnormal behaviors:

1. the burst of calls, which are $A - Numbers$ that produce enormous quantities of #$calls$ (above the $\overline{\#calls}$ of all $A - Numbers$) during a specific time window W. The size of this time window is typically small;
2. the repetitions, which are the repetition of some pattern (#$calls$) produced by a $A - Number$ during consecutive time windows W;
3. the mirror behaviors, which are two distinct $A - Numbers$ (typically these $A - Numbers$ are from the same country) that produces the same pattern of calls (#$calls$) during a time window W.

Algorithm 2 The Lossy Counting Algorithm.

1: **procedure** LossyCounting(S: A Sequence of Examples; ϵ: Error margin; α: fast forgetting parameter)
2:     $n \leftarrow 0; \Delta \leftarrow 0; T \leftarrow 0;$
3:     **for** example $e \in S$ **do**
4:         $n \leftarrow n+1$
5:         **if** e is monitored **then**
6:             Increment $Count_e$
7:         **else**
8:             $T \leftarrow T \cup \{e, 1 + \Delta\}$
9:         **end if**
10:        **if** $\lceil \frac{n}{\epsilon} \rceil \neq \Delta$ **then**
11:           $\Delta \leftarrow \frac{n}{\epsilon}$
12:        **end if**
13:        **for** all $j \in T$ **do**
14:           **if** $Count_j < \delta$ **then**
15:              $T \leftarrow T \backslash \{j\}$
16:           **end if**
17:        **end for**
18:     **end for**
19: **end procedure**

Algorithm 3 The Lossy Counting with Fast Forgetting Algorithm.

1: **procedure** LossyCounting(S: A Sequence of Examples; ϵ: Error margin; α: fast forgetting parameter)
2:     $n \leftarrow 0; \Delta \leftarrow 0; T \leftarrow 0;$
3:     **for** example $e \in S$ **do**
4:         $n \leftarrow n+1$
5:         **if** e is monitored **then**
6:             Increment $Count_e$
7:         **else**
8:             $T \leftarrow T \cup \{e, 1 + \Delta\}$
9:         **end if**
10:        **if** $\lceil \frac{n}{\epsilon} \rceil \neq \Delta$ **then**
11:           $\Delta \leftarrow \frac{n}{\epsilon}$
12:        **end if**
13:        **for** all $j \in T$ **do**
14:           $Count_j \leftarrow \alpha * Count_j$
15:           **if** $Count_j < \delta$ **then**
16:              $T \leftarrow T \backslash \{j\}$
17:           **end if**
18:        **end for**
19:     **end for**
20: **end procedure**

Figures 1 and 2 present the evolving top-10 most active phone numbers. The first Figure 1 presents the top-10 cumulative counts, while the Figure 2 presents the top-10 counts with forget.

3 Learning to Learn Hyperparameters

A hyperparameter is a parameter whose value is used to control the learning process. Hyperparameter optimization (or tuning) is the problem of choosing a set of optimal hyper-parameters for a learning algorithm. For this propose we adapt the Nelder-Mead algorithm [4] for the streaming context. This algorithm is a simplex search algorithm for multidimensional unconstrained optimization without derivatives. The vertexes of the simplex, which define a convex hull shape, are iteratively updated in order to sequentially discard the vertex associated with the largest cost function value.

The Nelder-Mead algorithm relies on four simple operations: *reflection*, *shrinkage*, *contraction* and *expansion*. Figure 3 illustrates the four corresponding Nelder-Mead operators R, S, C and E. Each vertex represents a model containing a set of hyper-parameters. The vertexes (models under optimisation) are ordered and named according to the root mean square error (RMSE) value: best (B), good (G), which is

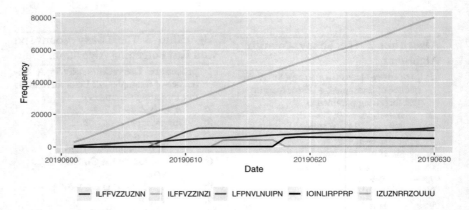

Fig. 1 Approximate Counts with Lossy Counting.

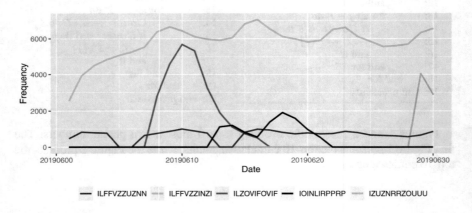

Fig. 2 Approximate Counts with Lossy Counting and Fast Forgetting.

the closest to the best vertex, and worst (W). M is a mid vertex (auxiliary model). The bottom panel in Figure 3 describe the four operations: Contraction, Reflexion, Expansion, and Shrink.

For each Nelder-Mead operation, it is necessary to compute an additional set of vertexes (midpoint M, reflection R, expansion E, contraction C and shrinkage S) and verify if the calculated vertexes belong to the search space. First, the algorithm computes the midpoint (M) of the best face of the shape as well as the reflection point (R). After this initial step, it determines whether to reflect or expand based on the set of heuristics.

The dynamic sample size, which is based on the RMSE metric, attempts to identify significant changes in the streamed data. Whenever such a change is detected, the Nelder-Mead compares the performance of the $n + 1$ models under analysis to choose the most promising model. The sample size S_{size} is given by Equation 1 where σ

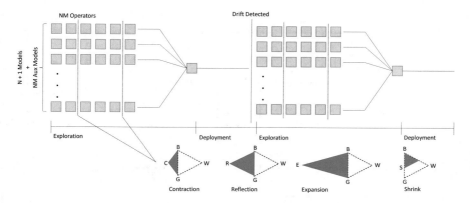

Fig. 3 SPT working modes: Exploration and Deployment. Bottom panel illustrates the Nelder & Mead operators.

represents the standard deviation of the RMSE and M the desired error margin. We use $M = 95\%$.

$$S_{size} = \frac{4\sigma^2}{M^2} \qquad (1)$$

However, to avoid using small samples, that imply error estimations with large variance, we defined a lower bound of 30 samples. The adaptation of the Nelder-Mead algorithm to on-line scenarios relies extensively on parallel processing. The main thread launches the $n+1$ model threads and starts a continuous event processing loop. This loop dispatches the incoming events to the model threads and, whenever it reaches the sample size interval, assesses the running models, and calculates the new sample size. The model assessment involves the ordering of the $n + 1$ models by RMSE value and the application of the Nelder-Mead algorithm to substitute the worst model. The Nelder-Mead parallel implementation creates a dedicated thread per Nelder-Mead operator, totaling seven threads. Each Nelder-Mead operator thread generates a new model and calculates the incremental RMSE using the instances of the last sample size interval. The worst model is substituted by the Nelder-Mead operator thread model with the lowest RMSE.

Figure 4 presents the critical difference diagram [2] of three hyper-parameter tuning algorithms: SPT, Grid search, default parameter values on four benchmark classification datasets. The diagram clearly illustrates the good performance of SPT.

4 Conclusions

This paper reviews our recent work in learning from data streams. The two works present different approaches to dealing with high-speed and time-evolving data: from applied research in fraud detection to fundamental research on hyperparameter

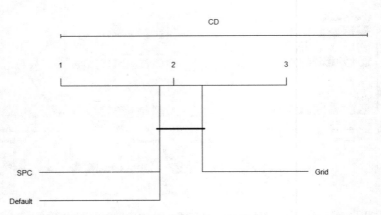

Fig. 4 Critical Difference Diagram comparing Self hyperparameter tuning, Grid hyperparameter tuning, and default parameters in 4 classification problems.

optimization for streaming algorithms. The first work identifies burst on the activity in phone calls, using approximate counting with forgetting. The last work presents a streaming optimization method to find the minimum of a function and its application in finding the hyper-parameter values that minimize the error. We believe that the two works reported here will have an impact on the work of other researchers.

Acknowledgements I would like to thank my collaborators Bruno Veloso and Rita P. Ribeiro that contribute to this work.

References

1. Ali, M. A., Azad, M. A., Centeno, M. P., Hao, F., van Moorsel, A.: Consumer-facing technology fraud: Economics, attack methods and potential solutions. Future Generation Computer Systems, **100**, 408–427 (2019)
2. Demšar, J.: Statistical comparisons of classifiers over multiple data sets. Journal of Machine Learning Research, **7**(Jan), 1–30 (2006)
3. Laleh, N., Azgomi, M. A.: A taxonomy of frauds and fraud detection techniques. In International Conference on Information Systems, Technology and Management, pp. 256–267. Springer (2009)
4. Veloso, B., Gama, Malheiro, J. B., Vinagre, J.: Hyperparameter self-tuning for data streams. Information Fusion, **76**, 75–86 (2021)
5. Veloso, B., Tabassum, S., Martins, C., Espanha, R., Azevedo, R., Gama, J.: Interconnect bypass fraud detection: a case study. Annals des Télécommunications, **75**(9), 583–596 (2020)

Old and New Constraints in Model Based Clustering

Luis A. García-Escudero, Agustín Mayo-Iscar, Gianluca Morelli, and Marco Riani

Abstract Model-based approaches to cluster analysis and mixture modeling often involve maximizing classification and mixture likelihoods. Without appropriate constrains on the scatter matrices of the components, these maximizations result in ill-posed problems. Moreover, without constrains, non-interesting or "spurious" clusters are often detected by the EM and CEM algorithms traditionally used for the maximization of the likelihood criteria. A useful approach to avoid spurious solutions is to restrict relative components scatter by a prespecified tuning constant. Recently new methodologies for constrained parsimonious model-based clustering have been introduced which include the 14 parsimonious models that are often applied in model-based clustering when assuming normal components as limit cases. In this paper we initially review the traditional approaches and illustrate through an example the benefits of the adoption of the new constraints.

Keywords: model based clustering, mixture modelling, constraints

L. A. García-Escudero
Department of Statistics and Operational Research and IMUVA, University of Valladolid, Spain,
e-mail: lagarcia@eio.uva.es

A. Mayo-Iscar
Department of Statistics and Operational Research and IMUVA, University of Valladolid, Spain,
e-mail: agustinm@eio.uva.es

G. Morelli
Department of Economics and Management and Interdepartmental Centre of Robust Statistics,
University of Parma, Italy, e-mail: gianluca.morelli@unipr.it

M. Riani (✉)
Department of Economics and Management and Interdepartmental Centre of Robust Statistics,
University of Parma, Italy, e-mail: mriani@unipr.it

© The Author(s) 2023   139
P. Brito et al. (eds.), *Classification and Data Science in the Digital Age*,
Studies in Classification, Data Analysis, and Knowledge Organization,
https://doi.org/10.1007/978-3-031-09034-9_16

1 Introduction

Given a sample of observations $\{x_1, ..., x_n\}$ in $\mathbb{R}^p$, a widely used method in unsupervised learning is to assume multivariate normal components and to adopt a maximum likelihood approach for clustering purposes. With this idea in mind, well-known classification and mixture likelihood approaches can be followed.

In this work, we use $\phi(\cdot; \mu, \Sigma)$ to denote the probability density function of a p-variate normal distribution with mean μ and covariance matrix Σ.

In the *classification likelihood* approach we search for a partition $\{H_1, ..., H_k\}$ of the indices $\{1, \cdots, n\}$, centres $\mu_1, \cdots, \mu_k$ in $\mathbb{R}^p$, symmetric positive semidefinite $p \times p$ scatter matrices $\Sigma_1, \cdots, \Sigma_k$ and positive weights $\pi_1, \cdots, \pi_k$ with $\sum_{j=1}^{k} \pi_j = 1$, which maximize

$$\sum_{j=1}^{k} \sum_{i \in H_j} \log\left(\pi_j \phi(x_i; \mu_j, \Sigma_j)\right). \tag{1}$$

On the other hand, in the *mixture likelihood* approach, we seek the maximization of

$$\sum_{i=1}^{n} \log\left(\sum_{j=1}^{k} \pi_j \phi(x_i; \mu_j, \Sigma_j)\right), \tag{2}$$

with similar notation and conditions on the parameters as above. In this second approach, a partition into k groups can be also obtained, from the fitted mixture model, by assigning each observation to the cluster-component with the highest posterior probability.

Unfortunately, it is well-known that the maximization of "log-likelihoods" like (1) and (2) without constraints on the Σ_j matrices is a mathematically ill-posed problem [1, 2]. To see this unboundedness issue, we can just take $\mu_1 = x_1$, $\pi_1 > 0$ and $|\Sigma_1| \to 0$ making (2) to diverge to infinity or (1) also to diverge with $H_1 = \{1\}$.

This lack of boundedness can be solved by just focusing on local maxima of the likelihood target functions. However, many local maxima are often found and it is difficult to know which are the most interesting ones. See [3] for a detailed discussion of this issue. In fact, non-interesting local maxima denoted as "spurious" solutions, which consist of a few, almost collinear, observations, are often detected by the Classification EM algorithm (CEM), traditionally applied when maximizing (1), and by the EM algorithm, traditionally applied when maximizing (2). A recent review of approaches for dealing with this lack of boundedness and for reducing the detection of spurious solutions can be found in [4].

It is also common to enforce constraints on the Σ_j scatter matrices when maximizing (1) or (2). Among them, the use of "parsimonious" models [5, 6] is one of the most popular and widely applied approaches in practice. These parsimonious models follow from a decomposition of the Σ_j scatter matrices as

$$\Sigma_j = \lambda_j \Omega_j \Gamma_j \Omega_j', \tag{3}$$

with $\lambda_j = |\Sigma_j|^{1/p}$ (volume parameters),

$$\Gamma_j = \text{diag}(\gamma_{j1}, ..., \gamma_{jl}, ..., \gamma_{jp}) \text{ with } \det(\Gamma_j) = \prod_{l=1}^{p} \gamma_{jl} = 1$$

(shape matrices), and Ω_j (rotation matrices) with $\Omega_j \Omega_j' = I_p$. Different constraints on the λ_j, Ω_j and Γ_j elements are considered across components to get 14 parsimonious models (which are coded with a combination of three letters). These models reduce notably the number of free parameters to be estimated, so improving efficiency and model interpretability. Moreover, many of them turn the constrained maximization of the likelihoods into well-defined problems and help to avoid spurious solutions. Unfortunately, the problems remain for models with unconstrained λ_j volume parameters, which are coded with the first letter as a V (V** models). Aside from relying on good initializations, it is common to consider the early stopping of iterations when approaching scatter matrices with very small eigenvalues or when detecting components accounting for a reduced number of observations. A not fully iterated solution (or no solution at all) is then returned in these cases. The idea is known, for instance, to be problematic when dealing with (well-separated) components made up of a few observations.

Starting from a seminal paper by [7], an alternative approach is to constrain the Σ_j scatter matrices by specifying some tuning constants that control the strength of the constraints. In this direction, the ratio between the largest and the smallest of the $k \times p$ eigenvalues of the Σ_j matrices was forced to be smaller than a given fixed constant $c^* \geq 1$ [8, 9, 10, 11, 12]. This means that the maximization of (1) and (2) is done under the (more simple) constraint:

$$\max_{jl} \lambda_l(\Sigma_j) / \min_{jl} \lambda_l(\Sigma_j) \leq c^*, \tag{4}$$

where $\{\lambda_l(\Sigma_j)\}_{l=1}^{p}$ are the set of eigenvalues of the Σ_j matrix, $j = 1, ..., k$.

With this eigenvalue-ratio approach, we need a very high c^* value to be close to affine equivalence. Unfortunately, such a high c^* value does not always successfully prevent us from incurring into spurious solutions.

2 The New Constraints

García-Escudero *et al.* [13] have recently introduced three different types of constraints on the Σ_j matrices which depend on three constants c_{det}, c_{shw} and c_{shb} all of them being greater than or equal to 1.

The first type of constraint serves to control the maximal ratio among determinants and, consequently, the maximum allowed difference between component volumes:

$$\text{"deter":} \qquad \frac{\max_{j=1,...,k} |\Sigma_j|}{\min_{j=1,...,k} |\Sigma_j|} = \frac{\max_{j=1,...,k} \lambda_j^p}{\min_{j=1,...,k} \lambda_j^p} \leq c_{\text{det}}. \tag{5}$$

The second type of constraint controls departures from sphericity "within" each component:

$$\text{shape-"within":} \qquad \frac{\max_{l=1,\dots,p} \gamma_{jl}}{\min_{l=1,\dots,p} \gamma_{jl}} \le c_{\text{shw}} \text{ for } j = 1, \dots, k. \qquad (6)$$

This provides a set of k constraints that in the most constrained case, $c_{\text{shw}} = 1$, imposes $\Gamma_1 = \dots = \Gamma_p = I_p$, where I_p is the identity matrix of size p, i.e., sphericity of components.

Note that the new determinant-and-shape constraints (based on $c_{\text{det}} > 1$ and $c_{\text{shw}} = 1$) in (4) allow us to deal with spherical "heteroscedastic" cases, whereas the eigenvalue ratio constraint with $c^* = 1$ can only handle the spherical "homoscedastic" case. Constraints (5) and (6) were the basis for the "deter-and-shape" constraints in [14]. These two constraints alone resulted in mathematically well-defined constrained maximizations of the likelihoods in (1) and (2). However, although highly operative in many cases, they do not include, as limit cases, all the already mentioned 14 parsimonious models. For instance, we may be interested in the same (or not very different) Γ_j or Σ_j matrices for all the mixture components and these cannot be obtained as limit cases from the "deter-and-shape" constraints.

The third constraint serves to control the maximum allowed difference between shape elements "between" components:

$$\text{shape-"between":} \qquad \frac{\max_{j=1,\dots,k} \gamma_{jl}}{\min_{j=1,\dots,k} \gamma_{jl}} \le c_{\text{shb}} \text{ for } l = 1, \dots, p. \qquad (7)$$

This new type of constraint allows us to impose "similar" shape matrices for the components and, consequently, enforce $\Gamma_1 = \dots = \Gamma_k$ in the most constrained $c_{\text{shb}} = 1$ case.

3 An Illustration Example of the New Constraints

Figure 1 shows an example based on three groups. The data have been generated imposing equal determinants $c_{\text{det}} = 1$, a sensible departure from sphericity "within" each component $c_{\text{shw}} = 40$ and a very moderate difference "between" shape elements components, $c_{\text{shb}} = 1.3$. No constraint has been imposed on the rotation matrices. Finally an average overlap of 0.10 has been imposed. The generation of these data sets has been done through the MixSim method of [15], as extended by [16] and incorporated into the FSDA Matlab toolbox [17]. The overlap is defined as a sum of pairwise misclassification probabilities. See more details in [16].

The application of traditional tclust approach with maximum ratio between eigenvalues (c^*) respectively equal to 128 and 10^{10} produces the classifications shown in the left panels of Figure 2. In fact, it could be seen that the results in the top left panel would be exactly the same one for any choice of c^* within the interval [16, 128]. This means that a higher value of c^* would be apparently needed to detect

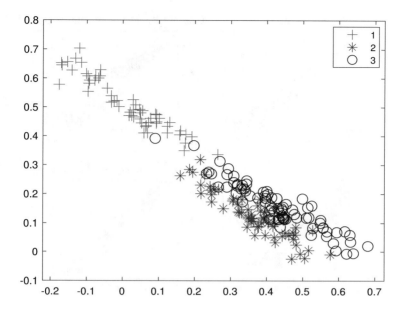

Fig. 1 An example with simulated data with 3 clusters in two dimensions. The average overlap is 0.10. The data have been generated using equal determinants, moderate difference between shape elements "between" components and sensible departure from sphericity "within" each component.

those two almost parallel clusters that were shown in Figure 1. However, choosing a value greater for c^* may destroy the desired protection against spurious solutions provided by the constraints. For example, we see in the lower left panel how the choice $c^* = 10^{10}$ results in the detection of a spurious group consisting of a single observation.

The panels on the right, on the other hand, show the partitions resulting from the 3 new constraints imposed on the components covariance matrices. The top right panel shows the result of applying the 3 new restrictions with values of the tuning constants very close to the real values used to generate the dataset. We can see that, in this case, it is possible to recover the real structure of the data generating process. Moreover, the real cluster structure is also recovered in the low right panel by choosing larger values of these tuning constants, but not too large just to avoid detection of spurious solutions. Some guidelines about how to choose these tuning constants can be found in [13].

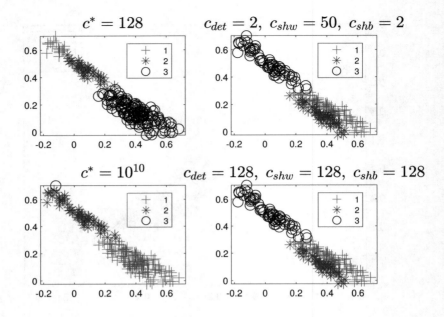

Fig. 2 Comparison between the traditional (left panels) and new tclust procedure (right panels).

References

1. Kiefer, J., Wolfowitz, J.: Consistency of the maximum likelihood estimator in the presence of infinitely many incidental parameters. Ann. Math. Stat. **27**, 887-906 (1956)
2. Day, N. E.: Estimating the components of a mixture of normal distributions. Biometrika, **56**, 463-474 (1969)
3. McLachlan, G., Peel, D. A.: Finite Mixture Models. Wiley Series in Probability and Statistics, New York (2000)
4. García-Escudero, L. A., Gordaliza, A., Greselin, F., Ingrassia, S., Mayo-Iscar, A.: Eigenvalues and constraints in mixture modeling: geometric and computational issues. Adv. Data Anal. Classif. **12**, 203-233 (2018)
5. Celeux, G., Govaert, G.: Gaussian parsimonious clustering models. Pattern Recogn. **28**, 781-793 (1995)
6. Banfield, J. D., Raftery, A. E.: Model-based Gaussian and non-Gaussian clustering. Biometrics **49**, 803-821 (1993)
7. Hathaway, R. J.: A constrained formulation of maximum likelihood estimation for normal mixture distributions. Ann. Stat. **13**, 795-800 (1985)
8. Ingrassia, S., Rocci, R.: Constrained monotone EM algorithms for finite mixture of multivariate Gaussians. Comput. Stat. Data Anal. **51**, 5339-5351 (2007)
9. García-Escudero, L. A., Gordaliza, A., Matrán, C., Mayo-Iscar, A.: A general trimming approach to robust cluster analysis. Ann. Stat. **36**, 1324-1345 (2008)
10. García-Escudero, L. A., Gordaliza, A., Matrán, C., Mayo-Iscar, A.: Exploring the number of groups in robust model-based clustering. Stat. Comput. **21**, 585-599 (2011)
11. García-Escudero, L. A., Gordaliza, A., Mayo-Iscar, A.: A constrained robust proposal for mixture modeling avoiding spurious solutions. Adv. Data Anal. Classif. **8**, 27-43 (2014)

12. García-Escudero, L. A., Gordaliza, A., Matrán, C., Mayo-Iscar, A.: Avoiding spurious local maximizers in mixture modeling. Stat. Comput. **25**, 619-633 (2015)
13. García-Escudero, L. A., Mayo-Iscar, A., Riani, M.: Constrained parsimonious model-based clustering. Stat. Comput. **32** (2022)
14. García-Escudero, L. A., Mayo-Iscar, A., Riani, M.: Model-based clustering with determinant-and-shape constraint. Stat. Comput. **25**, 1-18 (2020)
15. Maitra, R., Melnykov, V.: Simulating data to study performance of finite mixture modeling and clustering algorithms. J. Comput. Graph. Stat. **19**, 354-376 (2010)
16. Riani, M., Cerioli, A., Perrotta, D., Torti, F.: Simulating mixtures of multivariate data with fixed cluster overlap in FSDA library. Adv. Data Anal. Classif. **9**, 461-481 (2015)
17. Riani, M., Perrotta, D., Torti, F.: FSDA: a Matlab toolbox for robust analysis and interactive data exploration. Chemometr. Intell. Lab. Syst. **116**, 17-32 (2012)

Clustering Student Mobility Data in 3-way Networks

Vincenzo Giuseppe Genova, Giuseppe Giordano, Giancarlo Ragozini, and Maria Prosperina Vitale

Abstract The present contribution aims at introducing a network data reduction method for the analysis of 3-way networks in which classes of nodes of different types are linked. The proposed approach enables simplifying a 3-way network into a weighted two-mode network by considering the statistical concept of joint dependence in a multiway contingency table. Starting from a real application on student mobility data in Italian universities, a 3-way network is defined, where provinces of residence, universities and educational programmes are considered as the three sets of nodes, and occurrences of student exchanges represent the set of links between them. The Infomap community detection algorithm is then chosen for partitioning two-mode networks of students' cohorts to discover different network patterns.

Keywords: 3-way network, complex network, community detection, mobility data, tertiary education

Vincenzo Giuseppe Genova
Department of Economics, Business, and Statistics, University of Palermo, Italy,
e-mail: vincenzogiuseppe.genova@unipa.it

Giuseppe Giordano
Department of Political and Social Studies, University of Salerno, Italy,
e-mail: ggiordano@unisa.it

Giancarlo Ragozini
Department of Political Science, Federico II University of Naples, Italy,
e-mail: giragoz@unina.it

Maria Prosperina Vitale (✉)
Department of Political and Social Studies, University of Salerno, Italy,
e-mail: mvitale@unisa.it

P. Brito et al. (eds.), *Classification and Data Science in the Digital Age*,
Studies in Classification, Data Analysis, and Knowledge Organization,
https://doi.org/10.1007/978-3-031-09034-9_17

147

1 Introduction

Many complex relational data structures can be described as multimode or multiway networks in which nodes belonging to different modes are linked. The most common multimode network in social networks is represented by the affiliation network, where two-mode data, actors and events, form a bipartite graph divided into two groups [6]. In the case of tripartite networks, we deal with three types of nodes, and different graph structures can be defined.

Although only a few papers deal with methods for these networks, in recent years, a growing number of works have appeared –especially in bipartite and tripartite cases– to disentangle the inherent complexity of such kinds of data structures. Looking at clustering and community detection algorithms proposed to partition a network into groups, we can identify some strands, all deriving from generalizations of methods suited for one-mode [19] and two-mode networks [2]. A classical approach consists of applying the usual community detection algorithms on a unique supra-adjacency matrix defined by combining all the possible two-mode networks in a block matrix [11, 15]. Alternative methods rely on projecting each two-mode networks and on applying separately the usual community detection algorithms on these matrices [10]. In addition, there are methods adopting both an optimization procedure for 3-way networks [16, 17, 14] by extending the idea of bipartite modularity [2], and an indirect blockmodeling approach by deriving a dissimilarity measure based on structural equivalence concept [3].

In our opinion, approaches based on the analysis of the k-modes examined considering the collection of the $k(k-1)/2$ two-mode networks [10] cannot take into account statistical associations among all modes at same time. Hence, the aim of the contribution is to present a network data reduction method based on the concept of joint dependence in a multiway contingency table [1].

Starting from real applications on the Italian student mobility phenomenon in higher education [12, 21, 7, 8, 13, 22], a 3-way network is defined, where provinces of residence, universities and educational programmes are considered as the three modes. Student mobility flows, measured in terms of occurrences, represent the set of links between them. Assuming that the statistical dependency between the set of nodes provinces of residence and the other two sets of nodes can be captured by the joined pair of nodes (universities and educational programmes), the tripartite network is transformed into a bipartite network, where the two modes are given by Italian provinces of residence (first mode) and the set of nodes given by all possible pairs of universities and educational programmes (second mode). Thus, taking advantage of this approach of network simplification, network indexes and clustering techniques for bipartite networks are available. Hence, the Infomap community detection algorithm is adopted [9, 4] to partition the derived network.

The remainder of the paper is organized as follows. Section 2 presents the details of the proposed strategy of analysis, and the main results are reported from the analysis of student mobility data of Italian universities. Section 3 provides final remarks.

2 Simplification of 3-way Networks

In the present paper, the case of a tripartite network is considered as an example to show how the proposed network data simplification method works. In particular, we consider the real case study of student mobility paths in Italian universities. The MOBYSU.IT dataset[1] enables reconstruction of network data structures considering student mobility flows among territorial units and universities.

More formally, given $\mathcal{V}_P \equiv \{p_1, \ldots, p_i, \ldots, p_I\}$, the set of I provinces of residence; $\mathcal{V}_U \equiv \{u_1, \ldots, u_j, \ldots, u_J\}$, the set of J Italian universities, and $\mathcal{V}_E \equiv \{e_1, \ldots, e_k, \ldots, e_K\}$, the set of K educational programmes, a weighted tripartite 3-uniform hyper-graph $\mathcal{T}$ can be defined, consisting of a triple $(\mathcal{V}, \mathcal{L}, \mathcal{W})$, with $\mathcal{V} = \{\mathcal{V}_P, \mathcal{V}_U, \mathcal{V}_E\}$ the collection of three sets of vertices, one for each mode, and being $\mathcal{L} = \{\mathcal{L}_{PUE}\}$, $\mathcal{L}_{PUE} \subseteq \mathcal{V}_P \times \mathcal{V}_U \times \mathcal{V}_E$, the collection of hyper-edges, with generic term (p_i, u_j, e_k), which is the link joining the i-th province, the j-th university, and the k-th educational programme. Finally, $\mathcal{W}$ is the set of weights, obtained by the function $w : \mathcal{L}_{PUE} \to \mathbb{N}$, and $w(p_i, u_j, e_k) = w_{ijk}$ is the number of students moving from a province p_i towards a university u_j in an educational programme e_k. Such a network structure can be described as a three-way array $\mathbb{A} = (a_{ijk})$, with $a_{ijk} \equiv w_{ijk}$, and it has been called a 3-way network [3].

To deal with such a complex network structure and aiming at obtaining communities in which three modes are mixed, we wish to simplify the tripartite nature of the graph, without losing any significant information. In statistical terms, the array $\mathbb{A}$ can be interpreted as a 3-way contingency table, and then the statistical techniques to evaluate the association among variables (i.e. the modes) can be exploited [1]. Because a 3-way contingency table is a cross-classification of observations by the levels of three categorical variables, we are defining a network structure where the sets of nodes are the levels of the categorical variables. Specifically, we assume that if two modes are jointly associated –as are, for their own nature, universities and educational programmes– the tripartite network can be logically simplified into a bipartite one. In the student mobility network, we join the pair of nodes in $\mathcal{V}_U$ and in $\mathcal{V}_E$, and then we deal with the relationships between these *dyads* and the nodes in $\mathcal{V}_P$.

Following this assumption, the sets of nodes $\mathcal{V}_U$ and $\mathcal{V}_E$ are put together into a set of joint nodes, namely $\mathcal{V}_{UE}$. The tripartite network $\mathcal{T}$ can now be represented as a bipartite network $\mathcal{B}$ given by the triple $\{\mathcal{V}^*, \mathcal{L}^*, \mathcal{W}^*\}$, with $\mathcal{V}^* = \{\mathcal{V}_P, \mathcal{V}_{UE}\}$. The set of hyper-edges $\mathcal{L}$ is thus simplified into a set of edges $\mathcal{L}^* = \{\mathcal{L}_{P,UE}\}$, $\mathcal{L}_{P,UE} \subseteq \mathcal{V}_P \times \mathcal{V}_{UE}$. The new edges $(p_i, (u_j; e_k))$ connect a province p_i with an educational programme e_k running in a given university u_j. The weights $\mathcal{W}^*$ are the same as in the hyper-graph $\mathcal{T}$, i.e., $w^*_{ij,k} = w_{ijk}$. Note that the weights contained in the 3-way array $\mathbb{A}$ are preserved, but are now organized in a rectangular matrix $\mathbf{A}$ of I rows and $(J \times K)$ columns.

[1] Database MOBYSU.IT [Mobilità degli Studi Universitari in Italia], research protocol MUR - Universities of Cagliari, Palermo, Siena, Torino, Sassari, Firenze, Cattolica and Napoli Federico II, Scientific Coordinator Massimo Attanasio (UNIPA), Data Source ANS-MUR/CINECA.

Taking advantage of this method, we aim to analyse weighted bipartite graphs adopting clustering methods. Among others, we use the Infomap community detection algorithm [9, 4] to study the flows' patterns in network structures instead of modularity optimization proposed in topological approaches [18, 5]. Indeed, the rationale of this algorithm –*map equation*– takes advantage of the duality between finding communities and minimizing the length –*codelength*– of a random walker's movement on a network. The partition with the shortest path length is the one that best captures the community structure in the bipartite data. Formally, the algorithm defines a module partition **M** of n vertices into m modules such that each vertex is assigned to one and only one module. The Infomap algorithm looks for the best **M** partition that minimizes the expected *codelength*, $L(M)$, of a random walker, given by the following map equation:

$$L(M) = q_\curvearrowright H(\mathcal{Q}) + \sum_{i=1}^{m} p_\circlearrowright^i H(\mathcal{P}^i) \tag{1}$$

In equation (1), $q_\curvearrowright H(\mathcal{Q})$ represents the entropy of the movement between modules weighed for the probability that the random walker switches modules on any given step ($q_\curvearrowright$), and $\sum_{i=1}^{m} p_\circlearrowright^i H(\mathcal{P}^i)$ is the entropy of movements within modules weighed for the fraction of within-module movements that occur in module i, plus the probability of exiting module i ($p_\circlearrowright^i$), such that $\sum_{i=1}^{m} p_\circlearrowright^i = 1 + q_\curvearrowright$ [9].

In our case, the Infomap algorithm is adopted to discover communities of students characterized by similar mobility patterns. Indeed, to analyse mobility data, where links represent patterns of student movement among territorial units and universities, flow-based approaches are likely to identify the most important features. Finally, in our student mobility network, to focus only on relevant student flows, a filtering procedure is adopted by considering the Empirical Cumulative Density Function (ECDF) of links' weights distribution.

2.1 Main Findings

Students' cohorts enrolled in Italian universities in four academic years (a.y.) 2008–09, 2011–12, 2014–15, and 2017–18 are analysed. The number of nodes for the sets $\mathcal{V}_P$ (107 provinces), $\mathcal{V}_U$ (79-80 universities), and $\mathcal{V}_E$ (45 educational programmes), and the number of students involved in the four cohorts are quite stable over time (Table 1). Furthermore, the percentage of movers (i.e., students enrolled in a university outside of their region of residence) increased, from 16.4% in the a.y. 2008–09 to 20.6% in the a.y. 2017–18, and it is higher for males than females.

Table 1 Percentage of students according to their mobility status by cohort and gender.

Cohort	Gender		Mover status	
			Stayers%	Movers%
2008–09	F	136,381	84.2	15.8
	M	106,950	82.8	17.2
	Total	243,331	83.6	16.4
2011–12	F	126,606	81.7	18.3
	M	102,479	80.9	19.1
	Total	229,085	81.0	19.0
2014–15	F	121,121	80.5	19.5
	M	102,358	80.4	19.6
	Total	223,479	80.5	19.5
2017–18	F	134,315	79.1	20.9
	M	113,496	79.8	20.2
	Total	247,811	79.4	20.6

Following the network simplification approach, the tripartite networks –one for each cohort– are simplified into bipartite networks, and the four ECDFs of links' weights are considered to filter relevant flows. The distributions suggest that more than 50% of links between pairs of nodes have weights equal to 1 (i.e., flows of only one student), and about 95% of flows are characterized by flows not greater than a digit. Thus, networks holding links with a value greater or equal to 10 are further analysed.

To reveal groups of universities and educational programmes attracting students, the Infomap community detection algorithm is applied. Looking at Table 2, we notice a reduction of the number of communities from the first to the last student cohort, suggesting a sort of stabilization in the trajectories of movers towards brand universities of the center-north with also an increase in the north-north mobility [20], and a relevant dichotomy between scientific and humanistic educational programmes. Network visualizations by groups (Figures 1 and 2) confirm that the more attractive universities are located in the north of Italy, especially for educational programmes in economics and engineering (the Bocconi University, the Polytechnic of Turin and the Cattolica University).

Table 2 Number of communities, codelength, and relative saving codelength per cohort.

Cohort	Communities	Codelength	Relative saving codelength
2008–09	14	0.96	83%
2011–12	17	1.72	70%
2014–15	3	5.23	12%
2017–18	3	1.00	83%

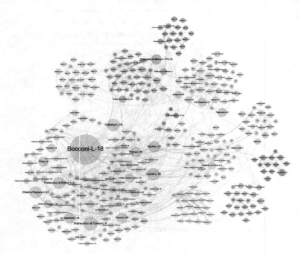

Fig. 1 Network visualization by groups, student cohort a.y. 2008–09.

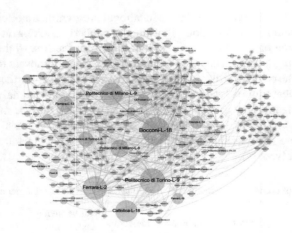

Fig. 2 Network visualization by groups, student cohort a.y. 2017–18.

3 Concluding Remarks

The proposed simplification network strategy on tripartite graphs defined for student mobility data provides interesting insights for the phenomenon under analysis. The main attractive destinations still remain the northern universities for educational programmes, such as engineering and business. Besides the well-known south-to-north route, other interregional routes in the northern area appear. In addition, the reduction in the number of communities suggests a sort of stabilization in terms of mobility roots of movers towards brand universities, highlighting student university destination choices close to the labor market demand.

Hyper-graphs and multipartite networks still remain very active areas for research and challenging tasks for scholars interested in discovering the complexities underlying these kinds of data. Specific tools for such complex network structures should be designed combining network analysis and other statistical techniques. As future lines of research, the comparison of community detection algorithms that better represent the structural constraints of the phenomena under analysis and the assessment of other backbone approaches to filter the significant links will be developed.

Acknowledgements The contribution has been supported from Italian Ministerial grant PRIN 2017 "From high school to job placement: micro-data life course analysis of university student mobility and its impact on the Italian North-South divide", n. 2017 HBTK5P - CUP B78D19000180001.

References

1. Agresti, A.: Categorical Data Analysis (Vol. 482). John Wiley & Sons, New York (2003)
2. Barber, M. J.: Modularity and community detection in bipartite networks. Phys. Rev. E, **76**, 066102 (2007)
3. Batagelj, V., Ferligoj, A., Doreian, P.: Indirect Blockmodeling of 3-Way Networks. In: Brito P., Cucumel G., Bertrand P., de Carvalho F. (eds) Selected Contributions in Data Analysis and Classification. Studies in Classification, Data Analysis, and Knowledge Organization, pp. 151–159. Springer, Berlin, Heidelberg (2007)
4. Blöcker, C., Rosvall, M.: Mapping flows on bipartite networks. Phys. Rev. E, **102**, 052305 (2020)
5. Blondel, V. D., Guillaume, J. L., Lambiotte, R., Lefebvre, E.: Fast unfolding of communities in large networks. J. Stat. Mech.-Theory E, **10**, P10008 (2008)
6. Borgatti, S. P., Everett, M. G.: Regular blockmodels of multiway, multimode matrices. Soc. Networks, **14**, 91–120 (1992)
7. Columbu, S., Porcu, M., Primerano, I., Sulis, I., Vitale, M.P.: Geography of Italian student mobility: A network analysis approach. Socio. Econ. Plan. Sci. **73**, 100918 (2021)
8. Columbu, S., Porcu, M., Primerano, I., Sulis, I., Vitale, M. P.: Analysing the determinants of Italian university student mobility pathways. Genus, **77**, 34 (2021)
9. Edler, D., Bohlin, L., Rosvall, M.: Mapping higher-order network flows in memory and multilayer networks with infomap. Algorithms, **10**, 112 (2017)
10. Everett, M. G., Borgatti, S.: Partitioning multimode networks. In: Doreian, P., Batagelj, V., Ferligoj, A. (eds.) Advances in Network Clustering and Blockmodeling, pp. 251-265, John Wiley & Sons, Hoboken, USA (2020)

11. Fararo, T. J., Doreian, P.: Tripartite structural analysis: Generalizing the Breiger-Wilson formalism. Soc. Networks, **6**, 141–175 (1984)
12. Genova, V. G., Tumminello, M., Aiello, F., Attanasio, M.: Student mobility in higher education: Sicilian outflow network and chain migrations. Electronic Journal of Applied Statistical Analysis, **12**, 774–800 (2019)
13. Genova, V. G., Tumminello, M., Aiello, F., Attanasio, M.: A network analysis of student mobility patterns from high school to master's. Stat. Method. Appl., **30**, 1445–1464 (2021)
14. Ikematsu, K., Murata, T.: A fast method for detecting communities from tripartite networks. In: International Conference on Social Informatics, pp. 192-205. Springer, Cham (2013)
15. Melamed, D., Breiger, R. L., West, A. J.: Community structure in multi-mode networks: Applying an eigenspectrum approach. Connections, **33**, 18–23 (2013)
16. Murata, T.: Detecting communities from tripartite networks. In: Proceedings of the 19th international conference on world wide web, pp. 1159-1160. (2010)
17. Neubauer, N., Obermayer, K.: Tripartite community structure in social bookmarking data. New Rev. Hypermedia M., **17**, 267-294 (2011)
18. Newman, M. E., Girvan, M.: Finding and evaluating community structure in networks. Phys. Rev. E, **69**, 026113 (2004)
19. Newman, M. E.: Modularity and community structure in networks. Proceedings of the National Academy of Sciences, **103**, 8577-8582 (2006)
20. Rizzi, L., Grassetti, L. Attanasio, M.: Moving from North to North: how are the students' university flows? Genus **77**, 1–22 (2021)
21. Santelli, F., Scolorato, C., Ragozini, G.: On the determinants of student mobility in an inter-regional perspective: A focus on Campania region. Statistica Applicata - Italian Journal of Applied Statistics, **31**, 119–142 (2019)
22. Santelli, F., Ragozini, G., Vitale, M. P.: Assessing the effects of local contexts on the mobility choices of university students in Campania region in Italy. Genus, **78**, 5 (2022)

Clustering Brain Connectomes Through a Density-peak Approach

Riccardo Giubilei

Abstract The density-peak (DP) algorithm is a mode-based clustering method that identifies cluster centers as data points being surrounded by neighbors with lower density and far away from points with higher density. Since its introduction in 2014, DP has reaped considerable success for its favorable properties. A striking advantage is that it does not require data to be embedded in vector spaces, potentially enabling applications to arbitrary data types. In this work, we propose improvements to overcome two main limitations of the original DP approach, i.e., the unstable density estimation and the absence of an automatic procedure for selecting cluster centers. Then, we apply the resulting method to the increasingly important task of graph clustering, here intended as gathering together similar graphs. Potential implications include grouping similar brain networks for ability assessment or disease prevention, as well as clustering different snapshots of the same network evolving over time to identify similar patterns or abrupt changes. We test our method in an empirical analysis whose goal is clustering brain connectomes to distinguish between patients affected by schizophrenia and healthy controls. Results show that, in the specific analysis, our method outperforms many existing competitors for graph clustering.

Keywords: nonparametric statistics, mode-based clustering, networks, graph clustering, kernel density estimation

1 Introduction

Clustering is the task of grouping elements from a set in such a way that elements in the same group, also defined as *cluster*, are in some sense similar to each other, and dissimilar to those from other groups. Mode-based clustering is a nonparametric approach that works by first estimating the density, and then identifying in some

Riccardo Giubilei (✉)
Luiss Guido Carli, Rome, Italy, e-mail: `rgiubilei@luiss.it`

© The Author(s) 2023
P. Brito et al. (eds.), *Classification and Data Science in the Digital Age*,
Studies in Classification, Data Analysis, and Knowledge Organization,
https://doi.org/10.1007/978-3-031-09034-9_18

way its modes and the corresponding clusters. An effective method to find modes and clusters is through the density-peak (DP) algorithm [12], which has drawn considerable attention since its introduction in 2014. One of the striking advantages of DP is that it does not require data to be embedded in vector spaces, implying that it can be applied to arbitrary data types, provided that a proper distance is defined. In this work, we focus on its application to clustering graph-structured data objects.

The expression *graph clustering* can refer either to *within-graph clustering* or to *between-graph clustering*. In the first case, the elements to be grouped are the vertices of a single graph; in the second, the objects are distinct graphs. Here, *graph clustering* is intended as *between-graph clustering*. Between-graph clustering is an emerging but increasingly important task due to the growing need of analyzing and comparing multiple graphs [10, 4]. Potential applications include clustering: brain networks of different people for ability assessment, disease prevention, or disease evaluation; online social ego networks of different users to find people with similar social structures; different snapshots of the same network evolving over time to identify similar patterns, cycles, or abrupt changes.

Heretofore, the task of between-graph clustering has not been exhaustively investigated in the literature, implying a substantial lack of well-established methods. The goal of this work is to improve and adapt the density-peak algorithm to define a fairly general method for between-graph clustering. For validation and comparison purposes, the resulting procedure and its main competitors are applied to grouping brain connectomes of different people to distinguish between patients affected by schizophrenia and healthy controls.

2 Related Work

Existing techniques for between-graph clustering can be divided into two main categories: 1) transforming graph-structured data objects into Euclidean feature vectors in order to apply standard clustering algorithms; 2) using the distances between the original graphs in distance-based clustering methods.

The most common technique within the first category is the use of classical clustering techniques on the vectorized adjacency matrices [10]. Nonetheless, more advanced numerical summaries have been proposed to better capture the structural properties of the graphs and to decrease feature dimensionality. Examples include: shell distribution [1], traces of powers of the adjacency matrix [10], and graph embeddings such as *graph2vec* [11]; see [4] for a longer list. Techniques from the first category share an important drawback: the transformation into feature vectors necessarily implies loss of information. Additionally, methods for extracting features may be domain-specific.

The second category features Partitioning Around Medoids (PAM) [7], or k-medoids, which finds representative observations by iteratively minimizing a cost function based on the distances between data objects, and assigns other observations to the closest medoid. PAM's main limitations are that it requires the number of clusters in advance and can only identify convex-shaped groups. Density-based spatial clustering of applications with noise [3], or DBSCAN, overcomes these two constraints by computing the density of data points starting from their distances, and defining clusters as samples of high density that are close to each other (and surrounded by areas of lower density). A similar approach is the DP, which is described in greater detail in Section 3.1. Alternatively, hierarchical clustering can be applied to distances between graphs, as in [13], where a spectral Laplacian-based distance is proposed and used. Finally, k-groups [8] is a clustering technique within the Energy Statistics framework [14] where the goal is minimizing the total within-cluster Energy distance, which is computed starting from the distances between original observations.

3 Methods

In this section, we first describe the original DP approach; then, we introduce the DP-KDE method, which is partly named after Kernel Density Estimation; finally, we discuss how to employ it for graph clustering.

3.1 Original DP

The density-peak algorithm [12] is based on a simple idea: since cluster centers are identified as the distribution's modes, they must be 1) surrounded by neighbors with lower density, and 2) at a relatively large distance from points with higher density. Consequently, two quantities are computed for each observation x_i: the local density ρ_i, and the minimum distance δ_i from other data points with higher density. The local density ρ_i of x_i is defined as:

$$\rho_i = \sum_j I_{(d_{ij} - d_c)}, \tag{1}$$

where $I_{(.)}$ is the indicator function, $d_{ij} = d(x_i, x_j)$ is the distance between x_i and x_j, and d_c is a cutoff distance. In simple terms, ρ_i is the number of points that are closer than d_c to x_i. The DP algorithm is robust with respect to d_c, at least with large datasets [12]. Once the density is computed, the definition of the minimum distance δ_i between point x_i and any other point x_j with higher density is straightforward:

$$\delta_i = \min_{j:\rho_j > \rho_i} (d_{ij}). \tag{2}$$

By convention, the point with highest density has $\delta_i = max_j (d_{ij})$. The interpretation of δ_i reflects the algorithm's core idea: data points that are not local or global maxima have their δ_i constrained by other points within the same cluster, hence cluster centers have large values of δ_i. However, this is not sufficient: they also need to have a large ρ_i

because otherwise the point could be merely distant from any other. After identifying cluster centers, other observations are assigned to the same cluster as their nearest neighbor of higher density.

The density-peak algorithm has many favorable properties: it manages to detect nonspherical clusters, it does not require the number of clusters in advance or data to be embedded in vector spaces, it is computationally fast because it does not maximize explicitly each data point's density field and it performs cluster assignment in a single step, it estimates a clear population quantity, and it has only one tuning parameter (the cutoff distance d_c).

3.2 DP-KDE

The density-peak approach also has drawbacks. Over the last few years, many articles have proposed improvements to overcome two main critical points: the unstable density estimation and the absence of an automatic procedure for selecting cluster centers. In this work, we explicitly tackle these two aspects.

The unstable density estimation induced by Equation (1) has been widely shown [9, 16, 15]. Although many solutions have been proposed, we espouse the research line suggesting the use of Kernel Density Estimation (KDE) to compute ρ_i [9, 15]:

$$\rho_i = \frac{1}{nh} \sum_{j=1}^{n} K\left(\frac{x_i - x_j}{h}\right). \tag{3}$$

In Equation (3), h is the *bandwidth*, which is a smoothing parameter, and $K(\cdot)$ is the *kernel*, which is a non-negative function weighting the contribution of each data point to the density of the i-th observation. We use the Epanechnikov kernel, which is normalized, symmetric, and optimal in the Mean Square Error sense [2]:

$$K(u) = \begin{cases} 3/4(1 - u^2), & |u| \le 1 \\ 0, & |u| > 1 \end{cases}. \tag{4}$$

Equation (4) implies a null contribution of observation j to the i-th density whenever $|(x_i - x_j)/h| \ge 1$, while, in the opposite case, it results in a positive weight depending quadratically on $(x_i - x_j)/h$. Consequently, h may be regarded as the cutoff distance for the DP-KDE method.

The automatic selection of cluster centers involves many aspects: the cutoff distance, the number of clusters, and which data points to select. In the following, we use a cutoff distance h such that the average number of neighbors is between 1 and 2% of the sample size, as suggested by [12]. The number of clusters k is here considered as a given parameter, leaving the search for its optimal value for future work. Finally, the method for selecting data points as cluster centers is obtained refining an intuition contained in [12], where candidates are observations with sufficiently large values of $\gamma_i = \delta_i \rho_i$. However, this quantity has two major drawbacks: first, if

δ_i and ρ_i are not defined over the same scale, results could be misleading; second, it implicitly assumes that δ_i and ρ_i shall be given the same weight in the decision. We overcome these two limitations by first normalizing both δ_i and ρ_i between 0 and 1, and then giving them different weights that are based on their informativeness. We measure the latter using the Gini coefficient of the two (normalized) quantities, under the assumption that the least concentrated distribution between the two is the most informative. Specifically, each observation is given a measure of importance that is defined as:

$$\gamma_i^G = \delta_{01,i}^{G(\delta_{01})} \rho_{01,i}^{G(\rho_{01})}, \tag{5}$$

where δ_{01} and ρ_{01} are the normalized versions of δ and ρ respectively, $\delta_{01,i}$ and $\rho_{01,i}$ are the corresponding i-th values, and $G(x)$ denotes the Gini coefficient of x. Then, the selected cluster centers are the top k observations in terms of γ_i^G. Assigning observations to the same cluster as their nearest neighbor of higher density is what concludes the DP-KDE method.

3.3 Graph Clustering

A *graph* is a mathematical object composed of a collection of *vertices* linked by *edges* between them. Formally, a graph is denoted with $\mathcal{G} = (V, E)$, where V is the set of vertices and E is the set of edges. If $e \in E$ joins vertices $u, v \in V$, i.e., $e = \{u, v\}$, then u and v are *adjacent* or *neighbors*. The number of edges incident with any vertex v is the *degree* of v. Each edge $e \in E$ is represented through a numerical value w_e called *edge weight*: if weights are equal to 1 for all and only the existent edges, and 0 for the others, $\mathcal{G}$ is *unweighted*; when existent edges have real-valued weights, $\mathcal{G}$ is *weighted*. If $w_{\{u,v\}} = w_{\{v,u\}}$ for all $u, v \in V$, the graph $\mathcal{G}$ is *undirected*; otherwise, it is *directed*. The entire information about $\mathcal{G}$'s connectivity is stored in a $|V| \times |V|$ *adjacency matrix* $\mathbf{A}$ whose generic entry in the u-th row and v-th column is w_e, where $e = \{u, v\}$ and $u, v \in V$.

The DP-KDE method can be used for graph clustering if a proper distance between graphs is defined. In this work, we employ the Edge Difference Distance [6], which is defined as the Frobenius norm of the difference between the two graphs' adjacency matrices. The choice is motivated by many factors: a flexible definition that can be directly applied also to directed and weighted graphs, the reasonable results it yields when node correspondence is a concern, and its limited computational time complexity. Formally, the Edge Difference Distance between two graphs x_i and x_j is defined as:

$$d_{ED}(x_i, x_j) = ||\mathbf{A}^i - \mathbf{A}^j||_F := \sqrt{\sum_p \sum_q |A_{pq}^i - A_{pq}^j|^2}, \tag{6}$$

where $\mathbf{A}^i$ and $\mathbf{A}^j$ are the adjacency matrices of x_i and x_j respectively, and $|| \cdot ||_F$ denotes the Frobenius norm.

Consequently, the two fundamental quantities of the DP-KDE method are computed in the following as:

$$\rho_i = \sum_{j=1}^{n} K\left(\frac{d_{ED}(x_i, x_j)}{h}\right), \tag{7}$$

where $K(\cdot)$ is the Epanechnikov kernel defined in Equation (4) and the normalizing constant is omitted because we are simply interested in the ranking between the densities, and:

$$\delta_i = \min_{j:\rho_j > \rho_i} (d_{ED}(x_i, x_j)). \tag{8}$$

Finally, cluster centers are selected as the observations with the largest values of γ_i^G, as defined in Equation (5), and other observations are assigned to the same cluster as their nearest neighbor in terms of δ_i.

4 Empirical Analysis

The DP-KDE method for graph clustering is employed in an unsupervised empirical analysis where the ground truth is known, and its performance is compared in terms of accuracy both with natural competitors and with a method treating the problem as supervised. The ultimate goal is clustering brain connectomes, one for each individual, correctly distinguishing between patients affected by schizophrenia (SZ) and healthy controls.

We use publicly available[1] data from a recent study [5] whose aim is finding relevant links between Regions of Interest (ROIs) for predicting schizophrenia from multimodal brain connectivity data. The cohort is composed of 27 schizophrenic patients and 27 age-matched healthy participants acting as control subjects. In the current work, we focus only on this cohort's functional Magnetic Resonance Imaging (fMRI) connectomes. Functional connectivity matrices have been computed starting from fMRI scans, treating them as time series, and computing Pearson's correlation coefficient between time series for distinct ROIs. The resulting matrices are weighted, undirected, and made of 83 nodes.

The aforementioned study [5] treats every functional connectivity matrix as a single multivariate realization of $(83 \cdot 82)/2 = 3403$ numeric variables, each representing a connection between two of the 83 ROIs. They reduce feature dimensionality by performing Recursive Feature Elimination based on Support Vector Machines (SVM-RFE), and tackle the classification problem as supervised using 20 repetitions of nested 5-fold cross-validation. When using only functional connectivity data, they achieve an average accuracy of 68.28%[2] over the resulting 100 test sets.

---

[1] https://doi.org/10.5281/zenodo.3758534.

[2] This exact figure is not included in the article, but the analysis is fully reproducible since the authors made their source code available at https://github.com/leoguti85/BiomarkersSCHZ.

The approach we adopt in this work is rather different. First, graphs are analyzed in their original form, without any simplification to numeric variables, resulting in only one graph-structured variable. Observations are 54, each one representing the functional connectome of a different individual. We tackle the problem with an unsupervised classification approach seeking to cluster connectomes into two groups: schizophrenic and healthy. To this end, we use the DP-KDE method for graph clustering described in Section 3.3. Starting from the 54 connectomes, each observation's local density ρ_i and minimum distance δ_i are computed using Equations (7) and (8), respectively. The centers of the two clusters are those whose γ_i^G is largest. Then, other observations are assigned to the same cluster as their nearest neighbor of higher density. Finally, the clustering performance is evaluated by comparing the algorithm's assignment to the ground truth. The DP-KDE method achieves an accuracy of 70.37%, which is more than 2% higher than the one obtained in [5].

Table 1 includes the performance in terms of accuracy of both the DP-KDE and the SVM-RFE methods, as well as that of other graph clustering competitors. Specifically, we consider: the classical DP algorithm on the original data objects, with the same cutoff distance as in DP-KDE and manually selected cluster centers; k-means clustering on the 3403 numeric variables obtained from vectorizing the adjacency matrices; DBSCAN on the original data objects, with parameters $\varepsilon = 20.2$ and 15 as the minimum number of points required to form a dense region; PAM and k-groups on the original data objects. In all these cases, the number of clusters has been kept fixed to $k = 2$. The method that yields the best accuracy in the specific problem is the DP-KDE.

Table 1 Accuracy for DP-KDE and some of its possible competitors.

Method	DP-KDE	SVM-RFE	DP	k-means	DBSCAN	PAM	k-groups
Accuracy	70.37	68.28	62.96	62.96	61.11	62.96	62.96

5 Concluding Remarks

After explaining the importance of graph clustering and briefly reviewing some existing methods to perform this task, we have considered the possibility of adopting a density-peak approach. We have improved the original DP algorithm by using a more robust definition of the density ρ_i, and by automatically selecting cluster centers based on the quantity γ_i^G we have introduced. We have also selected a proper distance between graphs, namely, the Edge Difference Distance. Finally, we have used the resulting method in an empirical analysis with the goal of clustering brain connectomes to distinguish between schizophrenic patients and healthy controls. Our method outperforms another one treating the specific task as supervised, and it is by far the best one with respect to many graph clustering competitors.

An initial idea for future work is the search for the optimal number of clusters. This may be achieved either by fixing a threshold for γ_i^G or by selecting all the data points after the largest increase in terms of γ_i^G. Also the cutoff distance could be tuned, possibly maximizing in some way the dispersion of points in the bivariate distribution of ρ and δ. Then, the DP-KDE method needs to be extended beyond the univariate case. Finally, other distances between graphs could be considered to better reflect alternative application-specific needs, e.g., when graphs are not defined over the same set of nodes.

Acknowledgements The author would like to thank Pierfancesco Alaimo Di Loro, Federico Carlini, Marco Perone Pacifico, and Marco Scarsini for several engaging and stimulating discussions.

References

1. Carmi, S., Havlin, S., Kirkpatrick, S., Shavitt, Y., Shir, E.: A model of Internet topology using k-shell decomposition. Proc. Natl. Acad. Sci. **104**, 11150–11154 (2007)
2. Epanechnikov, V.: Non-parametric estimation of a multivariate probability density. Theory Probab. Its Appl. **14**, 153–158 (1969)
3. Ester, M., Kriegel, H., Sander, J., Xu, X., et al.: A density-based algorithm for discovering clusters in large spatial databases with noise. KDD-96 **34** 226–231 (1996)
4. Gutiérrez-Gómez, L., Delvenne, J.: Unsupervised network embeddings with node identity awareness. Appl. Netw. Sci. **4**, 1–21 (2019)
5. Gutiérrez-Gómez, L., Vohryzek, J., Chiêm, B., Baumann, P., Conus, P., Do Cuenod, K., Hagmann, P., Delvenne, J.: Stable biomarker identification for predicting schizophrenia in the human connectome. NeuroImage Clin. **27** 102316 (2020)
6. Hammond, D., Gur, Y., Johnson, C.: Graph diffusion distance: A difference measure for weighted graphs based on the graph Laplacian exponential kernel. IEEE GlobalSIP 2013, pp. 419–422 (2013)
7. Kaufmann, L., Rousseeuw, P.: Clustering by means of medoids. Proc. of the Statistical Data Analysis based on the L1 Norm Conference, Neuchatel, Switzerland, pp. 405–416 (1987)
8. Li, S., Rizzo, M.: K-groups: A generalization of k-means clustering. ArXiv Preprint ArXiv:1711.04359 (2017)
9. Mehmood, R., Zhang, G., Bie, R., Dawood, H., Ahmad, H.: Clustering by fast search and find of density peaks via heat diffusion. Neurocomputing. **208**, 210–217 (2016)
10. Mukherjee, S., Sarkar, P., Lin, L.: On clustering network-valued data. NIPS2017, pp. 7074–7084 (2017)
11. Narayanan, A., Chandramohan, M., Venkatesan, R., Chen, L., Liu, Y., Jaiswal, S.: graph2vec: Learning distributed representations of graphs. ArXiv Preprint ArXiv:1707.05005 (2017)
12. Rodriguez, A., Laio, A.: Clustering by fast search and find of density peaks. Science **344**, 1492–1496 (2014)
13. Shimada, Y., Hirata, Y., Ikeguchi, T., Aihara, K.: Graph distance for complex networks. Sci. Rep. **6**, 1–6 (2016)
14. Székely, G., Rizzo, M.: The energy of data. Annu. Rev. Stat. Appl. **4**, 447–479 (2017)
15. Wang, X., Xu, Y.: Fast clustering using adaptive density peak detection. Stat. Methods Med. Res. **26**, 2800–2811 (2017)
16. Xie, J., Gao, H., Xie, W., Liu, X., Grant, P.: Robust clustering by detecting density peaks and assigning points based on fuzzy weighted K-nearest neighbors. Inf. Sci. **354**, 19–40 (2016)

Similarity Forest for Time Series Classification

Tomasz Górecki, Maciej Łuczak, and Paweł Piasecki

Abstract The idea of similarity forest comes from Sathe and Aggarwal [19] and is derived from random forest. Random forests, during already 20 years of existence, proved to be one of the most excellent methods, showing top performance across a vast array of domains, preserving simplicity, time efficiency, still being interpretable at the same time. However, its usage is limited to multidimensional data. Similarity forest does not require such representation – it is only needed to compute similarities between observations. Thus, it may be applied to data, for which multidimensional representation is not available. In this paper, we propose the implementation of similarity forest for time series classification. We investigate 2 distance measures: Euclidean and dynamic time warping (DTW) as the underlying measure for the algorithm. We compare the performance of similarity forest with 1-nearest neighbor and random forest on the UCR (University of California, Riverside) benchmark database. We show that similarity forest with DTW, taking into account mean ranks, outperforms other classifiers. The comparison is enriched with statistical analysis.

Keywords: time series, time series classification, random forest, similarity forest

Tomasz Górecki (✉)
Faculty of Mathematics and Computer Science, Adam Mickiewicz University, Uniwersytetu Poznańskiego 4, Poznań, Poland, e-mail: `tomasz.gorecki@amu.edu.pl`

Maciej Łuczak
Faculty of Mathematics and Computer Science, Adam Mickiewicz University, Uniwersytetu Poznańskiego 4, Poznań, Poland, e-mail: `maciej.luczak@amu.edu.pl`

Paweł Piasecki
Faculty of Mathematics and Computer Science, Adam Mickiewicz University, Uniwersytetu Poznańskiego 4, Poznań, Poland, e-mail: `pawel.piasecki@amu.edu.pl`

© The Author(s) 2023
P. Brito et al. (eds.), *Classification and Data Science in the Digital Age*,
Studies in Classification, Data Analysis, and Knowledge Organization,
https://doi.org/10.1007/978-3-031-09034-9_19

1 Introduction

Time series classification is a well-developing research field, that gained much attention from researchers and business during the last two decades apparently by the fact that more and more data around us seems to be located in the time domain – and thus fulfilling the definition of time series. Predictive maintenance [18], quality monitoring [22], stock market analysis [20] or sales forecasting [17] are just a few exemplar nowadays problems where time series are indeed present. The reason why we usually apply to time series different methods from regular (non-time series) data is the fact, that time series are ordered in time (or some other space with ordering) and it is beneficial to use the information conveyed by the ordering.

In recent years, one could observe many advances on the field of time series classification. In 2017, Bagnall et al. presented a comprehensive comparison of time series classification algorithms [2], showing that despite there are dozens of far more complex methods, 1-Nearest Neighbour (1NN) [6, 11] coupled with DTW [3] distance constitutes a good baseline. In fact, it has been outperformed by several classifiers, with Collective of Transformation Ensembles (COTE) [1] as the most efficient one. Furthermore, COTE was extended with Hierarchical Vote system, first to HIVE-COTE [13] and then finally to HIVE-COTE 2.0 [15] – a current state of the art classifier for time series. In general, the success of COTE-family classifiers is based on the observation, that in the case of time series it is highly beneficial to use different data representations. For example, HIVE-COTE 1.0 utilizes five ensembles based on different data transformation domains. However, a common criticism of such an approach is its time complexity. In the case of HIVE-COTE, it equals $O(n^2 l^4)$, where n is a number of observations and l is a length of series. Another drawback, especially significant for practitioners is the complex structure of the model ensembles that makes it hard to use HIVE-COTE without spending a decent amount of time studying its components beforehand.

As an alternative to such complex models may be trying to achieve possibly slightly worse performance in favour of model simplicity and reduced computation time. A group of classifiers that seems to hold a great potential are those inspired by Random Forest (RF) [4]. This already 20-years old algorithm remains in the classifiers' forefront, showing extremely good performance and robustness across multiple domains. Fernandez-Delgado et al. [10] performed a comparison of 179 classifiers on 121 non-time series data sets originated from UCI Machine Learning Repository [9], concluding RF to be the most accurate one. Unfortunately, the usage of RF is essentially limited to multidimensional data, as they sample features from original space while creating each node of decision trees.

In this paper, we propose a method for extending RF to work with time series using similarity forests (SF). We significantly extend the applicability of the RF method to time series data. Furthermore, the approach even outperforms traditional classifiers for time series. The main goal of this paper is to enrich the pool of time series classifiers by Similarity Forest for time series classification. SF was initially proposed by Sathe and Aggarwal in 2017 [19], as a method extending Random Forests to deal with arbitrary data sets, provided that we are able to compute similarities between observations. We would like to implement and tune the method to time series data. We investigate the performance of the model using two distance measures (the algorithm's hyper-parameter): Euclidean and DTW. Also, a comparison with other selected time series classifiers is provided. We compare its performance against 1NN-ED, 1NN-DTW and RF.

The rest of the paper is structured as follows. In Section 2, we provide details of similarity forest and we give more details about random forests. Additionally, we discuss how similarity forest is related to random forest. Section 3 describes data sets that we used and the comparison methodology. The corresponding results are presented in Section 4. Finally, in Section 5 we give a brief summary of our research.

2 Classification Methods Used in Comparison

In the paper, we compare the standard random forest and the similarity forest with the distance measure: ED (Euclidian distance) and DTW (dynamic time warping distance). As benchmark methods, we also use the nearest neighbor method (1NN) with distance measure ED and DTW. 1NN-ED and 1NN-DTW are very common classification methods for time series classification [2]. For a review of these methods refer to [14].

2.1 General Method of Random Forest Construction

Random forest consists of random decision trees. For the construction of a random forest we usually take decision trees as simple as possible — without special criteria for stopping, pruning, etc.

When building a decision tree, we start at a node N, which contains the entire data set (bootstrap sample). Then, according to an established criterion, we split the node N into two subnodes N_1 and N_2. In each subnode there are data subsets of the data set from node N. We make this split in a way that is optimal for a given split method. In each node, we write down how the split occurred. Then, proceeding recursively, we split next nodes into subnodes until the stop criterion occurs. In our case we take the simplest such criterion, namely we stop the split of a given node when only elements of the same class are included in a node. We call such a node a leaf and assign it a label which elements of the node (leaf) have.

Having built a tree, we can now use it (in the testing phase) to classify a new observation. We pass this observation through the trained tree — starting from the node N selecting each time one of the subnodes, according to the condition stored in the node. We do this until we reach one of the leaves, and then we assign the test observation to the class of the leaf.

Now, constructing the random forest, we collect a certain number of decision trees, train them independently according to the above method and, in the test phase, use each of the trees to test new observation. Thus, each tree assigns a label to the test observation. The final label (for the entire forest) we construct by voting, we choose the most frequently appearing label among the decision trees.

2.2 Classical Random Forest

To create a (classical) random tree and a random forest [4], we proceed as described above using the following node split method:

To obtain split conditions for a single tree, we select randomly a certain number of features ($\sqrt{k}$ for classification, k — number of features), and for each feature we create a feature vector (column, variable) made of all elements of the data set (bootstrap sample). For a given feature vector (variable), we determine a threshold vector. First, we sort values of the feature vector (uniquely — without repeating values). Let us name this sorted feature vector as $V = (V_1, V_2, \ldots)$. Then we take the values of the split as means of successive values of the vector V:

$$v_i = \frac{V_i + V_{i+1}}{2} \quad i = 1, 2, \ldots. \tag{1}$$

Each splitting value divides the data set in node N into two subsets — the one (left) in which we have elements with feature values smaller than v_i and the second (right) with other elements. Then we check the quality of such a split.

The splitting point is chosen such that it minimizes the Gini index of the children nodes. If $p_1, p_2 \ldots p_c$ are the fractions of data points belonging to the c different classes in node N, then the Gini index of that node is given by: $G(N) = 1 - \sum_{i=1}^{c} p_i^2$.

Then, if the node N is split into two children nodes N_1 and N_2, with n_1 and n_2 points, respectively, the Gini quality of the children nodes is given by:

$$GQ(N_1, N_2) = \frac{n_1 G(N_1) + n_2 G(N_2)}{n_1 + n_2}.$$

Quality of the split is given by: $GQ(N) = G(N) - GQ(N_1, N_2)$.

2.3 Similarity Forest

The similarity forest [19] differs from the ordinary (classical) random forest only in the way we split nodes of trees. Instead of selecting a certain number of features, we select randomly a pair of elements e_1, e_2 with different classes. Then, for each element e of the subset of elements in a given node, we calculate the difference of the squared distances to the elements e_1 and e_2:

$$w(e) = d(e, e_1)^2 - d(e, e_2)^2,$$

where d is any fixed distance measure of the elements of the data set. We sort the vector $\boldsymbol{w}$ uniquely (without duplicates) creating the vector $\boldsymbol{V}$ and continue as for the classical decision tree. We calculate values of the split v_i (1), calculate the quality of the node split using the Gini index (2.2) and choose the best split. In the learning phase, we remember in each node how the optimal split occurred (elements e_1, e_2, $w(e)$). In the learning phase, in each node we write down the optimal split — elements e_1, e_2, and value $w(e)$).

2.4 Random Forest vs Similarity Forest

The difference between a classical random tree and a similarity tree is that instead of selecting $\sqrt{k}$ of the features, we select only one pair of elements e_1, e_2. Generally, we have much fewer possible node splits, which has a very good effect on the computation time.

The second important difference is that in the similarity tree we use any distance measure between elements of the data set. Therefore, we can use distance measures specific to a data set. For example, for time series we can use the DTW distance, much better suited for calculating the distance between time series, instead of the Euclidean distance.

3 Experimental Setup

We investigated the performance of similarity forest on UCR time series repository [7] (128 data sets). The latest update of the UCR database introduced several data sets with missing observations and uneven sample lengths. However, the repository includes a standardized version of the database without these impediments, and that is the version we used.

All data sets are split into a training and testing subset, and all parameter optimization is conducted on the training set only. We combined both parts and in the next step, we used 100 random train/test splits.

4 Results

The error rates for each classifier can be found on the accompanying website[1]. In the Table 1 we show a short summary of results, including a number of wins (draw is not counted as a win) and mean ranks. Taking into account mean ranks, SF-DTW is the best classifier, sightly ahead of RF (mean ranks correspondingly equal 2.64

---

[1] https://github.com/ppias/similarity_forest_for_tsc

Table 1 Number of wins (clearly wins) and mean ranks for examined methods.

Method	1NN-ED	1NN-DTW	RF	SF-ED	SF-DTW
Wins	12	28	**38**	10	31
Mean rank	3.59	2.89	2.69	3.19	**2.64**

and 2.89). Figure 1 demonstrates comparison of error rates and ranks for classifiers. These results lead to a conclusion that even though there is no clear winner, the top efficient distances are dominated by RF and SF-based classifiers. Figure 2 shows scatter plots of errors for pairs of classifiers.

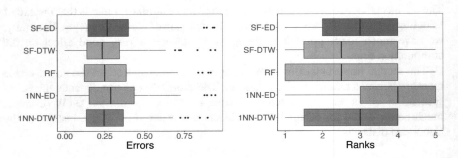

Fig. 1 Comparison of error rates and ranks.

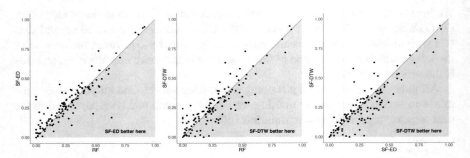

Fig. 2 Comparison of error rates.

To identify differences between the classifiers, we present a detailed statistical comparison. In the beginning, we test the null hypothesis that all classifiers perform the same and the observed differences are merely random. The Friedman test with Iman & Davenport extension is probably the most popular omnibus test, and it is usually a good choice when comparing different classifiers [12]. The p-value from this test is equal to 0. The obtained p-value indicates that we can safely reject the null hypothesis that all the algorithms perform the same. We can therefore proceed with the post-hoc tests in order to detect significant pairwise differences among all of the classifiers.

Demšar [8] proposes the use of the Nemenyi's test [16] that compares all the algorithms pair-wise. For a significance level α the test determines the critical difference (CD). If the difference between the average ranking of two algorithms is greater than CD the null hypothesis that the algorithms have the same performance is rejected. Additionally, Demšar [8] creates a plot to visually check the differences, the CD plot. In the plot, those algorithms that are not joined by a line can be regarded as different.

In our case, with a significance of $\alpha = 0.05$ any two algorithms with a difference in the mean rank above 0.54 will be regarded as non equal (Figure 3). We can see that we have three groups of methods. In the first group we have SF-DTW, RF and 1NN-DTW, in the second we have RF, 1NN-DTW and SF-ED and in the last group we have SF-ED and 1NN-ED. Unfortunately, groups are not disjoint. The first group is the group with the highest accuracy of classification. Hence, SF-DTW does not statistically outperform RF. However, we can recommend it over RF because of statistically the same quality and much better computational properties.

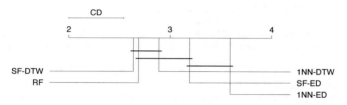

Fig. 3 Critical difference plot.

5 Conclusions

Our contribution is to implement similarity forest for time series classification using two distance measures: Euclidean and DTW. Comparison based on the recently updated UCR data repository (128 data sets) was presented. We showed that SF-DTW outperforms other classifiers, including 1NN-DTW which has been considered as a strong baseline hard to beat for years. The statistical comparison showed, that RF and SF-DTW are statistically insignificantly different, however taking into account mean ranks the latter one is the best one.

There are many improvements that could be applied to the implementation that we propose. For example, we could test other distance measures such as LCSS [21] or ERP [5] that were successfully used in time series tasks. Another idea could be to investigate the usage of boosting algorithm.

Acknowledgements The research work was supported by grant No. 2018/31/N/ST6/01209 of the National Science Centre.

References

1. Bagnall, A., Lines, J., Hills, J., Bostrom A.: Time-series classification with COTE: The collective of transformation-based ensembles. IEEE Trans. on Knowl. and Data Eng. **27**, 2522–2535 (2015)
2. Bagnall, A., Lines, J., Bostrom, A., Large J., Keogh, E.: The great time series classification bake off: a review and experimental evaluation of recent algorithmic advances. Data Min. and Knowl. Discov. **31**, 606–660 (2017)
3. Berndt, D. J., Clifford, J.: Using dynamic time warping to find patterns in time series. Proc. of the 3rd Int. Conf. on Knowl. Discov. and Data Min., pp. 359–370 (1994)
4. Brieman, L.: Random forests. J. Mach. Learn. Arch. **45**, 5–32 (2001)
5. Chen, L., Ng, R.: On the marriage of L_p-norms and edit distance. Proc. of the 30th Int. Conf. on Very Large Data Bases **30**, pp. 792–803 (2004)
6. Cover, T., Hart, P.: Nearest neighbor pattern classification. IEEE Trans. on Inf. Theor. **13**, 21–27 (1967)
7. Dau, H.A., Keogh, E., Kamgar, K., Yeh, Chin-Chia M., Zhu, Y.,Gharghabi, S., Ratanama-hatana, C.A., Yanping, C., Hu, B., Begum, N., Bagnall, A., Mueen, A., Batista, G., Hexagon-ML: The UCR time series classification archive (2019) `https://www.cs.ucr.edu/\str ing~eamonn/time\_series\_data\_2018`
8. Demšar, J.: Statistical comparisons of classifiers over multiple data sets. J. of Mach. Learn. Res. **7**, 1–30 (2006).
9. Du,a D., Graff, C.: UCI Machine Learning Repository. `http://archive.ics.uci.edu/ml`
10. Fernandez-Delgado, M., Cernadas, E., Barro, S., Amorim, D.: Do we need hundreds of classifiers to solve real world classification problems?. J. of Mach. Learn. Res. **15**, 3133–3181 (2014)
11. Fix, E, Hodges, J. L.: Discriminatory analysis: nonparametric discrimination, consistency properties. Techn. Rep. **4**, (1951)
12. García, S., Fernández, A., Luengo, J., Herrera, F.: Advanced nonparametric tests for multiple comparisons in the design of experiments in computational intelligence and data mining: Experimental Analysis of Power. Inf. Sci. **180**, 2044–2064 (2010)
13. Lines, J., Taylor S., Bagnall, A.: HIVE-COTE: The hierarchical vote collective of transfor-mation based ensembles for time series classification. IEEE Int. Conf. on Data Min., pp. 1041–1046 (2016)
14. Maharaj, E. A., D'Urso, P., Caiado, J.: Time Series Clustering and Classification. Chapman and Hall/CRC. (2019)
15. Middlehurst, M., Large, J., Flynn, M., Lines, J., Bostrom, A., Bagnall, A.: HIVE-COTE 2.0: a new meta ensemble for time series classification. (2021) `https://arxiv.org/abs/2104.07551`
16. Nemenyi, P.:Distribution-free multiple comparisons. PhD thesis at Princeton University (1963)
17. Pavlyshenko, B. M.: Machine-learning models for sales time series forecasting. Data **4**, 15 (2019)
18. Rastogi, V., Srivastava, S., Mishra, M., Thukral, R.: Predictive maintenance for SME in industry 4.0. 2020 Glob. Smart Ind. Conf., pp. 382–390 (2020)
19. Sathe, S., Aggarwal, C. C.: Similarity forests. Proc. of the 23rd ACM SIGKDD, pp. 395–403 (2017)
20. Tang, J., Chen, X.: Stock market prediction based on historic prices and news titles. Proc. of the 2018 Int. Conf. on Mach. Learn. Techn., pp. 29–34 (2018)
21. Vlachos, M., Kollios, G., Gunopulos, D.: Discovering similar multidimensional trajectories. Proc. 18th Int. Conf. on Data Eng., pp. 673–684 (2002)
22. Wuest, T., Irgens, C., Thoben, K. D.: An approach to quality monitoring in manufacturing using supervised machine learning on product state data. J. of Int. Man. **25**, 1167–1180 (2014)

Detection of the Biliary Atresia Using Deep Convolutional Neural Networks Based on Statistical Learning Weights via Optimal Similarity and Resampling Methods

Kuniyoshi Hayashi, Eri Hoshino, Mitsuyoshi Suzuki, Erika Nakanishi, Kotomi Sakai, and Masayuki Obatake

Abstract Recently, artificial intelligence methods have been applied in several fields, and their usefulness is attracting attention. These methods are techniques that correspond to models using batch and online processes. Because of advances in computational power, as represented by parallel computing, online techniques with several tuning parameters are widely accepted and demonstrate good results. Neural networks are representative online models for prediction and discrimination. Many online methods require large training data to attain sufficient convergence. Thus, online models may not converge effectively for low and noisy training datasets. For such cases, to realize effective learning convergence in online models, we introduce statistical insights into an existing method to set the initial weights of deep convolutional neural networks. Using an optimal similarity and resampling method, we proposed an initial weight configuration approach for neural networks. For a practice example, identification of biliary atresia (a rare disease), we verified the usefulness

Kuniyoshi Hayashi (✉)
Graduate School of Public Health, St. Luke's International University, 3-6 Tsukiji, Chuo-ku, Tokyo, Japan, 104-0045, e-mail: khayashi@slcn.ac.jp

Eri Hoshino · Kotomi Sakai
Research Organization of Science and Technology, Ritsumeikan University, 90-94 Chudoji Awatacho, Shimogyo Ward, Kyoto, Japan, 600-8815,
e-mail: erihoshino119@gmail.com;koto.sakai1227@gmail.com

Mitsuyoshi Suzuki
Department of Pediatrics, Juntendo University Graduate School of Medicine, 2-1-1 Hongo, Bunkyo-ku, Tokyo, Japan, 113-8421, e-mail: msuzuki@juntendo.ac.jp

Erika Nakanishi
Department of Palliative Nursing, Health Sciences, Tohoku University Graduate School of Medicine, 2-1 Seiryo-machi, Aoba-ku, Sendai, Japan, 980-8575,
e-mail: nakanishi.erika.q3@dc.tohoku.ac.jp

Masayuki Obatake
Department of Pediatric Surgery, Kochi Medical School, 185-1 Kohasu, Oko-cho, Nankoku-shi, Kochi, Japan, 783-8505, e-mail: mobatake@kochi-u.ac.jp

© The Author(s) 2023
P. Brito et al. (eds.), *Classification and Data Science in the Digital Age*,
Studies in Classification, Data Analysis, and Knowledge Organization,
https://doi.org/10.1007/978-3-031-09034-9_20

of the proposed method by comparing existing methods that also set initial weights
of neural networks.

Keywords: AUC, bootstrap method, leave-one-out cross-validation, projection matrix, rare disease, sensitivity and specificity

1 Introduction

The core technique in deep learning corresponds to neural networks, including the
convolutional process. Since 2012, deep learning architectures have been frequently
used for image classification [1, 2]. More so, deep convolution neural networks
(DCNN) are representative nonlinear classification methods for pattern recognition.
The DCNN technique is used as a powerful framework for the entirety of image
processing [3]. The clinical medicine field presents many opportunities to perform
diagnoses using imaging data from patients. Therefore, DCNN techniques are applied to enhance diagnostic quality, e.g., applying a DCNN to a chest X-ray dataset
to classify pneumonia [2] and detecting breast cancer [4]. However, DCNN architectures involve many parameters to be learned using training data. Therefore, effective
and efficient model development must realize effective learning convergence for
such parameters. Notably, it is important to set the initial parameter values to achieve
better learning convergence. Furthermore, several methods have been proposed to
set initial parameter values in the artificial intelligence (AI) field [5, 6]. However,
there are no clear guidelines for determining which existing methods should be used
in different situations. Thus, we propose an efficient initial weight approach using
existing methods from the viewpoints of optimal similarity and resampling methods.
Using a real-world clinical biliary atresia (BA) dataset, we evaluate the performance
of the proposed method compared with existing DCNNs. Additionally, we show the
usefulness of the proposed method in terms of learning convergence and prediction
accuracy.

2 Background

BA is a rare disease that occurs in children and is fatal unless treated early. Previous
studies have investigated models to identify BA by applying neural networks to patient data [7] and using an ensemble deep learning model to detect BA [8]. However,
these models were essentially for use in medical institutions, e.g., hospitals. Generally, certain stool colors in infants and children are highly correlated with BA [9]. In
Japan, the maternal and child health handbook includes a stool color card so parents
can compare their child's stool color to the information on the card. Such fecal color
cards are widely used to detect BA because of their easy accessibility outside the
clinical environments. However, this stool color card screening approach for BA is

subjective; thus, accurate and objective diagnoses are not always possible. Previously, we developed a mobile application to classify BA and non-BA stools using baby stool images captured using an iPhone [10]. Here, a batch type classification method was used, i.e., the subspace method, originating from the pattern recognition field. Since BA is a rare disease, the number of events in the case group is generally less. Thus, when we set the explanatory variables of the target observation as the pixel values of a target image, the number of explanatory variables is much higher than the number of observations, especially the disease group. With the subspace method, we can efficiently discriminate such high-dimensional small-sample data. For example, our previous study using the subspace method to classify BA and non-BA stools showed that BA could be discriminated with reasonable accuracy by applying the proposed method to image pixel data of the stool image data captured by a mobile phone [10]. This application was an automated version of the stool color card from the maternal and child health handbook. Unlike previous studies by [7, 8], the application is widely available outside hospital environments. As described previously, DCNNs are useful for image classification, including the automatic classification of stool images for early BA detection.

3 Proposed Method

Dimension reduction and discrimination processing can be realized using the subspace method and DCNN techniques. In DCNN, layers based on padding, convolution, and pooling correspond to the dimension reduction functions, and the affine layer performs the discrimination. The primary motivation of this study is to propose a method that properly sets the initial weights of the parameters in a DCNN using statistical approaches. Our secondary motivation is to apply the proposed method to real-world, high-dimensional, and small-sample clinical data.

3.1 Description of Related Procedures of the Convolution

For image discrimination in pattern recognition and machine learning fields, the pixel values of the image data are set as the explanatory variables for the target outcome. Here, the data to be classified correspond to a high-dimensional observation. To improve efficiency and demonstrate the feasibility of discriminant processing, the dimensionality must be reduced to a manageable size before classification. The most representative dimensionality reduction method is convolution in pattern recognition and machine learning, which involves padding, convolution, and pooling operations. After converting the input image to a pixel data matrix, the pixel data matrix is surrounded with a numeric value of 0. Using a convolution filter, we reconstruct the pixel data matrix while considering pixel adjacency information. Generally, the size and convolution filter type are parameters that need optimization to realize sufficient prediction accuracy. However, some representative convolution filters that exhibit good performance are known in the AI field, and we can essentially fix the size and type of the convolution filter. Finally, pooling is performed to reduce the size of the pixel data matrix after convolution. Here, we refer to the sequence of processing from padding to pooling as the layer for feature selection.

3.2 Setting Conditions Assumed in This Study

We denote the input pattern matrices comprising numerical pixel values in hue (H), saturation (S), and value (V) as $\mathbf{X}^H (\in \mathbb{R}^{p \times q})$, $\mathbf{X}^S (\in \mathbb{R}^{p \times q})$, and $\mathbf{X}^V (\in \mathbb{R}^{p \times q})$, respectively. First, we performed padding for the input pattern matrices in H, S, and V, respectively, and then, performed a convolution in each signal pattern matrix using a convolution filter. Next, we then applied max pooling to each pattern matrix after convolution. Here, we denote the pattern matrices after the padding, convolution, and max pooling as $\tilde{\mathbf{X}}^H (\in \mathbb{R}^{p' \times q'})$, $\tilde{\mathbf{X}}^S (\in \mathbb{R}^{p' \times q'})$, and $\tilde{\mathbf{X}}^V (\in \mathbb{R}^{p' \times q'})$, respectively, where p' and q' are less than p and q. Therefore, we combine the component values of each pattern matrix after padding, convolution, and max pooling into a single pattern matrix by simply adding them together. The combined pattern matrix after applying the feature selection layer is expressed as $\tilde{\mathbf{X}} (\in \mathbb{R}^{p' \times q'})$. Next, we applied convolution and max pooling to the combined pattern matrix k times. Additionally, the input vector after performing the convolution and max pooling k times is denoted by $\mathbf{x} (\in \mathbb{R}^{\ell \times 1})$, and the output of the DCNN and the label vectors are denoted $\mathbf{y} (\in \mathbb{R}^{1 \times 1})$ and $\mathbf{t} (\in \mathbb{R}^{1 \times 1})$, respectively. In this study, we evaluated the difference between $\mathbf{y}$ and $\mathbf{t}$ according to the mean square error function, i.e., $L(\mathbf{y}, \mathbf{t}) = \frac{1}{\ell} \parallel \mathbf{t} - \mathbf{y} \parallel_2^2$. Here, we consider a simple neural network with three layers. Concretely, between the first and second layers, we perform a linear transformation using $\mathbf{W}_1 (\in \mathbb{R}^{2 \times \ell})$ and $\mathbf{b}_1 (\in \mathbb{R}^{2 \times 1})$. Then, a linear transformation is performed using $\mathbf{W}_2 (\in \mathbb{R}^{1 \times 2})$ and $\mathbf{b}_2 (\in \mathbb{R}^{1 \times 1})$ between the second and third layers. Next, we defined $f_1(\mathbf{x})$ and $f_2(\mathbf{x})$ as $\mathbf{W}_1 \mathbf{x} + \mathbf{b}_1$ and $\mathbf{W}_2 f_1(\mathbf{x}) + \mathbf{b}_2$, respectively. Note that we assume η_2 is a nonlinear transformation between the second and third layers, and we calculated the output $\mathbf{y}$ as $\eta_2(f_2 \circ f_1(\mathbf{x}))$. Generally, $\mathbf{y}$ is calculated as a continuous value. For example, with classification and regression tree methods, we can determine the optimal cutoff point of $\mathbf{y}s$ from a prediction perspective.

3.3 General Approach to Update Parameters in CNNs

Here, we denote $f_1(\mathbf{x})$ and $f_2 \circ f_1(\mathbf{x})$ in the previous subsection as $\mathbf{u}_1$ and $\mathbf{u}_2$, respectively. By performing the partial derivative of $L(\mathbf{y}, \mathbf{t})$ with respect to $\mathbf{W}_2$, we obtain $\frac{\partial L}{\partial \mathbf{W}_2^T} = \frac{\partial L}{\partial \mathbf{y}} \frac{\partial \mathbf{y}}{\partial \mathbf{u}_2} \frac{\partial \mathbf{u}_2}{\partial \mathbf{W}_2^T}$ where $\frac{\partial L}{\partial \mathbf{y}} = -\frac{2}{\ell}(\mathbf{t} - \mathbf{y})$, $\frac{\partial \mathbf{y}}{\partial \mathbf{u}_2} = \frac{\partial \eta_2(\mathbf{u}_2)}{\partial \mathbf{u}_2}$, and $\frac{\partial \mathbf{u}_2}{\partial \mathbf{W}_2^T} = \mathbf{u}_1$. Additionally, we calculate $\eta_2(\mathbf{u}_2)$ as $1/(1 + \exp(-\mathbf{u}_2))$ using the representative sigmoid function. Then, $\frac{\partial \mathbf{y}}{\partial \mathbf{u}_2}$ is calculated as $\eta_2(\mathbf{u}_2)(1 - \eta_2(\mathbf{u}_2))$. Therefore, we obtain $\frac{\partial L}{\partial \mathbf{W}_2^T} = -\frac{2}{\ell}(\mathbf{t}-\mathbf{y})\eta_2(\mathbf{u}_2)(1-\eta_2(\mathbf{u}_2))\mathbf{u}_1$. With the learning coefficient of γ_2, we update $\mathbf{W}_2^T$ to $\mathbf{W}_2^T - \gamma_2 \frac{\partial L}{\partial \mathbf{W}_2^T}$. Then, when performing the partial derivative of $L(\mathbf{y}, \mathbf{t})$ with respect to $\mathbf{W}_1$, we can obtain $\frac{\partial L}{\partial \mathbf{W}_1} = \frac{\partial L}{\partial \mathbf{y}} \frac{\partial \mathbf{y}}{\partial \mathbf{u}_2} \frac{\partial \mathbf{u}_2}{\partial \mathbf{u}_1} \frac{\partial \mathbf{u}_1}{\partial \mathbf{W}_1}$ where $\frac{\partial L}{\partial \mathbf{y}} = -\frac{2}{\ell}(\mathbf{t} - \mathbf{y})$, $\frac{\partial \mathbf{y}}{\partial \mathbf{u}_2} = \eta_2(\mathbf{u}_2)(1 - \eta_2(\mathbf{u}_2))$, $\frac{\partial \mathbf{u}_2}{\partial \mathbf{u}_1} = \mathbf{W}_2^T$, and $\frac{\partial \mathbf{u}_1}{\partial \mathbf{W}_1} = 2\mathbf{x}^T$. Thus, we then obtain $\frac{\partial L}{\partial \mathbf{W}_1} = -\frac{4}{\ell}(\mathbf{t} - \mathbf{y})\eta_2(\mathbf{u}_2)(1 - \eta_2(\mathbf{u}_2))\mathbf{W}_2^T \mathbf{x}^T$. With the learning coefficient of γ_1, we update $\mathbf{W}_1$ to $\mathbf{W}_1 - \gamma_1 \frac{\partial L}{\partial \mathbf{W}_1}$.

3.4 Setting the Initial Weight Matrix in the Affine Layer

To ensure proper learning convergence in situations with limited training datasets, we proposed a method using optimal similarity and bootstrap methods. Here, the number of training data and the training dataset are denoted n and $S(\ni \mathbf{x}_j)$, respectively, where $\mathbf{x}_j$ is the j-th training observation (j takes values 1 to n). Additionally, we normalized each observation vector, such that its norm is one. By considering the discrimination problem of two groups whose outcomes are 0 and 1, respectively, we divided $\{\mathbf{x}_j\}$ into $\{\mathbf{x}_j|\mathbf{y}_j = 0\}$ and $\{\mathbf{x}_j|\mathbf{y}_j = 1\}$. Next, we defined $\{\mathbf{x}_j|\mathbf{y}_j = 0\}$ and $\{\mathbf{x}_j|\mathbf{y}_j = 1\}$ as S_0 and S_1, respectively. First, we calculated the autocorrelation matrix with the observations belonging to S_0. Then, using the eigenvalues ($\hat{\lambda}_{s_0}$) and eigenvectors ($\hat{\mathbf{u}}_{s_0}$) for the autocorrelation matrix, we calculated the following projection matrix:

$$\hat{P}_0 := \sum_{s_0=1}^{\ell'_0} \hat{\mathbf{u}}_{s_0} \hat{\mathbf{u}}_{s_0}^T, \tag{1}$$

where ℓ'_0 takes values 1 to ℓ in Equation (1). Similarly, we calculated the autocorrelation matrix with the observations belonging to S_1. Then, with eigenvalues ($\hat{\lambda}_{s_1}$) and eigenvectors ($\hat{\mathbf{u}}_{s_1}$) for the autocorrelation matrix, we calculate the following projection matrix:

$$\hat{P}_1 := \sum_{s_1=1}^{\ell'_1} \hat{\mathbf{u}}_{s_1} \hat{\mathbf{u}}_{s_1}^T, \tag{2}$$

where ℓ'_1 takes values 1 to ℓ in Equation (2). Here, if the value of $\mathbf{x}^T (\hat{P}_1 - \hat{P}_0)\mathbf{x} > 0$, we classify $\mathbf{x}$ into S_1; otherwise, we classify $\mathbf{x}$ into S_0.

From a prediction perspective, using the leave-one-out cross-validation [11], we determined the optimal $\hat{\ell}'_0$ and $\hat{\ell}'_1$, which are minimum values satisfying $\tau < (\sum_{s_0=1}^{\ell'_0} \hat{\lambda}_{s_0})/(\sum_{s_0=1}^{\ell} \hat{\lambda}_{s_0})$ and $\tau < (\sum_{s_1=1}^{\ell'_1} \hat{\lambda}_{s_1})/(\sum_{s_1=1}^{\ell} \hat{\lambda}_{s_1})$, respectively. Here, τ is a tuning parameter to be optimized using the leave-one-out cross-validation. In the second step, based on $\hat{P}_1$, we estimated $\hat{\mathbf{y}}_j$ as $\mathbf{x}_j^T \hat{P}_1 \mathbf{x}_j$. In the third step, using existing approaches [5, 6], we generated , we generated normal random numbers and set an initial matrix, vector, and scalar as $\hat{\mathbf{W}}_2$, $\hat{\mathbf{b}}_1$, and $\hat{\mathbf{b}}_2$, respectively. Next, we extracted

m observations randomly using the bootstrap method [12]. Using $\hat{\mathbf{W}}_2$, $\hat{\mathbf{b}}_1$, $\hat{\mathbf{b}}_2$, and a bootstrap sample of size m, we estimated $\mathbf{W}_2\mathbf{W}_1$ as follows:

$$\hat{\mathbf{W}}_2\hat{\mathbf{W}}_1 = \frac{1}{m} \sum_{i=1}^{m} (\eta_2^{-1}(\hat{\mathbf{y}}_i) - (\hat{\mathbf{W}}_2\hat{\mathbf{b}}_1 + \hat{\mathbf{b}}_2))\mathbf{x}_i^T (\mathbf{x}_i\mathbf{x}_i^T)^{-1}, \tag{3}$$

where we estimate the inverse of $\mathbf{x}_i\mathbf{x}_i^T$ in Equation (3) using the naive approach from the diagonal elements in $\mathbf{x}_i\mathbf{x}_i^T$. Additionally, using the generalized inverse approach, we obtained $\hat{\mathbf{W}}_1$ in the basis of $\hat{\mathbf{W}}_2$ and $\hat{\mathbf{W}}_2\hat{\mathbf{W}}_1$. Finally, $\hat{\mathbf{b}}_1$, $\hat{\mathbf{b}}_2$, $\hat{\mathbf{W}}_1$, and $\hat{\mathbf{W}}_2$ were used as initial vectors and matrices to update the parameters of the convolutional neural network.

4 Analysis Results on Real-world Data

In this paper, all analyses were performed using R version 4.1.2 (R Foundation for Statistical Computing). We applied the proposed method to a real BA dataset. Here, stool image data with objects, such as diapers partially photographed on the image were used. In this numeric experiment, we randomly divided 35 data into 15 training and 20 test data, respectively. Next, we compared the proposed and existing methods relative to the learning convergence and prediction accuracy on the training and test data, respectively. Here, we set the values of the learning coefficients γ_1 and γ_2 to 0.1, respectively. Also, we prepared a single feature selection layer and performed the convolution and max pooling process seven times. Each time an initial value was set randomly, learning was performed 1000 times using the 15 training data, and it was judged that learning converged when the value obtained by dividing the sum of the absolute values of the difference between $\hat{\mathbf{y}}_j$ and $\mathbf{t}_j$ by 1000 became less than 0.01. We repeated to randomly divide 35 data into 15 training and 20 test data five times. As a result, we created five datasets. For each dataset, the sensitivity, specificity, and AUC values of the training and test data were calculated using the parameters ($\hat{\mathbf{b}}_1$, $\hat{\mathbf{b}}_2$, $\hat{\mathbf{W}}_1$, and $\hat{\mathbf{W}}_2$) at the time the learning first converged in the existing and our proposed methods. Figure 1 shows the average of the five absolute values of the difference between the correct label and the predicted value at each step when learning was first converged for each method. We can observe that the error decreased steadily as the proposed method progressed compared to the existing methods. When the model was constructed using the weights at the learning convergence point and applied to 15 training data every time, the average values of sensitivity and specificity were 100.0%, and that of the AUC value was 1.000 for all methods. However, a difference was observed among the compared methods on the test data. For the method by [5], the average values of sensitivity, specificity, and AUC in the test data were 83.3%, 42.5%, and 0.629, respectively. Also, for that of [6], the average values of sensitivity, specificity, and AUC in the test data were 85.0%, 40.0%, and 0.625, respectively. With the proposed method, the average values of sensitivity, specificity, and AUC obtained on the test data were 85.0%, 67.5%, and 0.763, respectively.

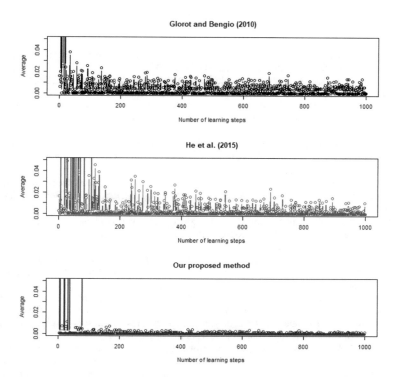

Fig. 1 Transition of learning in each method.

5 Conclusion and Limitations

In this paper, we considered a discrimination problem using a DCNN for high-dimensional small sample data and proposed a method by setting the initial weight matrix in the affine layer. In situations of limited learning data, although transfer learning can be used, we proposed an efficient learning method using the DCNN method. In terms of learning convergence and results obtained from the test data, we confirm that the proposed method is good. However, the results presented in this paper are limited and the proposed method needs to be examined in more detail. Therefore, in the future, through large-scale simulation studies and other real-world data applications, we plan to investigate the differences between the proposed method and existing methods by changing the number of feature selection layers and using different convolution filters. We also plan to investigate the proposed method by considering robustness and setting outliers on the simulation data.

Acknowledgements We thank Shinsuke Ito, Takashi Taguchi, Dr. Yusuke Yamane, Ms. Saeko Hishinuma, and Dr. Saeko Hirai for their advice. In addition, we acknowledge the biliary atresia patients' community (BA no kodomowo mamorukai) for their generous support of this project. This work was supported by the Mitsubishi Foundation.

References

1. Rawat, W., Wang, Z.: Deep convolutional neural networks for image classification: a comprehensive review. Neural Comput. **29**(9), 2352–2449 (2017)
2. Yadav, S. S., Jadhav, S.M.: Deep convolutional neural network based medical image classification for disease diagnosis. J. Big Data (2019) doi: 10.1186/s40537-019-0276-2
3. Huang, J., Xu, Z.: Cell detection with deep learning accelerated by sparse kernel. In: Lu, L. et al. (eds.) Advances in Computer Vision and Pattern Recognition, pp. 137-157. Springer, Switzerland (2017)
4. Abdelhafiz, D., Yang, C., Ammar, R., Nabavi, S.: Deep convolutional neural networks for mammography: advances, challenges and applications. BMC Bioinform. (2019) doi: 10.1186/s12859-019-2823-4
5. Glorot, X., Bengio, Y.: Understanding the difficulty of training deep feedforward neural networks. In: Proceedings of the Thirteenth International Conference on Artificial Intelligence and Statistics, pp. 249-256. (2010)
6. He, K., Zhang, X., Ren, S., Sun, J.: Delving deep into rectifiers: surpassing human-level performance on ImageNet classification. In: Proceedings of the IEEE International Conference on Computer Vision (ICCV), pp. 1026-1034. (2015)
7. Liu, J., Dai, S., Chen, G., Sun, S., Jiang, J., Zheng, S., Zheng, Y., Dong, R.: Diagnostic value and effectiveness of an artificial neural network in biliary atresia. Front. Pediatr. (2020) doi: 10.3389/fped.2020.00409
8. Zhou, W., Yang, Y., Yu, C., Liu, J., Duan, X., Weng, Z., Chen, D., Liang, Q., Fang, Q., Zhou, J., Ju, H., Luo, Z., Guo, W., Ma, X., Xie, X., Wang, R., Zhou, L.: Ensembled deep learning model outperforms human experts in diagnosing biliary atresia from sonographic gallbladder images. Nat. Commun. (2021) doi: 10.1038/s41467-021-21466-z
9. Gu, Y.H., Yokoyama, K., Mizuta, K., Tsuchioka, T., Kudo, T., Sasaki, H., Nio, M., Tang, J., Ohkubo, T., Matsui, A.: Stool color card screening for early detection of biliary atresia and long-term native liver survival: a 19-year cohort study in Japan. J. Pediatr. **166**(4), 897–902 (2015)
10. Hoshino, E., Hayashi, K., Suzuki, M., Obatake, M., Urayama, K.Y., Nakano, S., Taura, Y., Nio, M., Takahashi, O.: An iPhone application using a novel stool color detection algorithm for biliary atresia screening. Pediatr. Surg. Int. **33**(10), 1115–1121 (2017)
11. Hastie, T., Tibshirani, R., Friedman, J.: The Elements of Statistical Learning: Data Mining, Inference, and Prediction, Second Edition. Springer, New York (2009)
12. Efron, B.: Bootstrap methods: another look at the jackknife. Ann. Stat. **7**(1), 1–26 (1979)

Some Issues in Robust Clustering

Christian Hennig

Abstract Some key issues in robust clustering are discussed with focus on the Gaussian mixture model based clustering, namely the formal definition of outliers, ambiguity between groups of outliers and clusters, the interaction between robust clustering and the estimation of the number of clusters, the essential dependence of (not only) robust clustering on tuning decisions, and shortcomings of existing measurements of cluster stability when it comes to outliers.

Keywords: Gaussian mixture model, trimming, noise component, number of clusters, user tuning, cluster stability

1 Introduction

Cluster analysis is about finding groups in data. Robust statistics is about methods that are not affected strongly by deviations from the statistical model assumptions or moderate changes in a data set. Particular attention has been paid in the robustness literature to the effect of outliers. Outliers and other model deviations can have a strong effect on cluster analysis methods as well. There is now much work on robust cluster analysis, see [1, 19, 9] for overviews.

There are standard techniques of assessing robustness such as the influence function and the breakdown point [15] as well as simulations involving outliers, and these have been applied to robust clustering as well [19, 9].

Here I will argue that due to the nature of the cluster analysis problem, there are issues with the standard reasoning regarding robustness and outliers.

The starting point will be clustering based on the Gaussian mixture model, for details see [3]. For this approach, n observations are assumed i.i.d. with density

Christian Hennig (✉)
Dipartimento di Scienze Statistiche "Paolo Fortunati", University of Bologna, Via delle Belle Arti 41, 40126 Bologna, Italy, e-mail: `christian.hennig@unibo.it`

© The Author(s) 2023
P. Brito et al. (eds.), *Classification and Data Science in the Digital Age*,
Studies in Classification, Data Analysis, and Knowledge Organization,
https://doi.org/10.1007/978-3-031-09034-9_21

$$f_\eta(x) = \sum_{k=1}^{K} \pi_k \varphi_{\mu_k, \Sigma_k}(x),$$

$x \in \mathbb{R}^p$, with K mixture components with proportions π_k, $\varphi_{\mu_k, \Sigma_k}$ being the Gaussian density with mean vectors μ_k, covariance matrices Σ_k, $k = 1, \ldots, K$, η being a vector of all parameters. For given K, η can be estimated by maximum likelihood (ML) using the EM-algorithm, as implemented for example in the R-package "mclust". A standard approach to estimate K is the optimisation of the Bayesian Information Criterion (BIC). Normally, mixture components are interpreted as clusters, and observations x_i, $i = 1, \ldots, n$, can be assigned to clusters using the estimated posterior probability that x_i was generated by mixture component k. A problem with ML estimation is that the likelihood degenerates if all observations assigned to a mixture component lie on a lower dimensional hyperplane, i.e, a Σ_k has an eigenvalue of zero. This can be avoided by placing constraints on the eigenvalues of the covariance matrices [8]. Alternatively, a non-degenerate local optimum of the likelihood can be used, and if this cannot be found, constrained covariance matrix models (such as $\Sigma_1 = \ldots = \Sigma_K$) can be fitted instead, as is the default of mclust. Several issues with robustness that occur here are also relevant for other clustering approaches.

2 Outliers vs Clusters

It is well known that the sample mean and sample covariance matrix as estimators of the parameters of a single Gaussian distribution can be driven to breakdown by a single outlier [15]. Under a Gaussian mixture model with fixed K, an outlier must be assigned to a mixture component k and will break down the estimators of μ_k, Σ_k (which are weighted sample means and covariance matrices) for that component in the same manner; the same holds for a cluster mean in k-means clustering.

Addressing this issue, and dealing with more outliers in order to achieve a high breakdown point, is a starting point for robust clustering. Central ideas are trimming a proportion of observations [7], adding a "noise component" with constant density to catch the outliers [4, 3], mixtures with more robust component-wise estimators such as mixtures of heavy-tailed distributions (Sec. 7 of [18]).

But cluster analysis is essentially different from estimating a homogeneous population. Given a data set with K clear Gaussian clusters and standard ML-clustering, consider adding a single outlier that is far enough away from the clusters. Assuming a lower bound on covariance matrix eigenvalues, the outlier will form a one-point cluster, the mean of which will diverge with the added outlier, and the original clusters will be merged to form $K - 1$ clusters [10].

The same will happen with a group of several outliers being close together, once more added far enough away from the Gaussian clusters. "Breakdown" of an estimator it is usually understood as the estimator becoming useless. It is questionable that this is the case here. In fact, the "group of outliers" can well be interpreted as a cluster in its own right, and putting all these points together in a cluster could be seen as desirable behaviour of the ML estimator, at least if two of the original K clusters are close enough to each other that merging them will produce a cluster that is fairly well fitted by a single Gaussian distribution; note that the Gaussian mixture model does not assume strong separation between components, and a mixture of two Gaussians may be unimodal and in fact very similar to a single Gaussian. A breakdown point larger than a given α, $0 < \alpha < \frac{1}{2}$ may not be seen as desirable in cluster analysis if there can be clusters containing a proportion of less than α of the data, as a larger breakdown point will stop a method from taking such clusters (when added in large distance from the rest of the data) appropriately into account.

The core problem is that it is not clear what distinguishes a group of outliers from a legitimate cluster. I am not aware of any formal definition of outliers and clusters in the literature that allows this distinction. Even a one-point cluster is not necessarily invalid. Here are some possible and potentially conflicting aspects of such a distinction.

- A certain minimum size may be required for a cluster; smaller groups of points may be called outliers.
- Groups of points in low density areas of the data may be called outliers. Note that this particularly means that very widely spread Gaussian mixture components would also be defined as outliers, deviating from the standard interpretation of Gaussian mixture components as clusters.
- Members of non-Gaussian mixture components may be called outliers. This does not seem to be a good idea, because Gaussianity cannot be assessed for too small groups of observations, and furthermore in practice model assumptions are never perfectly fulfilled, and it may be desirable to interpret homogeneous or unimodal non-Gaussian parts of the data as "cluster" and fit them by a Gaussian component.
- The term "outlier" suggests that outliers lie far away from most other observations, so it may be required that outliers are farther away from the clusters than the clusters are from each other. But this would be in conflict with the intuition that strong separation is usually seen as a desirable feature for well interpretable clusters. It may only be reasonable in applications in which there is prior information that there is limited variation even between clusters, as is implied by certain Bayesian approaches to clustering [17].
- The term "cluster" may be seen as flexible enough that a definition of an outlier is not required. Clustering should accommodate whatever is "outlying" by fitting it by one or more further clusters, if necessary of size one (single linkage clustering can be useful for outlier detection, even though it is inappropriate for most clustering problems).

Most of these items require specific decisions that cannot be made in any objective and general manner, but only taking into account subject matter information, such as the minimum size of valid clusters or the density level below which observations are seen as outliers (potentially compared to density peaks in the distribution). This implies that an appropriate treatment of outliers in cluster analysis cannot be expected to be possible without user tuning.

3 Robustness and the Number of Clusters

The last item suggests that there is an interplay between outlier identification and the number of clusters, and that adding clusters might be a way of dealing with outliers; as long as clusters are assumed to be Gaussian, a single additional component may not be enough. More generally, concentrating robustness research on the case of fixed K may be seen as unrealistic, because K is rarely known, although estimating K is a notoriously difficult problem even without worrying about outliers [13].

The classical robustness concepts, breakdown point and influence function, assume parameters from $\mathbb{R}^q$ with fixed q. If K is not fixed, the number of parameters is not fixed either, and the classical concepts do not apply.

As an alternative to the breakdown point, [11] defined a "dissolution point". Dissolution is measured in terms of cluster memberships of points rather than in terms of parameters, and is therefore also applicable to nonparametric clustering methods. Furthermore, dissolution applies to individual clusters in a clustering; certain clusters may dissolve, i.e., there may be no sufficiently similar cluster in a new clustering computed after, e.g., adding an outlier; and others may not dissolve. This does not require K to be fixed; the definition is chosen so that if a clustering changes from K to $L < K$ clusters, at least $K - L$ clusters dissolve.

Hennig [10, 11] showed that when estimating K using the BIC and standard ML estimation, reasonably well separated clusters do not dissolve when adding possibly even a large percentage of outliers (this does not hold for every method to estimate the number of clusters, see [11]). Furthermore, [11] showed that no method with fixed K can be robust for data in which K is misspecified - already [7] had found that robustness features in clustering generally depend on the data.

An implication of these results is that even in the fixed K problem, the standard ML method can be a valid competitor regarding robustness if it comes with a rule that allows to add one or possibly more clusters that can then be used to fit the outliers (this is rarely explored in the literature, but [18], Sec. 7.7, show an example in which adding a single component does not work very well).

An issue with adding clusters to accommodate outliers is that in many applications it is appropriate to distinguish between meaningful clusters, and observations that cannot be assigned to such clusters (often referred to as "noise"). Even though adding clusters of outliers can formally prevent the dissolution of existing clusters, it may be misleading to interpret the resulting clusters as meaningful, and a classification as outliers or noise can be more useful. This is provided by the trimming and noise component approaches to robust clustering. Also some other clustering methods such as the density-based DBSCAN [5] provide such a distinction. On the other hand, modelling clusters by heavy-tailed distributions such as in mixtures of t-distributions will implicitly assign outlying observations to clusters that potentially are quite far away. For this reason, [18], Sec. 7.7, provide an additional outlier identification rule on top of the mixture fit. [6] even distinguish between "mild" outliers that are modelled as having a larger variance around the same mean, and "gross" outliers to be trimmed. The variety of approaches can be connected to the different meanings that outliers can have in applications. They can be erroneous, they can be irrelevant noise, but they can also be caused by unobserved but relevant special conditions (and would as such qualify as meaningful clusters), or they could be valid observations legitimately belonging to a meaningful cluster that regularly produces observations further away from the centre than modelled by a Gaussian distribution.

Even though currently there is no formal robustness property that requires both the estimation of K and an identification or downweighting of outliers, there is demand for a method that can do both.

Estimating K comes with an additional difficulty that is relevant in connection with robustness. As mentioned before, in clustering based on the Gaussian mixture model normally every mixture component will be interpreted as a cluster. In reality, however, meaningful clusters are not perfectly Gaussian. Gaussian mixtures are very flexible for approximating non-Gaussian distributions. Using a consistent method for estimating K means that for large enough n a non-Gaussian cluster will be approximated by several Gaussian mixture components. The estimated K will be fine for producing a Gaussian mixture density that fits the data well, but it will overestimate the number of interpretable clusters. The estimation of K, if interpreted as the number of clusters, relies on precise Gaussianity of the clusters, and is as such itself riddled with a robustness problem; in fact slightly non-Gaussian clusters may even drive the estimated $K \to \infty$ if $n \to \infty$ [12, 14].

This is connected with the more fundamental problem that there is no unique definition of a cluster either. The cluster analysis user needs to specify the cluster concept of interest even before robustness considerations, and arguably different clustering methods imply different cluster concepts [13]. A Gaussian mixture model defines clusters by the Gaussian distributional shape (unless mixture components are merged to form clusters [12]). Although this can be motivated in some real situations, robustness considerations require that distributional shapes fairly close to the Gaussian should be accepted as clusters as well, but this requires another specification, namely how far from a Gaussian a cluster is allowed to be, or alternatively how separated Gaussian components have to be in order to count as separated clusters. A similar problem can also occur in nonparametric clustering; if clusters are associated with density modes or level sets, the cluster concept depends on how weak a mode or gap between high level density sets is allowed to be to be treated as meaningful.

Hennig and Coretto [14] propose a parametric bootstrap approach to simultaneously estimate K and assign outliers to a noise component. This requires two basic tuning decisions. The first one regards the minimum percentage of observations so that a researcher is willing to add another cluster if the noise component can be reduced by this amount. The second one specifies a tolerance that allows a data subset to count as a cluster even though it deviates to some extent from what is expected under a perfectly Gaussian distribution. There is a third tuning parameter that is in effect for fixed K and tunes how much of the tails of a non-Gaussian cluster can be assigned to the noise in order to improve the Gaussian appearance of the cluster. One could even see the required constraints on covariance matrix eigenvalues as a further tuning decision. Default values can be provided, but situations in which matters can be improved deviating from default values are easy to construct.

4 More on User Tuning

User tuning is not popular, as it is often difficult to make appropriate tuning decisions. Many scientists believe that subjective user decisions threaten scientific objectivity, and also background knowledge dependent choices cannot be made when investigating a method's performance by theory and simulations. The reason why user tuning is indispensable in robust cluster analysis is that it is required in order to make the problem well defined. The distinction between clusters and outliers is an interpretative one that no automatic method can make based on the data alone. Regarding the number of clusters, imagine two well separated clusters (according to whatever cluster concept of interest), and then imagine them to be moved closer and closer together. Below what distance are they to be considered a single cluster? This is essentially a tuning decision that the data cannot make on their own.

There are methods that do not require user tuning. Consider the mclust implementation of Gaussian mixture model based clustering. The number of clusters is by default estimated by the BIC. As seen above, this is not really appropriate for large data sets, but its derivation is essentially asymptotic, so that there is no theoretical justification for it for small data sets either. Empirically it often but not always works well, and there is little investigation of whether it tends to make the "right" decision in ambiguous situations where it is not clear without user tuning what it even means to be "right". Covariance matrix constraints in mclust are not governed by a tuning of eigenvalues or their ratios to be specified by the user. Rather the BIC decides between different covariance matrix models, but this can be erratic and unstable, as it depends on whether the EM-algorithm gets caught in a degenerate likelihood maximum or not, and in situations where two or more covariance matrix models have similar BIC values (which happens quite often), a tiny change in the data can result in a different covariance matrix model being selected, and substantial changes in the clustering. A tunable eigenvalue condition can result in much smoother behaviour. When it comes to outlier identification, mclust offers the addition of a uniform "noise" mixture component governed by the range of the data, again supposedly without user tuning. This starts from an initial noise estimation that requires tuning (Sec. 3.1.2 of [3]) and is less robust in terms of breakdown and dissolution than trimming and the improper noise component, both of which require tuning [10, 11]. The ICL, an alternative to the BIC (Sec. 2.6 of [3]), on the other hand, is known to merge different Gaussian mixture components already at a distance at which they intuitively still seem to be separated clusters. Similar comments apply to the mixture of t-distributions; it requires user tuning for identifying outliers, scatter matrix constraints, and it has the same issues with BIC and ICL as the Gaussian mixture.

Summarising, both the identification of and robustness against outliers and the estimation of the number of clusters require tuning in order to be well defined problems; user tuning can only be avoided by taking tuning decisions out of the user's hands and making them internally, which will work in some situations and fail in others, and the impression of automatic data driven decision making that a user may have is rather an illusion. This, however, does not free method designers from the necessity to provide default tunings for experimentation and cases in which the users do not feel able to make the decisions themselves, and tuning guidance for situations in which more information is available. A decision regarding the smallest valid size of a cluster is rather well interpretable; a decision regarding admissible covariance matrix eigenvalues is rather difficult and abstract.

5 Stability Measurement

Robustness is closely connected to stability. Both experimental and theoretical investigation of the stability of clusterings require formal stability measurements, usually comparing two clusterings on the same data (potentially modified by replacing or adding observations). Not assuming any parametric model, proximity measures such as the Adjusted Rand Index (ARI; [16]), the Hamming distance (HD; [2]), or the Jaccard distance between individual clusters [11] can be used. Note that [2], standard reference on cluster stability in the machine learning community, state that stability and instability are caused in the first place by ambiguities in the cluster structure of the data, rather than by a method's robustness or lack of it. Although the outlier problem is ignored in that paper, it is true that cluster analysis can have other stability issues that are as serious as or worse than gross outliers.

To my knowledge, none of the measures currently in use allow for a special treatment of a set of outliers or noise; either these have to be ignored, or treated just as any other cluster. Both ARI and HD, comparing clusterings C_1 and C_2, consider pairs of observations x_i, x_j and check whether those that are in the same cluster in C_1 are also in the same cluster in C_2. An appropriate treatment of noise sets $N_1 \in C_1, N_2 \in C_2$ would require that $x_i, x_j \in N_1$ are not just in the same cluster in C_2 but rather in N_2, i.e., whereas the numberings of the regular clusters do not have to be matched (which is appropriate because cluster numbering is meaningless), N_1 has to be matched to N_2. Corresponding re-definitions of these proximities will be useful to robustness studies.

6 Conclusion

Key practical implications of the above discussions are:

- Outliers can be treated as forming their own clusters, or be collected in outlier/noise or trimmed sets, or be integrated in clusters of non-outliers. Which of these is appropriate depends on the nature of outliers in a given application.
- Methods that do not identify outliers but add clusters in order to accommodate them are valid competitors of robust clustering methods, as are nonparametric density-based methods.
- Cluster analysis involving estimating the number of clusters and robustness require tuning in order to define the problem they are meant to solve well. Method

developers need to provide sensible defaults, but also to guide the users regarding a meaningful interpretation of the tuning decisions.

References

1. Banerjee, A., Davé, R. N.: Robust clustering. WIREs Data Mining Knowl. Discov. **2**, 29–59 (2012)
2. Ben-David, S., von Luxburg, U., Pál, D.: A sober look at clustering stability. In: Proceedings of the 19th annual conference on Learning Theory (COLT'06), pp. 5–19, Springer, Berlin (2006)
3. Bouveyron, C., Celeux, G., Murphy, T. B., Raftery, A. E.: Model-based clustering and classification for data science. Cambridge University Press, Cambridge MA (2019)
4. Coretto, P., Hennig, C.: Consistency, breakdown robustness, and algorithms for robust improper maximum likelihood clustering. J. Mach. Learn. Res.**18**, 1–39 (2017)
5. Ester, M., Kriegel, H. P., Sander, J., Xu, X.: A density-based algorithm for discovering clusters in large spatial databases with noise. In: Proceedings of the 2nd International Conference on Knowledge Discovery and Data Mining, pp. 226-231, AAAI Press, Portland OR (1996)
6. Farcomeni, A., Punzo, A.: Robust model-based clustering with mild and gross outliers. TEST **29**, 989–1007 (2020)
7. García-Escudero, L. A., Gordaliza, A.: Robustness properties of k-means and trimmed k-means. J. Am. Stat. Assoc. **94**, 956–969 (1999)
8. García-Escudero, L. A., Gordaliza, A., Greselin, F., Ingrassia, S., Mayo-Iscar, A.: Eigenvalues and constraints in mixture modeling: geometric and computational issues. Adv. Data Anal. Classi. **12**, 203–233 (2018)
9. García-Escudero, L. A., Gordaliza, A., Matrán, C., Mayo-Iscar, A., Hennig, C.: Robustness and outliers. In: Hennig, C., Meila, M., Murtagh, F., Rocci, R. (eds.) Handbook of Cluster Analysis, pp. 653–678. Chapman & Hall/CRC, Boca Raton FL (2016)
10. Hennig, C.: Breakdown points for maximum likelihood estimators of location-scale mixtures. Ann. Stat. **32**, 1313–1340 (2004)
11. Hennig, C.: Dissolution point and isolation robustness: robustness criteria for general cluster analysis methods. J. Multivariate Anal. **99**, 1154–1176 (2008)
12. Hennig, C.: Methods for merging Gaussian mixture components. Adv. Data Anal. Classi. **4**, 3–34 (2010)
13. Hennig, C.: Clustering strategy and method selection. In: Hennig, C., Meila, M., Murtagh, F., Rocci, R. (eds.) Handbook of Cluster Analysis, pp. 703–730. Chapman & Hall/CRC, Boca Raton FL (2016)
14. Hennig, C., Coretto, P.: An adequacy approach for deciding the number of clusters for OTRIMLE robust Gaussian mixture-based clustering. Aust. N. Z. J. Stat. (2021) doi: 10.1111/anzs.12338
15. Huber, P. J., Ronchetti, E. M.: Robust Statistics (2nd ed.). Wiley, Hoboken NJ (2009)
16. Hubert, L., Arabie, P.: Comparing partitions. J. Classif. **2**, 193–218 (1985)
17. Malsiner-Walli, G., Frühwirth-Schnatter, S., Grün, B.: Identifying mixtures of mixtures using Bayesian estimation. J. Comput. Graph. Stat. **26**, 285–295 (2017)
18. McLachlan, G. J., Peel, D.: Finite Mixture Models. Wiley, New York (2000)
19. Ritter, G.: Robust cluster analysis and variable selection. Chapman & Hall/CRC, Boca Raton FL (2015)

Robustness Aspects of Optimized Centroids

Jan Kalina and Patrik Janáček

Abstract Centroids are often used for object localization tasks, supervised segmentation in medical image analysis, or classification in other specific tasks. This paper starts by contributing to the theory of centroids by evaluating the effect of modified illumination on the weighted correlation coefficient. Further, robustness of various centroid-based tools is investigated in experiments related to mouth localization in non-standardized facial images or classification of high-dimensional data in a matched pairs design. The most robust results are obtained if the sparse centroid-based method for supervised learning is accompanied with an intrinsic variable selection. Robustness, sparsity, and energy-efficient computation turn out not to contradict the requirement on the optimal performance of the centroids.

Keywords: image processing, optimized centroids, robustness, sparsity, low-energy replacements

1 Introduction

Methods based on centroids (templates, prototypes) are simple yet widely used for object localization or supervised segmentation in image analysis tasks and also within other supervised or unsupervised methods of machine learning. This is true e.g. in various biomedical imaging tasks [1], where researchers typically cannot afford a too large number of available images [3]. Biomedical applications also benefit from the interpretability (comprehensibility) of centroids [11].

This paper is focused on the question how are centroid-based methods influenced by data contamination. Section 2 recalls the main approaches to centroid-based object localization in images, as well as a recently proposed method of [6] for op-

Jan Kalina (✉) · Patrik Janáček
The Czech Academy of Sciences, Institute of Computer Science, Pod Vodárenskou věží 2, 182 07 Prague 8, Czech Republic, e-mail: `kalina@cs.cas.cz`;`janacekpatrik@gmail.com`

© The Author(s) 2023
P. Brito et al. (eds.), *Classification and Data Science in the Digital Age*,
Studies in Classification, Data Analysis, and Knowledge Organization,
https://doi.org/10.1007/978-3-031-09034-9_22

timizing centroids and their weights. The performance of these methods to data contamination (non-standard conditions) has not been however sufficiently investigated. Particularly, we are interested in the performance of low-energy replacements of the optimal centroids and in the effect of posterior variable selection (pixel selection). Section 2.1 presents novel expressions for images with a changed illumination. Numerical experiments are presented in Section 3. These are devoted to mouth localization over raw facial images as well as over artificially modified images; other experiments are devoted to high-dimensional data in a matched pairs design. The optimized centroids of [6] and especially their modification proposed here turn out to have remarkable robustness properties. Section 4 brings conclusions.

2 Centroid-based Classification (Object Localization)

Commonly used centroid-based approaches to object localization (template matching) in images construct the centroid simply as the average of the positive examples and typically use Pearson product-moment correlation coefficient r as the most common measure of similarity between a centroid $\mathbf{c}$ and a candidate part of the image (say $\mathbf{x}$). While the centroid and candidate areas are matrices of size (say) $I \times J$ pixels, they are used in computations after being transformed to vectors of length $d := IJ$. This allows us to use the notation $\mathbf{c} = (c_1, \ldots, c_d)^T$ and $\mathbf{x} = (x_1, \ldots, x_d)^T$.

Assumptions $\mathcal{A}$: We assume the whole image to have size $N_R \times N_C$ pixels. We assume the centroid $\mathbf{c} = (c)_{i,j}$ with $i = 1, \ldots, I$ and $j = 1, \ldots, J$ to be a matrix of size $I \times J$ pixels. A candidate area $\mathbf{x}$ and nonnegative weights $\mathbf{w}$ with $\sum_i \sum_j w_{ij} = 1$ are assumed to be matrices of the same size as $\mathbf{c}$.

For a given image, E will denote the set of its rectangular candidate areas of size $I \times J$. The candidate area fulfilling

$$\arg\max_{\mathbf{x} \in \mathsf{E}} r(\mathbf{x}, \mathbf{c}) \tag{1}$$

or (less frequently)

$$\arg\min_{\mathbf{x} \in \mathsf{E}} ||\mathbf{x} - \mathbf{c}||_2 \tag{2}$$

are classified to correspond to the object (e.g. mouth).

Let us consider here replacing r by the weighted correlation coefficient r_w

$$\arg\max_{\mathbf{x} \in \mathsf{E}} r_w(\mathbf{x}, \mathbf{c}; \mathbf{w}) \tag{3}$$

with given non-negative weights $\mathbf{w} = (w_1, \ldots, w_d)^T \in \mathcal{R}^p$ with $\sum_{i=1}^{n} w_i = 1$, where $\mathcal{R}$ denotes the set of all real numbers. Let us further use the notation $\bar{x}_w = \sum_{j=1}^{d} w_j x_j = \mathbf{w}^T \mathbf{x}$ and $\bar{c}_w = \mathbf{w}^T \mathbf{c}$. We may recall r_w between $\mathbf{x}$ and $\mathbf{c}$ to be defined as

$$r_W(\mathbf{x}, \mathbf{c}; \mathbf{w}) = \frac{\sum_{i=1}^{d} w_i (x_i - \bar{x}_w)(c_i - \bar{c}_w)}{\sqrt{\sum_{i=1}^{d} [w_i (x_i - \bar{x}_w)^2] \sum_{i=1}^{d} [w_i (c_i - \bar{c}_w)^2]}}. \tag{4}$$

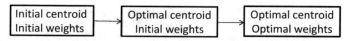

Fig. 1 The workflow of the optimization procedure of [6].

A detailed study of [2] investigated theoretical foundations of centroid-based classification, however for the rare situation when (1) is replaced by

The sophisticated centroid optimization method of [6], outlined in Figure 1, requires to minimize a nonlinear loss function corresponding to a regularized margin-like distance (exploiting r_w) evaluated for the worst pair from the worst image over the training database (i.e. the worst with respect to the loss function). Subsequently, optimization of the weights may be also performed, ensuring many pixels to obtain zero weights (i.e. yielding a sparse solution). The optimal centroid may be used as such, even without any weights at all; still, optimization of the weights leads to a further improvement of the classification performance. In the current paper, we always consider a linear (i.e. approximate) approach to centroid optimization, although a nonlinear optimization is also successful as revealed in the comparisons in [6].

2.1 Centroid-Based Object Localization: Asymmetric Modification of the Candidate Area

In the context of object localization as described above, our aim is to express $r_w(\mathbf{x}^*, \mathbf{c}; \mathbf{w})$ under modified candidate areas (say $\mathbf{x}^*$) of the image $\mathbf{x}$; we stress that the considered modification of the image does not allow to modify the centroid $\mathbf{c}$ and weights $\mathbf{w}$. These considerations are useful for centroid-based object localization, when asymmetric illumination is present in the whole image or its part. The weighted variance $S_w^2(\mathbf{x}; \mathbf{w})$ of $\mathbf{x}$ with weights $\mathbf{w}$ and the weighted covariance $S_w(\mathbf{x}, \mathbf{c})$ between $\mathbf{x}$ and $\mathbf{c}$ are denoted as

$$S_w^2(\mathbf{x}) = \sum_{i,j} w_{ij}(x_{ij} - \bar{x}_w)^2, \quad S_w(\mathbf{x}, \mathbf{c}) = \sum_{i,j} w_{ij}(x_{ij} - \bar{x}_w)(c_{ij} - \bar{c}_w). \quad (5)$$

Further, the notation $\mathbf{x} + a$ with $\mathbf{x} = (x_{ij})_{i,j}$ is used to denote the matrix $(x_{ij} + a)_{i,j}$ for a given $a \in \mathcal{R}$. We also use the following notation. The image $\mathbf{x}$ is divided to two parts $\mathbf{x} = (\mathbf{x}_1, \mathbf{x}_2)^T \in \mathcal{R}^d$, where $\sum_I$ or $\sum_{II}$ denote the sum over the pixels of the first or second part, respectively.

Theorem 1 *Under Assumptions $\mathcal{A}$, the following statements hold.*

1. *For $\mathbf{x}^* = \mathbf{x} + \varepsilon$, it holds $r_w(\mathbf{x}^*, \mathbf{c}) = r_w(\mathbf{x}, \mathbf{c})$ for $\varepsilon > 0$.*
2. *For $\mathbf{x}^* = k\mathbf{x}$ with $k > 0$, it holds $r_w(\mathbf{x}^*, \mathbf{c}) = r_w(\mathbf{x}, \mathbf{c})$.*
3. *For $\mathbf{x} = (\mathbf{x}_1, \mathbf{x}_2)^T$ and $\mathbf{x}^* = (\mathbf{x}_1, \mathbf{x}_2 + \varepsilon)^T$, it holds $r_w(\mathbf{x}^*, \mathbf{c}) =$*

$$= \frac{S_w(\mathbf{x}, \mathbf{c}) + \varepsilon \sum_{II} w_{ij} c_{ij} - \varepsilon v_2 \bar{c}_w}{S_w(\mathbf{c}) \sqrt{S_w^2(\mathbf{x}) + v_2(1 - v_2)\varepsilon^2 + 2\varepsilon(2v_2 - 1)(\sum_{II} w_{ij} x_{ij} - v_2 \bar{x}_w)}}, \tag{6}$$

where $v_2 = \sum_{II} w_{ij}$ and $\varepsilon \in \mathcal{R}$.

4. For $\mathbf{x} = (\mathbf{x}_1, \mathbf{x}_2)^T$ and $\mathbf{x}^* = (\mathbf{x}_1, k\mathbf{x}_2)^T$ with $k > 0$, it holds

$$r_w(\mathbf{x}^*, \mathbf{c}) = r_w(\mathbf{x}, \mathbf{c}) \frac{S_w(\mathbf{x})}{S_w^*(\mathbf{x})} + \frac{(k - 1) \sum_{II} w_{ij} x_{ij}(c_{ij} - \bar{c}_w)}{S_w(\mathbf{c}) S_w^*(\mathbf{x})}, \tag{7}$$

where

$$\left(S_w^*(\mathbf{x})\right)^2 = S_w^2(\mathbf{x}) + (k^2 - 1) \sum_{II} w_{ij} x_{ij}^2 - \frac{k^2 - 1}{n} \left(\sum_{II} w_{ij} x_{ij}\right)^2 -$$

$$- \frac{2}{n}(k - 1) \left(\sum_{I} w_{ij} x_{ij}\right) \left(\sum_{II} w_{ij} x_{ij}\right). \tag{8}$$

The proofs of the formulas are technical but straightforward exploiting known properties of r_w. The theorem reveals r_w to be vulnerable to the modified illumination, i.e. all the methods based on centroids of Section 2 may be too influenced by the data modification.

3 Experiments

3.1 Data

Three datasets are considered in the experiments. In the first dataset, the task is to localize the mouth in the database containing 212 grey-scale 2D facial images of faces of healthy individuals of size 192×256 pixels. The database previously analyzed in [6] was acquired at the Institute of Human Genetics, University of Duisburg-Essen, within research of genetic syndrome diagnostics based on facial images [1] under the projects BO 1955/2-1 and WU 314/2-1 of the German Research Council (DFG). We consider the training dataset to consist of the first 124 images, while the remaining 88 images represent an independent test set acquired later but still under the same standardized conditions fulfilling assumptions of unbiased evaluation. The centroid described below is used with $I = 26$ and $J = 56$.

Using always raw training images, the methods are applied not only to the raw test set, but also to the test set after being artificially modified using models inspired by Section 2.1. On the whole, five different versions of the test database are considered; the modifications required that we first manually localized the mouths in the test images:

1. Raw images.

2. Illumination. If we consider a pixel $[i, j]$ with intensity x_{ij} in an image (say) f, then the grey-scale intensity f_{ij} will be

$$f_{ij}^* = f_{ij} + \lambda|j - j_0|, \quad i = 1, \ldots, I, \quad j = 1, \ldots, J, \quad (9)$$

where $[i_0, j_0]$ are the coordinates of the mouth and $\lambda = 0.002$.
3. A more severe version of the modification (ii) with $\lambda = 0.004$.
4. Asymmetry. In every test image, each true mouth $\mathbf{x}$ of size 26×56 pixels with intensities x_{ij} is replaced by

$$x_{ij}^* = \begin{cases} x_{ij} + 0.2, & i = 1, \ldots, 26, \ j = 1, \ldots, 15, \\ x_{ij}, & i = 1, \ldots, 26, \ j = 16, \ldots, 41, \\ x_{ij} + 0.1, & i = 1, \ldots, 26, \ j = 42, \ldots, 56. \end{cases} \quad (10)$$

5. Rotation. Such candidate area is classified as the mouth in the given image, which maximizes the loss (1) or (3) over the three versions of the image, namely after rotations by +5, 0, and −5 degrees.
6. Image denoising (for raw images). The LWS-filter [5], replacing each grey value by the least weighted squares estimate [7] computed from a circular neighborhood with radius 4 pixels, was applied to each test image.

The optimized centroids were explained in [6] to be applicable also to classification tasks for other data than images, if they follow a matched pairs design. We use two datasets from [6] in the experiments and their classification accuracies are reported in a 10-fold cross validation.

- AMI. The gene expressions of 4000 genes over 92 individuals in two versions (raw or contaminated by outliers). The aim is to learn a classification rule allowing to assign a new individual to one of the two given groups (controls or patients with acute myocardial infarction (AMI)).
- Simulated data. The design mimics a 1:1 matched case-control study with 2500 variables over 60 individuals in two versions (raw or contaminated by outliers) and the aim is again to classify between two given groups (patients and controls).

Fig. 2 The average centroid used as the initial choice for the centroid optimization.

3.2 Methods

The following methods are compared in the experiments; standard methods are computed using R software and we use our own C++ implementation of centroid-based methods. The average centroid is obtained as the average of all mouths of the training set, or the average across all patients. The centroid optimization starts with the average centroid as the initial one, and the optimization of weights starts with equal weights as the initial ones:

A.   Centroid-based method (2).
B.   Centroid-based method (1) with average centroid (Figure 2) and equal weights.
C.   Centroid-based method (1) with average centroid, replacing r_w by cosine similarity defined for $\mathbf{x} \in \mathcal{R}^d$ and $\mathbf{y} \in \mathcal{R}^d$ as

$$\cos \theta = \frac{\mathbf{x}^T \mathbf{y}}{||\mathbf{x}||_2 ||\mathbf{y}||_2} = \frac{\sum_{i=1}^{d} x_i y_i}{\left(\sum_{i=1}^{d} x_i^2\right)^{1/2} \left(\sum_{j=1}^{d} y_j^2\right)^{1/2}}. \tag{11}$$

D.   Centroid-based method (1) with optimal centroid and equal weights [6].
E.   Centroid-based method (1) with optimal centroid and optimal weights as in [6] (optimizing the centroid and only after that the weights), i.e. with posterior variable selection (pixel selection).
F.   Centroid-based method (1) as in [6], where however the weights are optimized first, and then the centroid is optimized.
G.   Centroid-based method (1) as in [6], where however each step of centroid optimization is immediately followed by optimization of the weights; this method performs (in contrary to [6]) intrinsic variable selection.
H.   Centroid-based method (1) as in [6], where however each optimization step proceeds over 10 worst images (instead of the very worst image).
I.   Centroid-based method (1) with average centroid, where r_w is used as r_{LWS} [7] with weight function

$$\psi_1(t) = \exp\left\{-\frac{t^2}{2\tau^2}\right\} 1 \left[t < \frac{3}{4}\right], \quad t \in [0, 1], \tag{12}$$

corresponding to a (trimmed) density of the Gaussian $N(0, 1)$ distribution; 1 denotes an indicator function. To explain, the computation of $r_{\mathsf{LWS}}(x, y)$ starts by fitting the LWS estimator in the linear regression of y as the response of x, and r_w is used with the weights determined by the LWS estimator.
J.   The method (I) with the weight function $\psi_2(t) = 1 \left[t < \frac{3}{4}\right]$ for $t \in [0, 1]$.
K.   The approach of [12] that is meaningful however only for the mouth localization dataset.

Table 1 Classification accuracy for three datasets. For the mouth localization data, modifications of the test images are described in Section 3: (i) None (raw images); (ii) Illumination; (iii) Asymmetry; (iv) Rotation; (v) Image denoising. A detailed description of the methods is given in Section 3.2.

				Dataset						
			Mouth localization				AMI		Simul.	
Method	(i)	(ii)	(iii)	(iv)	(v)	(vi)	Raw	Cont.	Raw	Cont.
A	0.90	0.86	0.81	0.88	0.81	0.93	0.73	0.66	0.71	0.67
B	0.93	0.90	0.86	0.92	0.86	0.95	0.76	0.70	0.77	0.70
C	0.89	0.84	0.74	0.89	0.84	0.93	0.72	0.61	0.70	0.64
D	1.00	0.98	0.95	0.99	0.93	0.98	0.85	0.83	0.80	0.77
E	1.00	1.00	0.98	1.00	0.95	0.98	0.87	0.85	0.83	0.80
F	1.00	0.98	0.96	1.00	0.89	0.97	0.86	0.82	0.79	0.73
G	1.00	0.96	0.95	1.00	0.93	0.99	0.88	0.85	0.86	0.82
H	1.00	1.00	0.98	1.00	0.92	0.96	0.86	0.83	0.84	0.79
I	0.96	0.96	0.93	0.99	0.94	0.96	0.77	0.72	0.75	0.71
J	0.94	0.93	0.89	0.95	0.89	0.93	0.74	0.69	0.72	0.66
K	1.00	1.00	0.97	0.95	0.97	0.96	Not meaningful			

3.3 Results

The results as ratios of correctly classified cases are presented in Table 1. For the mouth localization, the optimized centroids of methods D, F, and H turn out to outperform simple centroids (A, B, and C); the novel modifications E and G performing intrinsic variable selection yield the best results. Simple standard centroids (A, B, and C) are non-robust to data contamination; this follows from Section 2.1 and from analogous considerations for other types of contaminating the images. On the other hand, the robustness of optimized centroids is achieved by their optimization (but not by using r_w as such). Methods E and G are even able to overcome methods I and J based on r_{LWS}. We recall that r_{LWS} is globally robust in terms of the breakdown point [4]), is computationally very demanding, and does not seem to allow any feasible optimization. Other results reported previously in [6] revealed that also numerous standard machine learning methods are too vulnerable (non-robust) with respect to data contamination, if measuring the similarity by r or r_w.

For the AMI dataset, methods E and G with variable selection perform the best results for raw as well as contaminated datasets. For the simulated data, the method G yields the best results and the method E stays only slightly behind as the second best method.

4 Conclusions

Understanding the robustness of centroids represents a crucial question in image processing with applications for convolutional neural networks (CNNs), because centroids are very versatile tools that may be based on deep features learned by deep learning. We focus on small datasets, for which CNNs cannot be used [10]. This paper is interested in performance of centroid-based object localization over small databases with non-standardized images, which commonly appear e.g. in medical image analysis.

The requirements on robustness with respect to modifications of the images turn out not to contradict the requirements on optimality of the centroids. The method G applying an intrinsic variable selection on the optimal centroid and weights [6] can be interpreted within a broader framework of robust dimensionality reduction (see [8] for an overview) or low-energy approximate computation. Additional results not presented here reveal the method based on optimized centroids to be robust also to small shift. Neither the theoretical part of this paper nor the experiments exploit any specific properties of faces. The presented robust method has potential also for various other applications, e.g. for deep fake detection by centroids, robust template matching by CNNs [9], or applying filters in convolutional layers of CNNs.

Acknowledgements The research was supported by the grant 22-02067S of the Czech Science Foundation.

References

1. Böhringer, S., de Jong, M. A.: Quantification of facial traits. Frontiers in Genetics **10**, 397 (2019)
2. Delaigle, A., Hall, P.: Achieving near perfect classification for functional data. Journal of the Royal Statistical Society **74**, 267–286 (2012)
3. Gao, B., Spratling, M. W.: Robust template matching via hierarchical convolutional features from a shape biased CNN. ArXiv:2007.15817 (2021)
4. Jurečková, J., Picek, J., Schindler, M.: Robust statistical methods with R. 2nd edn. CRC Press, Boca Raton (2019)
5. Kalina, J.: A robust pre-processing of BeadChip microarray images. Biocybernetics and Biomedical Engineering **38**, 556–563 (2018)
6. Kalina, J., Matonoha, C.: A sparse pair-preserving centroid-based supervised learning method for high-dimensional biomedical data or images. Biocybernetics and Biomedical Engineering **40**, 774–786 (2020)
7. Kalina, J., Schlenker, A.: A robust supervised variable selection for noisy high-dimensional data. BioMed Research International **2015**, 320385 (2015)
8. Rousseeuw, P. J., Hubert, M.: Anomaly detection by robust statistics. WIREs Data Mining and Knowledge Discovery **8**, e1236 (2018)
9. Sun, L., Sun, H., Wang, J., Wu, S., Zhao, Y., Xu, Y.: Breast mass detection in mammography based on image template matching and CNN. Sensors **2021**, 2855 (2021)
10. Sze, V., Chen, Y. H., Yang, T. J., Emer, J. S.: Efficient processing of deep neural networks. Morgan & Claypool Publishers, San Rafael (2020)
11. Watanuki, S.: Watershed brain regions for characterizing brand equity-related mental processes. Brain Sciences **11**, 1619 (2021)
12. Zhu, X., Ramanan, D.: Face detection, pose estimation, and landmark localization in the wild. IEEE Conference on Computer Vision and Pattern Recognition 2012. IEEE, New York, pp. 2879–2886 (2012)

Data Clustering and Representation Learning Based on Networked Data

Lazhar Labiod and Mohamed Nadif

Abstract To deal simultaneously with both, the attributed network embedding and clustering, we propose a new model exploiting both content and structure information. The proposed model relies on the approximation of the relaxed continuous embedding solution by the true discrete clustering. Thereby, we show that incorporating an embedding representation provides simpler and easier interpretable solutions. Experiment results demonstrate that the proposed algorithm performs better, in terms of clustering, than the state-of-art algorithms, including deep learning methods devoted to similar tasks.

Keywords: networked data, clustering, representation learning, spectral rotation

1 Introduction

In recent years, *Networks* [4] and *Attributed Networks* (AN) [8] have been used to model a large variety of real-world networks, such as academic and health care networks where both node links and attributes/features are available for analysis. Unlike plain networks in which only node links and dependencies are observed, with AN, each node is associated with a valuable set of features. In other words, we have $\mathbf{X}$ and $\mathbf{W}$ obtained/available independently of $\mathbf{X}$. More recently, the learning representation has received a significant amount of attention as an important aim in many applications including social networks, academic citation networks and protein-protein interaction networks. Hence, *Attributed network Embedding* (ANE) [2] aims to seek a continuous low-dimensional matrix representation for nodes in a network, such that original network topological structure and node attribute proximity can be preserved in the new low-dimensional embedding.

Although, many approaches have emerged with *Network Embedding* (NE), the research on ANE (Attributed Network Embedding) still remains to be explored

Lazhar Labiod (✉) · Mohamed Nadif
Centre Borelli UMR9010, Université Paris Cité, 75006-Paris, France,
e-mail: lazhar.labiod@u-paris.fr, e-mail: mohamed.nadif@u-paris.fr

© The Author(s) 2023
P. Brito et al. (eds.), *Classification and Data Science in the Digital Age*,
Studies in Classification, Data Analysis, and Knowledge Organization,
https://doi.org/10.1007/978-3-031-09034-9_23

[3]. Unlike NE that learns from plain networks, ANE aims to capitalize both the proximity information of the network and the affinity of node attributes. Note that, due to the heterogeneity of the two information sources, it is difficult for the existing NE algorithms to be directly applied to ANE. To sum up, the learned representation has been shown to be helpful in many learning tasks such as network clustering [13], Therefore ANE is a challenging research problem due to the high-dimensionality, sparsity and non-linearity of the graph data.

The paper is organized as follows. In Section 2 we formulate the objective function to be optimized, describe the different matrices used, and present a *Simultaneous Attributed Network Embedding and Clustering* (SANEC) framework for embedding and clustering. Section 3 is devoted to numerical experiments. Finally, the conclusion summarizes the advantages of our contribution.

2 Proposed Method

In this section, we describe the SANEC method. We will present the formulation of an objective function and an effective algorithm for data embedding and clustering. But first, we show how to construct two matrices $\mathbf{S}$ and $\mathbf{M}$ integrating both types of information –content and structure information– to reach our goal.

2.1 Content and Structure Information

An attributed network $\mathcal{G} = (\mathcal{V}, \mathcal{E}, \mathcal{X})$ consists of $\mathcal{V}$ the set of nodes, $E \subseteq \mathcal{V} \times \mathcal{V}$ the set of links, and $\mathbf{X} = [\mathbf{x}_1, \mathbf{x}_2, \ldots, \mathbf{x}_n]$ where $n = |\mathcal{V}|$ and $\mathbf{x}_i \in \mathbb{R}^d$ is the feature/attribute vector of the node v_i. Formally, the graph can be represented by two types of information, the content information $\mathbf{X} \in \mathbb{R}^{n \times d}$ and the structure information $\mathbf{A} \in \mathbb{R}^{n \times n}$, where $\mathbf{A}$ is an adjacency matrix of G and $a_{ij} = 1$ if $e_{ij} \in E$ otherwise 0; we consider that each node is a neighbor of itself, then we set $a_{ii} = 1$ for all nodes. Thereby, we model the nodes proximity by an $(n \times n)$ transition matrix $\mathbf{W}$ given by $\mathbf{W} = \mathbf{D}^{-1}\mathbf{A}$, where $\mathbf{D}$ is the degree matrix of $\mathbf{A}$ defined by $d_{ii} = \sum_{i'=1}^{n} a_{i'i}$.

In order to exploit additional information about nodes similarity from $\mathbf{X}$, we preprocessed the above dataset $\mathbf{X}$ to produce similarity graph input $\mathbf{W_X}$ of size $(n \times n)$; we construct a K-Nearest-Neighbor (KNN) graph. To this end, we use the heat kernel and L_2 distance, KNN neighborhood mode with $K = 15$ and we set the width of the neighborhood $\sigma = 1$. Note that any appropriate distance or dissimilarity measure can be used. Finally we combine in an $(n \times n)$ matrix $\mathbf{S}$, nodes proximity from both content information $\mathbf{X}$ and structure information $\mathbf{W}$. In this way, we intend to perturb the similarity $\mathbf{W}$ by adding the similarity from $\mathbf{W_X}$; we choose to take $\mathbf{S}$ defined by $\mathbf{S} = \mathbf{W} + \mathbf{W_X}$ (Figure 1).

As we aim to perform clustering, we propose to integrate it in the formulation of a new data representation by assuming that nodes with the same label tend to have

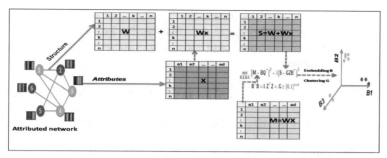

Fig. 1 Model and objective function of SANEC.

similar social relations and similar node attributes. This idea is inspired by the fact that, the labels are strongly influenced by both content and structure information and inherently correlated to both these information sources. Thereby the new data representation referred to as $\mathbf{M} = (m_{ij})$ of size $(n \times d)$ can be considered as a multiplicative integration of both $\mathbf{W}$ and $\mathbf{X}$ by replacing each node by the centroid of their neighborhood (barycenter): i.e, $m_{ij} = \sum_{k=1}^{n} \mathbf{w}_{ik}\mathbf{x}_{kj}, \forall i, j$ or $\mathbf{M} = \mathbf{W}\mathbf{X}$. In this way, given a graph G, a graph clustering aims to partition the nodes in G into k disjoint clusters $\{C_1, C_2, \ldots, C_k\}$, so that: (1) nodes within the same cluster are close to each other while nodes in different clusters are distant in terms of graph structure; and (2) the nodes within the same cluster are more likely to have similar attribute values.

2.2 Model, Optimization and Algorithm

Let k be the number of clusters and the number of components into which the data is embedded. With $\mathbf{M}$ and $\mathbf{S}$, the SANEC method that we propose aims to obtain the maximally informative embedding according to the clustering structure in the attributed network data. Therefore, we propose to optimize

$$\min_{\mathbf{B},\mathbf{Z},\mathbf{Q},\mathbf{G}} \left\|\mathbf{M} - \mathbf{B}\mathbf{Q}^{\top}\right\|^2 + \lambda\left\|\mathbf{S} - \mathbf{G}\mathbf{Z}\mathbf{B}^{\top}\right\|^2 \quad \mathbf{B}^{\top}\mathbf{B} = \mathbf{I}, \mathbf{Z}^{\top}\mathbf{Z} = \mathbf{I}, \mathbf{G} \in \{0, 1\}^{n \times k} \quad (1)$$

where $\mathbf{G} = (g_{ij})$ of size $(n \times k)$ is a cluster membership matrix, $\mathbf{B} = (b_{ij})$ of size $(n \times k)$ is the embedding matrix and $\mathbf{Z} = (z_{ij})$ of size $(k \times k)$ is an orthonormal rotation matrix which most closely maps $\mathbf{B}$ to $\mathbf{G} \in \{0, 1\}^{n \times k}$. $\mathbf{Q} \in \mathbb{R}^{d \times k}$ is the features embedding matrix. Finally, The parameter λ is a non-negative value and can be viewed as a regularization parameter. The intuition behind the factorization of $\mathbf{M}$ and $\mathbf{S}$ is to encourage the nodes with similar proximity, those with higher similarity in both matrices, to have closer representations in the latent space given by $\mathbf{B}$. In doing so, the optimisation of (1) leads to a clustering of the nodes into k clusters given by $\mathbf{G}$. Note that, both tasks –embedding and clustering– are performed simultaneously and supported by $\mathbf{Z}$; it is the key to attaining good embedding while taking into account the clustering structure. To infer the latent factor matrices $\mathbf{Z}$, $\mathbf{B}$, $\mathbf{Q}$ and $\mathbf{G}$, we shall derive an alternating optimization algorithm. To this end, we rely on the following proposition.

Proposition 1. Let be $\mathbf{S} \in \mathbb{R}^{n \times n}$, $\mathbf{G} \in \{0, 1\}^{n \times k}$, $\mathbf{Z} \in \mathbb{R}^{k \times k}$, $\mathbf{B} \in \mathbb{R}^{n \times k}$, we have

$$\left\| \mathbf{S} - \mathbf{GZB}^\top \right\|^2 = \left\| \mathbf{S} - \mathbf{BB}^\top \mathbf{S} \right\|^2 + \left\| \mathbf{SB} - \mathbf{GZ} \right\|^2 \tag{2}$$

proof. We first expand the matrix norm of the left term of (2)

$$\left\| \mathbf{S} - \mathbf{GZB}^T \right\|^2 = \|\mathbf{S}\|^2 + \left\| \mathbf{GZB}^\top \right\|^2 - 2Tr(\mathbf{SGZB}^\top) \tag{3}$$

In a similar way, we obtain from the two terms of the right term of (2)

$$\left\| \mathbf{S} - \mathbf{SBB}^T \right\|^2 = \|\mathbf{S}\|^2 - \|\mathbf{SB}\|^2 \quad \text{due to } \mathbf{B}^\top \mathbf{B} = \mathbf{I} \tag{4}$$

and $\quad \|\mathbf{SB} - \mathbf{GZ}\|^2 = \|\mathbf{SB}\|^2 + \|\mathbf{GZ}\|^2 - 2Tr(\mathbf{SBZG}^\top).$

Due also to $\mathbf{B}^\top \mathbf{B} = \mathbf{I}$, we have

$$\|\mathbf{SB} - \mathbf{GZ}\|^2 = \|\mathbf{SB}\|^2 + \|\mathbf{GZB}^\top\|^2 - 2Tr(\mathbf{SGZB}^\top) \tag{5}$$

Summing the two terms of (4) and (5) leads to the left term of (2).

$$\|\mathbf{S}\|^2 + \|\mathbf{GZ}\|^2 - 2Tr(\mathbf{SGZB}^\top) = \left\| \mathbf{S} - \mathbf{GZB}^T \right\|^2 \text{ due to } \|\mathbf{GZ}\|^2 = \left\| \mathbf{GZB}^\top \right\|^2$$

Compute Z. Fixing $\mathbf{G}$ and $\mathbf{B}$ the problem which arises in (1) is equivalent to $\min_{\mathbf{Z}} \left\| \mathbf{S} - \mathbf{GZB}^\top \right\|^2$. From Proposition 1, we deduce that

$$\min_{\mathbf{Z}} \left\| \mathbf{S} - \mathbf{GZB}^\top \right\|^2 \Leftrightarrow \min_{\mathbf{Z}} \left\| \mathbf{S} - \mathbf{BB}^\top \mathbf{S} \right\|^2 + \|\mathbf{SB} - \mathbf{GZ}\|^2 \tag{6}$$

which can be reduced to $\max_{\mathbf{Z}} Tr(\mathbf{G}^\top \mathbf{SBZ})$ s.t. $\mathbf{Z}^\top \mathbf{Z} = \mathbf{I}$. As proved in page 29 of [1], let $\mathbf{U\Sigma V}^\top$ be the SVD for $\mathbf{G}^\top \mathbf{SB}$, then $\mathbf{Z} = \mathbf{UV}^\top$.

Compute Q. Given $\mathbf{G}$, $\mathbf{Z}$ and $\mathbf{B}$, the opimization problem (1) is equivalent to $\min_{\mathbf{Q}} \|\mathbf{M} - \mathbf{BQ}^\top\|^2$, and we get

$$\mathbf{Q} = \mathbf{M}^\top \mathbf{B}. \tag{7}$$

Thereby $\mathbf{Q}$ is somewhere an embedding of attributes.

Compute B. Given $\mathbf{G}$, $\mathbf{Q}$ and $\mathbf{Z}$, the problem (1) is equivalent to

$$\max_{\mathbf{B}} Tr((\mathbf{M}^\top \mathbf{Q} + \lambda \mathbf{SGZ})\mathbf{B}^\top) \quad \text{s.t.} \quad \mathbf{B}^\top \mathbf{B} = \mathbf{I}.$$

In the same manner for the computation of $\mathbf{Z}$, let $\hat{\mathbf{U}}\hat{\mathbf{\Sigma}}\hat{\mathbf{V}}^\top$ be the SVD for $(\mathbf{M}^\top\mathbf{Q} + \lambda\mathbf{SGZ})$, we get

$$\mathbf{B} = \hat{\mathbf{U}}\hat{\mathbf{V}}^\top. \tag{8}$$

It is important to emphasize that, at each step, $\mathbf{B}$ exploits the information from the matrices $\mathbf{Q}$, $\mathbf{G}$, and $\mathbf{Z}$. This highlights one of the aspects of the simultaneity of embedding and clustering.

Compute $\mathbf{G}$: Finally, given $\mathbf{B}$, $\mathbf{Q}$ and $\mathbf{Z}$, the problem (1) is equivalent to $\min_\mathbf{G} \|\mathbf{SB} - \mathbf{GZ}\|^2$. As $\mathbf{G}$ is a cluster membership matrix, its computation is done as follows: We fix $\mathbf{Q}$, $\mathbf{Z}$, $\mathbf{B}$. Let $\tilde{\mathbf{B}} = \mathbf{SB}$ and calculate

$$g_{ik} = 1 \text{ if } k = \arg\min_{k'} \|\tilde{\mathbf{b}}_i - \mathbf{z}_{k'}\|^2 \text{ and } 0 \text{ otherwise} . \tag{9}$$

In summary, the steps of the SANEC algorithm relying on $\mathbf{S}$ referred to as SANEC$_\mathbf{S}$ can be deduced in Algorithm 1. The convergence of SANEC$_\mathbf{S}$ is guaranteed and depends on the initialization to reach only a local optima. Hence, we start the algorithm several times and select the best result which minimizes the objective function (1).

Algorithm 1 : SANEC$_\mathbf{S}$ algorithm

Input: $\mathbf{M}$ and $\mathbf{S}$ from structure matrix $\mathbf{W}$ and content matrix $\mathbf{X}$;
Initialize: $\mathbf{B}$, $\mathbf{Q}$ and $\mathbf{Z}$ with arbitrary orthonormal matrix;
repeat
    (a) - Compute $\mathbf{G}$ using (9)
    (b) - Compute $\mathbf{B}$ using (8)
    (c) - Compute $\mathbf{Q}$ using (7)
    (d) - Compute $\mathbf{Z}$ using (6)
until convergence
Output: $\mathbf{G}$: clustering matrix, $\mathbf{Z}$: rotation matrix, $\mathbf{B}$: nodes embedding and $\mathbf{Q}$: attributes embedding

---

3 Numerical Experiments

In the following, we compare SANEC with some competitive methods described later. The performances of all clustering methods are evaluated using challenging real-world datasets commonly tested with ANE where the clusters are known. Specifically, we consider three public citation network data sets, Citeseer, Cora and Wiki, which contain sparse bag-of-words feature vector for each document and a list of citation links between documents. Each document has a class label. We treat documents as nodes and the citation links as the edges. The characteristics of the used datasets are summarized in Table 1. The balance coefficient is defined as the ratio of the number of documents in the smallest class to the number of documents in the largest class while *nz* denotes the percentage of sparsity.

Table 1 Description of datasets (#: the cardinality).

datasets	# Nodes	# Attributes	# Edges	#Classes	nz(%)	Balance
Cora	2708	1433	5294	7	98.73	0.22
Citeseer	3312	3703	4732	6	99.14	0.35
Wiki	2405	4973	17981	17	86.46	0.02

In our comparison we include standard methods and also recent deep learning methods; these differ in the way they use available information. Some of them (such as K-means) use only $\mathbf{X}$ as the baseline, while others use more recent algorithms based on $\mathbf{X}$ and $\mathbf{W}$. All the compared methods are: TADW [14], DeepWalk [7] and Spectral Clustering [11]. Using $\mathbf{X}$ and $\mathbf{W}$ we evaluated GAE and VGAE [5], ARVGA [6], AGC [15] and DAEGC [12].

With the SANEC model, the parameter λ controls the role of the second term $||\mathbf{S} - \mathbf{GZB}^\top||^2$ in (1). To measure its impact on the clustering performance of SANEC$_S$, we vary λ in $\{0, 10^{-6}, 10^{-3}, 10^{-1}, 10^0, 10^1, 10^3\}$. Through, many experiments, as illustrated in Figure 2 we choose to take $\lambda = 10^{-3}$. The choice of λ warrants in-depth evaluation.

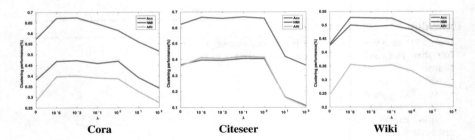

Fig. 2 Sensitivity analysis of λ using ACC, NMI and ARI.

Compared to the true available clusters, in our experiments the clustering performance is assessed by *accuracy* (ACC), *normalized mutual information* (NMI) and *adjusted rand index* (ARI). We repeat the experiments 50 times, with different random initialization and the averages (mean) are reported in Table 2; the best performance for each dataset is highlighted in bold.

First, we observe the high performances of methods integrating information from $\mathbf{W}$. For instance, RTM and RMSC are better than classical methods using only either $\mathbf{X}$ or $\mathbf{W}$. On the other hand, all methods including deep learning algorithms relying on $\mathbf{X}$ and $\mathbf{W}$ are better yet. However, regarding SANEC with both versions relying on $\mathbf{W}$, referred to as SANEC$_W$ or $\mathbf{S}$ referred to as SANEC$_S$, we note high performances for all the datasets and with SANEC$_S$, we remark the impact of $\mathbf{W_X}$; it learns low-dimensional representations while suits the clustering structure.

To go further in our investigation and given the sparsity of $\mathbf{X}$ we proceeded to standardization tf-idf followed by L_2, as it is often used to process document-term matrices; see e.g, [9, 10], while in the construction of $\mathbf{W_X}$ we used the cosine metric. In Figure 3 are reported the results where we observe a slight improvement.

Table 2 Clustering performances (ACC % , NMI % and ARI %).

Methods	Input	Cora ACC	Cora NMI	Cora ARI	Citeseer ACC	Citeseer NMI	Citeseer ARI	Wiki ACC	Wiki NMI	Wiki ARI
K-means	$\mathbf{X}$	49.22	32.10	22.96	54.01	30.54	27.86	41.72	44.02	15.07
Spectral	$\mathbf{W}$	36.72	12.67	03.11	23.89	05.57	01.00	22.04	18.17	01.46
DeepWalk	$\mathbf{W}$	48.40	32.70	24.27	33.65	08.78	09.22	38.46	32.38	17.03
RTM	$\mathbf{X, W}$	43.96	23.01	16.91	45.09	23.93	20.26	43.64	44.95	13.84
RMSC	$\mathbf{X, W}$	40.66	25.51	08.95	29.50	13.87	04.88	39.76	41.50	11.16
TAWD	$\mathbf{X, W}$	56.03	44.11	33.20	45.48	29.14	22.81	30.96	27.13	04.54
VGAE	$\mathbf{X, W}$	50.20	32.92	25.47	46.70	26.05	20.56	45.09	46.76	26.34
ARGE	$\mathbf{X, W}$	64.0	44.9	35.2	57.3	35.0	34.1	47.34	47.02	28.16
ARVGE	$\mathbf{X, W}$	63.8	45.0	37.74	54.4	26.1	24.5	46.45	47.8	29.65
SANEC$_\mathbf{W}$	$\mathbf{X, W}$	64.47	43.30	36.19	64.71	38.61	39.20	46.21	42.83	28.30
SANEC$_\mathbf{S}$	$\mathbf{X, S}$	**67.38**	**47.14**	**39.88**	**66.77**	**40.60**	**41.78**	**52.80**	**50.02**	**35.57**

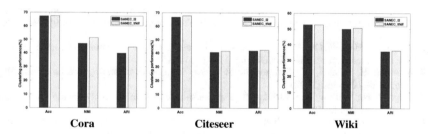

| Cora | Citeseer | Wiki |

Fig. 3 Evaluation of SANEC$_\mathbf{S}$ using tf-idf normalization of $\mathbf{X}$ and cosine metric for $\mathbf{W_X}$.

4 Conclusion

In this paper, we proposed a novel matrix decomposition framework for simultaneous attributed network data embedding and clustering. Unlike known methods that combine the objective function of AN embedding and the objective function of clustering separately, we proposed a new single framework to perform SANEC$_\mathbf{S}$ for AN embedding and nodes clustering. We showed that the optimized objective function can be decomposed into three terms, the first is the objective function of a kind of PCA applied to $\mathbf{X}$, the second is the graph embedding criterion in a low-dimensional space, and the third is the clustering criterion. We also integrated a discrete rotation functionality, which allows a smooth transformation from the relaxed continuous embedding to a discrete solution, and guarantees a tractable optimization problem with a discrete solution. Thereby, we developed an effective algorithm capitalizing on learning representation and clustering. The obtained results show the advantages of combining both tasks over other approaches. SANEC$_S$ outperforms all recent methods devoted to the same tasks including deep learning methods which require deep models pretraining. However, there are other points that warrant in-depth evaluation, such as the choice of λ and the complexity of the algorithm in terms of network size. The proposed framework offers several perspectives and investigations. We have noted that the construction of $\mathbf{M}$ and $\mathbf{S}$ is important, it highlights the introduction of $\mathbf{W}$. As for the $\mathbf{W_X}$ we have observed that it is fundamental as it makes possible to link the information from $\mathbf{X}$ to the network; this has been verified by many experiments. First, we would like to be able to measure the impact of each matrix $\mathbf{W}$ and $\mathbf{W_X}$ in the construction of $\mathbf{S}$ by considering two different weights for $\mathbf{W}$ and $\mathbf{W_X}$ as follows: $\mathbf{S} = \alpha\mathbf{W} + \beta\mathbf{W_X}$. Finally, as we have stressed that $\mathbf{Q}$ is an embedding of attributes, this suggests to consider also a simultaneously ANE and co-clustering.

References

1. Ten Berge, J. M. F: Least Squares Optimization in Multivariate Analysis. DSWO Press, Leiden University Leiden, (1993)
2. Cai, H. Y., Zheng, V. W., Chang, K. C. C.: A comprehensive survey of graph embedding: Problems, techniques, and applications. IEEE Trans. Knowl. Data Eng. **30**(9), 1616-1637 (2018)
3. Chang, S., Han, W., Qi, G. J., Aggarwal, C. C., Huang, T.S.: Heterogeneous network embedding via deep architectures. In SIGKDD, pp. 119–128 (2015)
4. Doreian, P., Batagelj, V., Ferligoj, A.: Advances in network clustering and blockmodeling. John Wiley & Sons (2020)
5. Kipf, T. N., Welling, M.: Variational graph auto-encoders. In NIPS Workshop on Bayesian Deep Learning, (2016)
6. Pan, S., Hu, R., Long, G., Jiang, J., Yao, L., Zhang, C.: Adversarially regularized graph autoencoder for graph embedding. In IJCAI, pp. 2609-2615, (2018)
7. Perozzi, B., Al-Rfou, R., Skiena, S.: Deepwalk: Online learning of social representations. In SIGKDD, pp. 701-710 (2014)
8. Qi, G.J., Aggarwal, C. C., Tian, Q., Ji, H., Huang, T. S.: Exploring context and content links in social media: A latent space method. IEEE Trans. Pattern Anal. Mach. Intell. **34**(5), 850-862 (2012)
9. Salah, A., Nadif, M.: Model-based von Mises-Fisher co-clustering with a conscience. In SDM, pp. 246–254. SIAM (2017)
10. Salah, A., Nadif, M.: Directional co-clustering. Data Analysis and Classification. **13**(3), 591-620 (2019)
11. Tang, L., Liu, H.: Leveraging social media networks for classification. Data mining and knowledge discovery. **23**(3), 447-478 (2011)
12. Wang, C., Pan, S., Hu, R., Long, G., Jiang, J., Zhang, C.: Attributed graph clustering: A deep attentional embedding approach. arXiv preprint arXiv:1906.06532 (2019) Available via . https://arxiv.org/pdf/1906.06532.pdf
13. Wang, C., Pan, S., Long, G., Zhu,X., Jiang, J.: Mgae: Marginalized graph autoencoder for graph clustering. In CIKM, pp. 889-898, (2017)
14. Yang, C., Liu, Z., Zhao, D., Sun, M., Chang, E. Y.: Network representation learning with rich text information. In IJCAI, pp. 2111-2117 (2015)
15. Zhang, X., Liu, H., Li, Q., Wu, X. M.: Attributed graph clustering via adaptive graph convolution. arXiv preprint arXiv:1906.01210, (2019) Available via . https://arxiv.org/pdf/1906.01210.pdf?ref=https://githubhelp.com

Towards a Bi-stochastic Matrix Approximation of k-means and Some Variants

Lazhar Labiod and Mohamed Nadif

Abstract The k-means algorithm and some k-means variants have been shown to be useful and effective to tackle the clustering problem. In this paper we embed k-means variants in a bi-stochastic matrix approximation (BMA) framework. Then we derive from the k-means objective function a new formulation of the criterion. In particular, we show that some k-means variants are equivalent to algebraic problem of bi-stochastic matrix approximation under some suitable constraints. For optimizing the derived objective function, we develop two algorithms; the first one consists in learning a bi-stochastic similarity matrix while the second seeks for the optimal partition which is the equilibrium state of a Markov chain process. Numerical experiments on real data-sets demonstrate the interest of our approach.

Keywords: k-means, reduced k-means, factorial k-means, bi-stochastic matrix

1 Introduction

These last decades unsupervised learning and specifically clustering, have received a significant amount of attention as an important problem with many application in data science. Let $A = (a_{ij})$ be a $n \times m$ continuous data matrix where the set of rows (objects, individuals) is denoted by I and the set of columns (attributes, features) by J. Many clustering methods such as hierarchical or not aim to construct an optimal partition of I or, sometimes of J.

In this paper we show how some k-means variants can be presented as a bi-stochastic matrix approximation problem under some suitable constraints generated by the properties of the reached solution. To reach this goal, we first demonstrate that some variants of k-means are equivalent to learning a bi-stochastic similarity matrix having a diagonal block structure. Based on this formulation, referred to as BMA, we derive two iterative algorithms, the first algorithm learns a bi-stochastic $n \times n$ similarity matrix while the second directly seeks an optimal clustering solution.

Our main contribution is to establish the theoretical connection of the conventional k-means and some of its variants to BMA framework. The implications of the reformulation of k-means as a BMA problem are multi-folds:

Lazhar Labiod (✉) · Mohamed Nadif
Centre Borelli UMR9010, Université Paris Cité, 75006-Paris, France,
e-mail: lazhar.labiod@u-paris.fr, e-mail: mohamed.nadif@u-paris.fr

© The Author(s) 2023
P. Brito et al. (eds.), *Classification and Data Science in the Digital Age*,
Studies in Classification, Data Analysis, and Knowledge Organization,
https://doi.org/10.1007/978-3-031-09034-9_24

- It makes connections with recent clustering methods like spectral clustering and subspace clustering.
- It learns a well normalized (bi-stochastic normalization) similarity matrix, beneficial for spectral clustering [12].
- Unlike existing spectral and subspace methods which combine in a sequential way, the steps of similarity learning and clustering derivation, our proposed method jointly learns a block diagonal bi-stochastic affinity matrix which naturally expresses a clustering structure.

The rest of paper is organized as follows. Section 2 introduces some variants of k-means. Section 3 provides *Matrix Factorization* (MF) and BMA formulations of k-means variants. Section 4 discusses the BMA clustering algorithm and section 5 is devoted to numerical experiments. Finally, the conclusion summarizes the interest of our contribution.

2 Variants of k-Means

Given a data matrix $A = (a_{ij}) \in R^{n \times m}$, the aim of clustering is to cluster the rows or the columns of A, so as to optimize the difference between $A = (a_{ij})$ and the clustered matrix revealing significant block structure. More formally, we seek to partition the set of rows $I = \{1, \ldots, n\}$ into k clusters $C = \{C_1, \ldots, C_l, \ldots, C_k\}$. The partitioning naturally induce clustering index matrix $R = (r_{il}) \in \mathbb{R}^{n \times k}$, defined as binary classification matrix such as we have $r_{il} = 1$, if the row $a_i \in C_l$, and 0 otherwise. On the other hand, we note $S \in \mathbb{R}^{m \times k}$ a reduced matrix specifying the cluster representation. The detection of homogeneous clusters of objects can be reached by looking for the two matrices R and S minimizing the total squared residue measure

$$\mathcal{J}_{KM}(R, S) = ||A - RS^\top||^2 \tag{1}$$

The term $RS^\top$ characterizes the information of A that can be described by the clusters structure. The clustering problem can be formulated as a matrix approximation problem where the clustering aims to minimize the approximation error between the original data A and the reconstructed matrix based on the cluster structures.

Factorial k-means analysis (FKM) [9] and Reduced k-means analysis (RKM) [1] are clustering methods that aim at simultaneously achieving a clustering of the objects and a dimension reduction of the features. The advantage of these methods is that both clustering of objects and low-dimensional subspace capturing the cluster structure are simultaneously obtained. To achieve this objective, RKM is defined by the minimizing problem of the following criterion

$$\mathcal{J}_{RKM}(R, S, Q) = ||A - RS^\top Q^\top||^2 \tag{2}$$

and FKM is defined by the minimizing problem of the following criterion

$$\mathcal{J}_{FKM}(R, S, Q) = ||AQ - RS^\top||^2 \tag{3}$$

where $S \in \mathbb{R}^{p \times k}$ with RKM and FKM, and Q is an m by p column-wise orthonormal loading matrix.

3 Bi-stochastic Matrix Approximation of k-Means Variants

3.1 Low-rank Matrix Factorization (MF)

By considering k-means as a lower rank matrix factorization with constraints, rather than a clustering method, we can formulate constraints to impose on MF formulation. Let $D_r^{-1} \in \mathcal{R}^{k \times k}$ be diagonal matrix defined as follow $D_r^{-1} = Diag(r_1^{-1}, \ldots, r_k^{-1})$. Using the matrices D_r, A and R, the matrix summary S can be expressed as $S^T = D_r^{-1} R^T A$. Plugging S into the objective function in equation, (1) leads to optimize $||A - R(D_r^{-1} R^T A)||^2$ equal to

$$\mathcal{J}_{MF-KM}(\mathbf{R}) = ||A - \mathbf{R}\mathbf{R}^T A||^2, \text{ where } \mathbf{R} = RD_r^{-0.5}. \tag{4}$$

On the other hand, it is easy to verify that the approximation $\mathbf{R}\mathbf{R}^T A$ of A is formed by the same value in each block $A_{l,(l=1,\ldots,k)}$. Specifically, the matrix $\mathbf{R}^T A$, equal to S^T, plays the role of a summary of A and absorbs the different scales of A and $\mathbf{R}$. Finally $\mathbf{R}\mathbf{R}^T A$ gives the row clusters mean vectors. Note that it is easy to show that $\mathbf{R}$ verifies the following properties

$$\mathbf{R} \geq 0, \mathbf{R}^T \mathbf{R} = I_k, \mathbf{R}\mathbf{R}^T \mathbb{1} = \mathbb{1}, Trace(\mathbf{R}\mathbf{R}^T) = k, (\mathbf{R}\mathbf{R}^T)^2 = \mathbf{R}\mathbf{R}^T \tag{5}$$

Next, in similar way, we can derive a MF formulation of FKM,

$$\mathcal{J}_{MF-FKM}(\mathbf{R}) = ||AQ - \mathbf{R}\mathbf{R}^T AQ||^2, \tag{6}$$

and of RKM, $\quad \mathcal{J}_{MF-RKM}(\mathbf{R}) = ||A - \mathbf{R}\mathbf{R}^T AQQ^T||^2. \tag{7}$

3.2 BMA Formulation

Let $\mathbf{\Pi} = \mathbf{R}\mathbf{R}^T$ be a bi-stochastic similarity matrix, before giving the BMA formulation of k-means variants, we need first to spell out the good properties of $\mathbf{\Pi}$. Indeed, by construction from $\mathbf{R}$, $\mathbf{\Pi}$ has at least the following properties reported below that can be easily proven.

$$\mathbf{\Pi} \geq 0, \mathbf{\Pi}^T = \mathbf{\Pi}, \mathbf{\Pi}\mathbb{1} = \mathbb{1}, Trace(\mathbf{\Pi}) = k, \mathbf{\Pi}\mathbf{\Pi}^T = \mathbf{\Pi}, Rank(\mathbf{\Pi}) = k \tag{8}$$

Given a data matrix A and k row clusters, we can hope to discover the cluster structure of A from $\mathbf{\Pi}$. Notice that from (8) $\mathbf{\Pi}$ is nonnegative, symmetric, bi-stochastic (doubly stochastic) and idempotent. By setting the k means in the BMA framework, the problem of clustering is reformulated as the learning of a structured bi-stochastic similarity matrix $\mathbf{\Pi}$ by minimizing the following k-means variants objective,

$$\mathcal{J}_{BMA-kM}(\mathbf{\Pi}) = ||A - \mathbf{\Pi}A||^2, \tag{9}$$

$$\mathcal{J}_{BMA-FKM}(\mathbf{\Pi}) = ||AQ - \mathbf{\Pi}AQ||^2, \tag{10}$$

$$\mathcal{J}_{BMA-RKM}(\mathbf{\Pi}) = ||A - \mathbf{\Pi}AQQ^\top||^2, \tag{11}$$

with respect to the following constraints on $\mathbf{\Pi}$

$$\mathbf{\Pi} \geq 0, \mathbf{\Pi} = \mathbf{\Pi}^\top, \mathbf{\Pi}\mathbb{1} = \mathbb{1}, Tr(\mathbf{\Pi}) = k, \mathbf{\Pi}\mathbf{\Pi}^\top = \mathbf{\Pi} \tag{12}$$

$$\text{and } Q^\top Q = I \quad \text{for equations (10) and (11).}$$

In the rest of the paper, we will consider only non-negativity, symmetry and bi-stochastic constraints.

3.3 The Equivalence Between BMA and k-Means

The theorem below demonstrates that the optimization of the k-means objective and the BMA objective under some suitable constraints are equivalent. The equation (13) establishes the equivalence between k-means and the BMA formulation. Then, solving the BMA objective function (9) is equivalent to finding a global solution of the k-means criterion (1).

Theorem 1

$$\arg\min_{R,S} ||A - RS^\top||^2 \Leftrightarrow \arg\min_{\{\mathbf{\Pi} \geq 0, \mathbf{\Pi} = \mathbf{\Pi}^\top, \mathbf{\Pi}\mathbb{1} = \mathbb{1}, Tr(\mathbf{\Pi}) = k, \mathbf{\Pi}\mathbf{\Pi}^\top = \mathbf{\Pi}\}} ||A - \mathbf{\Pi}A||^2 \tag{13}$$

The proof of this equivalence is given in the appendix. Note that this new formulation gives some interesting highlights on k-means and its variants:

- First, this shows that k-means is equivalent to learning a structured bi-stochastic similarity matrix which is normalized bi-stochastic matrix with block diagonal structure.
- Secondly, it establishes very interesting connections of k-means to many state-of-the-art subspace clustering methods [10, 5]. Moreover, this formulation combines the traditional two-step process used by subspace clustering methods, which consist in first constructing an affinity matrix between data points and then applying spectral clustering to this affinity. This allows joint learning of a similarity matrix that better reflects the clustering structure by its block diagonal shape.
- Finally, it allows to apply the spirit of k-means for graph or similarity data.

4 BMA Clustering Algorithm

First, we establish the relationship between our objective function and that used in
[12, 11]. From $||A - \mathbf{\Pi}A||^2 = Trace(AA^\top) + Trace(\mathbf{\Pi}AA^\top\mathbf{\Pi}) - 2Trace(AA^\top\mathbf{\Pi})$
and by using the idempotent property, $\mathbf{\Pi}\mathbf{\Pi}^\top = \mathbf{\Pi}$, we can show that

$$\arg\min_{\mathbf{\Pi}} ||A - \mathbf{\Pi}A||^2 \Leftrightarrow \arg\min_{\mathbf{\Pi}} ||AA^\top - \mathbf{\Pi}||^2 \Leftrightarrow \arg\max_{\mathbf{\Pi}} Trace(AA^\top\mathbf{\Pi}).$$

The algorithm for learning similarity matrix is summarized in Algorithm 1 as in
[12, 11]. Once the bi-stochastic similarity matrix $\mathbf{\Pi}$ is obtained, the basic idea of
BMA is based on the following steps:

Algorithm 1 : Learning similarity matrix

Input: data A
Output: similarity matrix $\mathbf{\Pi}$
Initialize: $t = 0$ and $\mathbf{\Pi}^{(0)} = AA^\top$
repeat
$\quad \mathbf{\Pi}^{(t+1)} \leftarrow [\mathbf{\Pi}^{(t)} + \frac{1}{n}(I - \mathbf{\Pi}^{(t)} + \frac{\mathbf{1}\mathbf{1}^\top\mathbf{\Pi}^{(t)}}{n})\mathbf{1}\mathbf{1}^\top - \frac{1}{n}\mathbf{1}\mathbf{1}^\top\mathbf{\Pi}^{(t)}]$
until Satisfied convergence condition

---

1. Estimating iteratively A by applying at each time the matrix $\mathbf{\Pi}$ on the current
   A using the following update $\hat{A}^{(t+1)} = \mathbf{\Pi}A^{(t)}$. This process converges to an
   equilibrium (steady) state. Let k be the multiplicity of the eigenvalue of matrix
   $\mathbf{\Pi}$ equal to 1, $\hat{A}$ is composed of $k << n$ quasi-similar rows, where each row is
   represented by its prototype.
2. Extracting the first left singular vectors π of $\hat{A}$ using the Power method [4];
   it is a well-known technique used for computing the largest left eigenvector of
   data matrix. The numerical computation of the leading left singular vector of $\hat{A}$,
   consists in starting with an arbitrary vector $\pi^{(0)}$, repeatedly performing updates
   of π until stabilization of π as follow: $\pi^{(t+1)} = \hat{A}\hat{A}^\top\pi^{(t)}$ and $\pi^{(t)} \leftarrow \frac{\pi^{(t)}}{||\pi^{(t)}||}$. We
   stop the Power method if, $|\gamma^{(t+1)} - \gamma^{(t)}| \simeq \epsilon$ where $\gamma^{(t+1)} \leftarrow ||\pi^{(t+1)} - \pi^{(t)}||$.

Why does this work? At first glance, this process might seem uninteresting since it
eventually leads to a vector with all rows and columns coincide for any starting vector.
However our practical experience shows that, first the vectors π very quickly collapse
into rows blocks and these blocks move towards each other relatively slowly. If we
stop the Power method iteration at this point, the algorithm would have a potential
application for data visualization and clustering. The structure of π during short-run
stabilization makes the discovery of rows data ordering straightforward. The key is
to look for values of π that are approximately equal and reordering rows and columns
data accordingly. The BMA algorithm involves a reorganization of the rows of data
matrix $\hat{A}$ according to sorted π. It also allows to locate the points corresponding to
an abrupt change in the curve of the first left singular vector π, and then assess the
number of clusters and the rows belonging to each cluster.

5 Experiments Analysis

In this subsection we first ran our algorithm on two real world data set, the 16 townships data which consists of the characteristics (rows) of 16 townships (columns), each cell indicates the presence 1 or absence 0 of a characteristic on a township . This example has been used by Niermann [7] for data ordering task and the author aims to reveal a block diagonal form. The second data called Mero data, comes from archaeological data on Merovingian buckles found in north eastern France. This data matrix consists of 59 buckles characterized by 26 attributes of description (see Marcotorchino for more details [6]). Figure 1 shows in order, A, $\hat{A}$, $SR = AA^T$ reorganized according to the sorted π and the sorted π plot for both data sets. We also evaluated

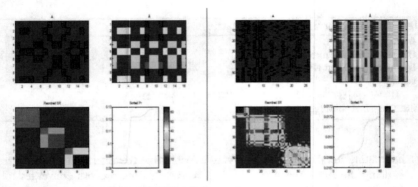

Fig. 1 left: 16 Townships data - right: Mero data.

the performances of BMA on some real challenging datasets described in Table1. We compared the performance of BMA with the spectral co-clustering (SpecCo) [2], Non-negative Matrix Factorization (NMF) and Orthgogonal Non-negative Matrix Tri-Factorization (NMTF) [3] by using two evaluation metrics: accuracy (ACC) corresponding to the percentage of well-classified elements and the normalized mutual information (NMI) [8]. In Table 1, we observe that BMA outperforms all compared algorithms for all tested datasets.

Table 1 Clustering Accuracy and Normalized Mutual Information (%).

datasets	# samples	# features	# classes	per	k-means	NMF	ONMTF	SpecCO	BMA
Classic3	3891	4303	3	ACC	88.6	73.33	70.10	97.89	98.30
				NMI	74.9	51.46	51.46	91.17	91.91
CSTR	476	1000	4	ACC	76.3	75.30	77.41	80.21	90.73
				NMI	65.4	66.40	67.30	66.36	77.86
Webkb4	4199	1000	4	ACC	60.10	66.30	67.10	61.68	68.8
				NMI	45.7	42.70	45.36	48.64	49
Leukemia	38	5000	3	ACC	72.2	89.21	90.32	94.73	97.36
				NMI	19.4	75.42	80.50	82	90.69

6 Conclusion

In this paper we have presented a new reformulation of some variants of k-means as a unified BMA framework and established the equivalence between k-means and BMA under suitable constraints. By doing so, k-means leads to learning a structured bi-stochastic matrix which is beneficial for clustering task. The proposed approach, not only learns a similarity matrix from data matrix, but uses this matrix in an iterative process that converges to a matrix $\hat{A}$ in which each row is represented by its prototype. The clustering solution is given by the first left eigenvector of $\hat{A}$ while overcoming the knowledge of the number of clusters. We expect for future work to integrate the idempotent and trace constraints on Π to make the approximate similarity matrix fits the best the case of a block diagonal structure.

Appendix

From the BMA's formulation, we know that one can easily construct a feasible solution for k-means from a feasible solution of BMA's formulation. Therefore, it remains to show that from a global solution of BMA's formulation, we can obtain a feasible solution of k-means. In order to show the equivalence between the optimization of k-means formulation and the BMA formulation, we first consider the following lemma.

Lemma If Π is a symmetric and positive semi-definite matrix, then we have

$$
\begin{cases}
(a)\pi_{ii'} \leq \sqrt{\pi_{ii}\pi_{i'i'}} & \text{(geometric mean)} \ \forall i, i' \\
(b)\pi_{ii'} \leq \frac{1}{2}(\pi_{ii} + \pi_{i'i'}) & \text{(arithmetic mean)} \ \forall i, i' \\
(c) \max_{ii'} \pi_{ii'} = \max_i \pi_{ii} \\
(d)\pi_{ii} = 0 \Rightarrow \pi_{ii'} = \pi_{i'i} = 0 \ \forall i, i'
\end{cases}
$$

Proposition. Any positive semi-definite matrix Π satisfying the constraints:

$$
\begin{cases}
\pi_{ii'} = \pi_{i'i} \ \forall i, i' & \text{(symmetry)} \\
\pi_{ii'} = \sum_{i''} \pi_{ii''}\pi_{i'i''} \ \forall i, i' & \text{(idempotence)} \\
\sum_{i'} \pi_{ii'} = 1 \ \forall i \\
\sum_i \pi_{ii} = k
\end{cases}
$$

is a matrix partitioned into k blocks $\Pi = diag(\Pi^1, \ldots, \Pi^l, \ldots, \Pi^k)$ with $\Pi^l = \frac{1}{n_l}\mathbb{1}_l\mathbb{1}_l^t$, $trace(\Pi^l) = 1 \ \forall l$ and $\sum_{l=1}^k n_l = n$; $\mathbb{1}_l$ denotes the vector of appropriate dimension with all its values are 1.

Proof. Since Π is idempotent ($\Pi^2 = \Pi$), we have: $\forall i; \ \pi_{ii} = \sum_{i'} \pi_{ii'}^2$. From the Lemma above, we know that there exist; $i^0 \in \{1, 2, \ldots, n\}$ such as $\max_{ii'} \pi_{i'i} = \pi_{i^0 i^0} > 0$. Consider the set A_{i^0} defined by $A_{i^0} = \{i | \pi_{i^0 i} > 0\}$, we can rewrite; $\forall i \in A_{i^0}; \ \pi_{ii} = \sum_{i' \in A_{i^0}} \pi_{i'i}^2$

$$
\forall i \in A_{i^0}; \quad \sum_{i' \in A_{i^0}} \pi_{i'i} = \sum_{i' \in I} \pi_{i'i} = 1 \tag{14}
$$

and,

$$\sum_{i' \in A_{i^0}} \sum_{i \in A_{i^0}} \pi_{i'i} = \sum_{i \in A_{i^0}} \pi_{i.} = \sum_{i \in A_{i^0}} 1 = |A_{i^0}| \tag{15}$$

$$\forall i \ \pi_{ii} = \sum_{i'} \pi_{ii'}^2 \Rightarrow \forall i \in A_{i^0}; \quad \sum_{i' \in A_{i^0}} \frac{\pi_{ii'}^2}{\pi_{ii}} = \sum_{i' \in A_{i^0}} (\frac{\pi_{ii'}}{\pi_{ii}})\pi_{ii'} = 1. \tag{16}$$

From (14) and (16), we deduce that $\forall i \in A_{i^0}$; $\sum_{i' \in A_{i^0}} \pi_{i'i} = \sum_{i' \in A_{i^0}} (\frac{\pi_{ii'}}{\pi_{ii}})\pi_{ii'}$, implying that: $\pi_{ii'} = \pi_{ii}$, $\forall i, i' \in A_{i^0}$. Substituting in (15) $\pi_{ii'}$ by π_{ii} for all $i, i' \in A_{i^0}$ leads to $\sum_{i' \in A_{i^0}} \pi_{ii'} = \sum_{i' \in A_{i^0}} \pi_{ii} = |A_{i^0}|\pi_{ii} = 1$, $\forall i \in A_{i^0}$. From this we can deduce that $\pi_{ii} = \pi_{ii'} = \frac{1}{|A_{i^0}|}$, $\forall i, i' \in A_{i^0}$. We can therefore rewrite the matrix Π in the form of a block diagonal matrix $\Pi = \begin{pmatrix} \Pi^0 & 0 \\ 0 & \bar{\Pi}^0 \end{pmatrix}$ where Π^0 is a block matrix whose general term is defined by $\Pi_{ii'}^0 = \frac{1}{|A_{i^0}|}$, $\forall i, i' \in A_{i^0}$ and $trace(\Pi^0) = 1$. The matrix $\bar{\Pi}^0$ is a positive semi-definite matrix which also verified the constraints $(\bar{\Pi}^0)^t = \bar{\Pi}^0$, $\bar{\Pi}^0 \mathbb{1} = \mathbb{1}$, $(\bar{\Pi}^0)^2 = \bar{\Pi}^0$ and $trace(\bar{\Pi}^0) = k - 1$.
By repeating the same process $k - 1$ times, we get the block diagonal form of Π.
$\Pi = diag(\Pi^0, \Pi^1, \ldots, \Pi^l, \ldots, \Pi^{k-1})$ with, $\Pi^l = \frac{1}{n_l} \mathbb{1}_l \mathbb{1}_l'$, $trace(\Pi^l) = 1 \forall l$ and $\sum_{l=0}^{k-1} n_l = n$.

References

1. De Soete, G., Carroll, J. D.: K-means clustering in a low-dimensional euclidean space. In: E. Diday et al. (eds.) New Approaches in Classification and Data Analysis, pp. 212–219. Springer-Verlag Berlin (1994)
2. Dhillon, I. S.: Co-clustering documents and words using bipartite spectral graph partitioning. In SIGKDD, pp. 269-274 (2001)
3. Ding, C., Li, T., Peng, W., Park, H.: Orthogonal nonnegative matrix trifactorizations for clustering. In SIGKDD, pp. 126-135 (2006)
4. Golub, G. H., van Loan, C. F.: Matrix Computations (3rd ed.). Johns Hopkins University Press (1996)
5. Lim, D., Vidal, R., Haeffele, B. D.: Doubly stochastic subspace clustering. ArXiv, abs/2011.14859, 2020. Available via ArXiv. https://arxiv.org/abs/2011.14859
6. Marcotorchino, J. F.: Seriation problems: an overview. Appl. Stoch. Model. D. A., 7(2), 139–151 (1991)
7. Niermann, S.: Optimizing the ordering of tables with evolutionary computation. American Statistician. 59(1), 41-46 (2005)
8. Strehl, A., Ghosh, J.: Cluster ensembles—a knowledge reuse framework for combining multiple partitions. Journal of Machine Learning Research, 3, 583-617 (2002)
9. Vichi, M., Kiers, H. A.: Factorial k-means analysis for two-way data. CSDA, 37(1), 49-64 (2001)
10. Vidal, R.: Subspace clustering. IEEE Signal Processing Magazine 28(2), 52-68 (2011)
11. Wang, F., Li, P., König, A. C.: Improving clustering by learning a bi-stochastic data similarity matrix. Knowl. Inf. Syst. 32(2), 351-382 (2012)
12. Zass, R., Shashua, A.: A unifying approach to hard and probabilistic clustering. In ICCV, pp. 294-301 (2005)

Clustering Adolescent Female Physical Activity Levels with an Infinite Mixture Model on Random Effects

Amy LaLonde, Tanzy Love, Deborah R. Young, and Tongtong Wu

Abstract Physical activity trajectories from the Trial of Activity in Adolescent Girls (TAAG) capture the various exercise habits over female adolescence. Previous analyses of this longitudinal data from the University of Maryland field site, examined the effect of various individual-, social-, and environmental-level factors impacting the change in physical activity levels over 14 to 23 years of age. We aimed to understand the differences in physical activity levels after controlling for these factors. Using a Bayesian linear mixed model incorporating a model-based clustering procedure for random deviations that does not specify the number of groups *a priori*, we find that physical activity levels are starkly different for about 5% of the study sample. These young girls are exercising on average 23 more minutes per day.

Keywords: Bayesian methodology, Markov chain Monte Carlo, mixture model, reversible jump, split-merge procedures

1 Introduction

Physical activity and diet are arguably the two main controllable factors having the greatest impact on our health. Whereas we have little to no control over factors like our genetic predisposition to disease or exposure to environmental toxins, we have

Amy LaLonde
University of Rochester, NY, USA, e-mail: amylalonde2@gmail.com

Tanzy Love (✉)
University of Rochester, NY, USA, e-mail: tanzy_love@urmc.rochester.edu

Deborah Rohm Young
University of Maryland, MD, USA, e-mail: dryoung@umd.edu

Tongtong Wu
University of Rochester, NY, USA, e-mail: tongtong_wu@urmc.rochester.edu

© The Author(s) 2023
P. Brito et al. (eds.), *Classification and Data Science in the Digital Age*,
Studies in Classification, Data Analysis, and Knowledge Organization,
https://doi.org/10.1007/978-3-031-09034-9_25

much greater control over our diet and activity levels. Despite our ability to choose to engage in healthy behaviors such as exercising and eating a healthy diet, these choices are plagued with the complexity of human psychology and the modern demands and distractions that pervade our lives today. Several factors influence levels of physical activity; we explore the factors impacting female adolescents using longitudinal data.

The University of Maryland, one of the six initial university field centers of the Trial of Activity in Adolescent Girls (TAAG), selected to follow its 2006 8th grade cohort for two additional time points over adolescence: 11th grade and 23 years of age. The females were therefore measured roughly at ages 14, 17, and 23. In these waves, there was no intervention as this observational longitudinal study aimed at exploring the patterns of physical activity levels and associated factors over time.

The model presented in Wu et al. [1] motivates the current work. We fit a similar linear mixed model controlling for the same variables. Rather than cluster the raw physical activity trajectories to identify groups, we cluster the females within the model-fitting procedure based on the values of the subject-specific deviations from the adjusted physical activity levels. Fitting a Bayesian linear mixed model, we simultaneously explore the subject groups through the use of reversible jump Markov chain Monte Carlo (MCMC) applied to the random effects. Bayesian model-based clustering methods have been applied within linear mixed models to identify groups by clustering the fitted values of the dependent variable. For example, [2] fits cluster-specific linear mixed models to the gene expression outcome using an EM algorithm and [3] clusters gene expression in a similar fashion, except using Bayesian methods. In contrast, we perform the clustering on the random effects, which allows us to investigate the variability that is unexplained by the covariates of interest. This methodology is advantageous because of its ability to jointly estimate all effects, while also exploring the infinite space of group arrangements.

2 Bayesian Mixture Models for Heterogeneity of Random Effects

Let $\mathbf{y}_i = (y_{i,1}, \ldots, y_{i,T})$ be the ith subject's average daily moderate-to-vigorous physical activity (MVPA) at each of the $T = 3$ time points. The MVPA was collected from ActiGraph accelerometers (Manufacturing Technologies Inc. Health Systems, Model 7164, Shalimar, FL) worn for seven consecutive days. Accelerometers offered a great alternative to self-report for tracking physical activity levels, and measuring over seven days helped to account for differences in activity patterns during weekdays and weekends. Wu et al. [1] analyzed this cohort using mixed models that accounted for the subject-specific variability. We let $\mathbf{X}_i$ represent the ith subject's values for covariates.

Furthermore, let $\mathbf{r} = (r_1, \ldots, r_n)$ represent the subject-specific random effects for the n subjects. The simple linear mixed model is written in terms of each subject as

$$\mathbf{y}_i = \mathbf{X}_i\boldsymbol{\beta} + r_i\mathbf{1}_T + \boldsymbol{\epsilon}_i \tag{1}$$

where β represents the coefficients for the covariate effects and $\epsilon_i = (\epsilon_{i,1}, \ldots, \epsilon_{i,T})$ are the residuals. We assume independence and normality in the residuals and the random effects; hence, $r_i \sim N(0, \sigma_r^2)$ and $\epsilon_i \sim N(\mathbf{0}, \sigma_\epsilon^2 \mathbf{I}_T)$ for $i = 1, \ldots, n$.

Fitting the mixed model demonstrates substantial heterogeneity in the residuals, the variability increases as the fitted values increase. A traditional approach to fixing this violation would re-fit the model to the log-transformed MVPA values. Plots of residuals versus fitted values in this model approach also exhibited evidence of heterogeneity in the model; thus, still violating a core assumption of the regression framework. Given the changes adolescents experience as they grow into young adults, we expect to see heterogeneity in the physical activity patterns across this duration of follow-up time. However, the inability of the model to capture such changes over time at these higher levels of physical activity suggests the need for model improvements. The purpose of this analysis is to present our adjustments to previous analyses in order to investigate underlying characteristics across different groups of females formed based on deviations from adjusted physical activity levels.

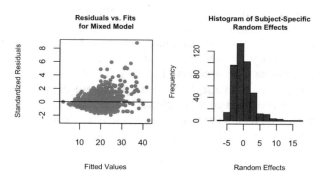

Fig. 1 The plot on the left depicts the residuals versus fitted values for the linear mixed model in Eq. (1); they demonstrate severe heteroscedasticity. The variance increases as the fitted values increase. The plot on the right depicts the distribution of the random effects.

We fit the mixed model in Eq. (1) to the sample of female adolescents. The heteroscedasticity depicted in Figure 1 reveals an increase in variance with predicted minutes of moderate-to-vigorous physical activity, which we would expect. The plot on the right in Figure 1 demonstrates that the distribution of the random effects do not appear to follow our assumption of normally distributed and centered around zero. The random effects do appear to follow a normal distribution over the lower range of deviations with a subset of the subjects having larger positive deviations from the estimated adjusted physical activity levels.

To capture the heterogeneity and allow the random effects to follow a non-normal distribution, we assign the random effects a Gaussian mixture distribution. Before introducing the model for heterogeneity, we note the likelihood distribution for the observed outcomes, $\mathbf{Y} = (\mathbf{y}_1, \ldots, \mathbf{y}_T)'$. The moderate-to-vigorous physical activity distribution is

$$p(\mathbf{Y}|\boldsymbol{\beta}, \mathbf{r}, \sigma_\epsilon^2) = \prod_{i=1}^{n} \prod_{t=1}^{T} \left(2\pi\sigma_\epsilon^2\right)^{-\frac{1}{2}} \exp\left\{-\frac{1}{2\sigma_\epsilon^2}(y_{i,t} - X_{i,t}\boldsymbol{\beta} - r_i)^2\right\}. \qquad (2)$$

Then to account for the heterogeneity across subjects, the probability density for the subject-specific deviations in physical activity is expressed as a mixture of one-dimensional normal densities,

$$p(r_i|\boldsymbol{\mu}, \sigma_r^2) = \sum_{g=1}^{G} \pi_g \left(2\pi\sigma_{r,g}^2\right)^{-\frac{1}{2}} \exp\left\{-\frac{1}{2\sigma_{r,g}^2}(r_i - \mu_g)^2\right\}. \qquad (3)$$

Here, $\boldsymbol{\mu} = (\mu_1, \ldots, \mu_G)'$ defines the group-specific mean deviations, $\sigma_r^2 = (\sigma_{r,1}^2, \ldots, \sigma_{r,G}^2)'$ characterizes the variances of the group-specific deviations, and $\pi = (\pi_1, \ldots, \pi_G)'$ is the probability of membership in each group g.

The model in Eqs. (2) and (3) can be fit using either an EM or Bayesian MCMC procedures. Both require specification of a fixed number of G-groups. While we may hypothesize that there are only two groups—one that is normally distributed and centered at zero and another that is normally distributed and centered at a larger mean—the assumption hinges on what we have seen from plots like those in Figure 1. The random effects in the aforementioned histogram, however, are being shrunk towards zero by assumption; while a mixture model will allow the data to more accurately depict the deviations observed in the girl's physical activity levels. The assumption of G groups can strongly influence the results of our model fitting. To circumvent the issues associated with selecting G in either an EM algorithm or a Bayesian finite mixture model framework, we implement a Bayesian mixture model that incorporates G as an additional unknown parameter.

2.1 Bayesian Mixed Models With Clustering

Richardson and Green [4] adapts the reversible jump methodology to univariate normal mixture models. In addition to being able to characterize the distribution of G, this Bayesian framework has the ability to simultaneously explore the posterior distribution for the covariate effects of interest. Furthermore, we will have the posterior distributions of the group-defining parameters rather than just point estimates. Since we are interested in the physical activity differences in subjects when controlling for these covariates, we use Eq. (1) as the basis of our model.

The foundation of our clustering model is a finite mixture model on the random effects, r_i, as shown in Eq. (3). For all $i = 1, \ldots, n$ and $g = 1, \ldots, G$, $r_i|c_i, \boldsymbol{\mu} \sim F_r(\mu_{c_i}, \sigma_{r,c_i}^2)$, $(c_i = g)|\boldsymbol{\pi}, G \sim \text{Categorical}(\pi_1, \ldots, \pi_G)$, $\mu_g|\tau \sim N(\mu_0, \tau)$, $\sigma_{r,g}^2|c, \delta \sim IG(c, \delta)$, $\boldsymbol{\pi}|G \sim \text{Dirichlet}(\alpha, \ldots, \alpha)$, $G \sim \text{Uniform}[1, G_{max}]$, where c_i is the latent grouping variable tracking the assignment of r_i into any one of the G clusters. The *likelihood function* for these subject-specific deviations, given the group assignment, c_i, is simply $p(r_i|c_i = g, \mu_g, \sigma_{r,g}^2) = \left(2\pi\sigma_{r,g}^2\right)^{-\frac{1}{2}} \exp\left\{-\frac{1}{2\sigma_{r,g}^2}(r_i - \mu_g)^2\right\}.$

This replaces the typical independent and identically distributed assumption of $r_i \sim N(0, \sigma_r^2)$ for all i with a normal distribution that is now conditional on group assignment. The remainder of the model formulation follows closely to the framework constructed in [4], except we have an additional layer of unknown parameters defining the linear mixed model in Eq. (1).

We select conjugate priors so that the the posterior distributions of the unknown parameters are analytically tractable. The prior on the mixing probabilities, π, is a symmetric Dirichlet distribution, reflecting the prior belief that belonging to any one cluster is equally likely. To use the sampling methods of [4], we select a discrete uniform prior on G that reflects our uncertainty on the number of groups, and impose an a priori ordering of the μ_g, such that for any given value G, $\mu_1 < \mu_2 < \cdots < \mu_G$, to remove label switching. Thus, in the prior for the clustering parameters,

$$p(\mu) = G! \prod_{g=1}^{G} \sqrt{(2\pi\tau)^{-\frac{1}{2}}} \exp\left\{-\frac{1}{2\tau}(\mu_g - \mu_0)^2\right\}$$

$$p(\sigma_{r,g}^2) = \frac{\delta^c}{\Gamma(c)}(\sigma_{r,g}^2)^{-c-1} \exp\left\{-\frac{\delta}{\sigma_{r,g}^2}\right\}$$

$$p(G) = \frac{1}{G_{max}} \mathbf{1}\{G \in [1, G_{max}]\},$$

where G_{max} is set to be reasonably large and $\mathbf{1}\{G \in [1, G_{max}]\}$ is a discrete indicator function, equal to 1 on the interval $[1, G_{max}]$ and 0 elsewhere.

The capacity of our sampler to move between dimensions is essential to our ability to explore the grouping of the observations while simultaneously exploring the parameters describing the relationships between the covariates and the outcome. This means that we can allow the number of components of our mixture model on the random effects to increase or decrease at each state of our MCMC chain. Such changes impact the dimension of the parameters of the mixture model, $\theta = (\mu, \sigma_r^2, G, \pi, \mathbf{c})$.

Let θ denote the current state of the parameters $(\mu, \sigma_r^2, G, \pi, \mathbf{c})$ when proposing move m where $m \in \{S, M, B, D\}$ corresponds to a split, merge, birth and death, respectively. Given the current state, θ, and move m, we propose a new state, θ^m, under move m. The acceptance probability is written as $acc_m(\theta^m, \theta) = \min\left[1, \frac{p(\theta^m|\mathbf{r})q(\theta^m|m^{-1})}{p(\theta|\mathbf{r})q(\theta|m)}|J|\right]$ where $p(\cdot)$ and $q(\cdot)$ denote the target and proposal distribution, respectively. In our case, the target distribution is the posterior distribution of our group-specific parameters, $(\mu, \sigma_r^2, \pi, \mathbf{c})$, given the data, $\mathbf{r}$, which are the random effects. Each proposed move changes the dimension of the parameters in θ by 1, adding or deleting group-specific parameters. The ratio $q(\theta^m|m^{-1})/q(\theta|m)$ ensures "dimension balancing", as explained in [4]. For moves increasing in dimension, the Jacobian, $|J|$, is computed as $|\delta\theta^m/\delta(\theta, \mathbf{u})|$ because moving from θ to θ^m will require additional parameters, $\mathbf{u}$ to appropriately match dimensions. The opposite is true for moves decreasing in dimension. This is what we refer to as the reversible jump mechanism; each time a split is proposed, we must also design the reversible move that would result in the currently merged component, and vice versa.

Split and merge moves are implemented for our model. These moves update π, μ, and σ for two adjacent groups or create two adjacent groups using three Beta-distributed additional parameters, u, for dimension balancing in a similar way to [4]. Within our context of random effects, births and deaths are not appropriate. A singleton causes issues of identifiability because the r_i is no longer defined as random. We do not allow for birth and death moves in our reversible jump methods.

3 Trial of Activity in Adolescent Girls (TAAG) and Model Results

Our analysis focuses only on these girls from the University of Maryland site of the TAAG study who were measured at all three follow-up time points, beginning in 2006. After excluding girls with missing outcomes, the final sample consisted of 428 girls measured in 2006, 2009, and 2014. Missing covariate values were imputed for four subjects using the values from the nearest time point.

We determine the group assignments using an MCMC sampler having 10,000 iterations, with a burn-in of 500 draws. The posterior distribution for G was extremely peaked at $G = 2$. Summarization of the posterior distribution of the group assignments via the least squares clustering method delivers the final arrangement, $\hat{\mathbf{c}}_{LS}$, of girls into two groups describing their physical activity levels [5]. Since our sampler explores several models for which group assignments and G can vary, we sample additional draws from the posterior distribution of the remaining parameters of interest using an MCMC sampler with the model specification of Eq. (1) with groups fixed at our posterior assignment, $\hat{\mathbf{c}}_{LS}$, for the subject-specific random effects. This additional chain was run for 10,000 iterations with a burn-in 500 draws, yielding the results summarized below. Convergence diagnostics indicated that 10,000 iterations sufficiently met the effective sample size threshold for estimating the coefficients for the covariate effects, β, and the group-specific means, μ, describing the deviations of the girls' physical activity levels [6].

After controlling for covariates believed to best describe the variation in the physical activity levels of females, our method finds that there is a small subset of the females who are much more active than the remainder of the sample. Every subject in the more active group has fitted trajectories above the recommended 30 minutes of exercise. Most of the population does not get the recommended allowance of daily physical activity and this is well-supported in our analysis. All but two subjects in the less active group have fitted trajectories that never pass the recommended 30 minutes of exercise. The random effects from this model better fit a normal distribution (not centered at 0) for each of the two groups and do not show as much heteroscedasticity over time as the one group model depicted in Figure 1.

Given these differences are observed even after controlling for the aforementioned variables, we would like to further examine the characteristics that may set these highly active females apart from the rest of the girls in our sample. To do this, we look at a number of other covariates that were either excluded during the variable selection process or were not measured at all time points. We use simple Wilcoxon tests on the available time points of the additional variables and on all time points for covariates we adjusted for in the initial model.

We first note that the median BMI of the subset of highly active girls is significantly lower than that of the remaining girls consistently at each TAAG wave. Similarly, mother's education level is also consistently significant at each time point. These values are measured at each time point to reflect changes as the mother pursues additional education, or as the girls become more aware of their mother's education. The majority of the highly active girls have mother's who have completed college or higher (75% or higher at each time point); whereas, the remainder of the sample has mother's with a range of education levels (less than high school through college or more). The number of parks within a one-mile radius of the home is significantly different among the high and low groups in the middle school and high school years, when the girls are likely to be living at home. This variable may be an indicator of socioeconomic status as families with more money may live in neighborhoods nearer to parks. Finally, in the high school and college-aged years, the self-management strategies among the highly active girls are significantly higher rated than the remainder of the population.

In high school, the subset of highly active girls tend to have better self-described health, participate in more sports teams, have access to more physical education classes, and have been older at the time of their first menstrual period. At the college age, these girls still have higher self-described health; however, the higher levels of the global physical activity score and self-esteem scores are now significantly improved in the subset of highly active females.

4 Discussion

We extended the mixed models of [1] with the application still focused on the same 428 girls from the TAAG, TAAG 2, and TAAG 3 studies. Within the Bayesian linear mixed model, we implemented a clustering procedure aimed at clustering girls into groups based on deviations from the adjusted physical activity levels. These groups reflected the tendency for small subsets of females to be highly active. Not surprisingly, only 24 girls (5% of our sample) were classified as highly active.

This group of highly active girls differs in several ways. These girls are more active, and thus we expect that the age at first menstrual period will be higher. We may also expect that the highly active girls are involved in more sports teams and that they will have higher global physical activity scores. Some other interesting characteristics of these girls, however, is their increased self-management strategies, self-esteem scores, and self-described health. This may suggest that interventions focusing on time management and emphasizing self-efficacy could impact adolescent female physical activity levels. In doing so, we could aim to increase self-esteem and self-described health.

The ability to account for heterogeneity in the subject-specific deviations from an adjusted model allows us to keep the outcome on the original scale while still improving model assumptions. Our model estimates model parameters while identifying groups of observations with differing activity levels. In contrast, a frequentist approach could be taken using EM algorithm; however, we would lose the ability for the data to give statistical inference on the appropriate number of groups and to incorporate posterior samples with different numbers of groups into the estimated class label.

The current analysis looks only at identifying groups based on deviations from the overall adjusted minutes of MVPA for the females. A natural extension would be to look at clustering on the slope for time to begin to understand the various patterns we observe among adolescent females over time. Furthermore, we may want to incorporate a variable selection procedure into the fixed portion of the model. The groups we find by either clustering on subject-specific intercepts and/or slopes would be sensitive to the covariates selected, depending on the variability captured by this fixed portion of the model. Physical activity, like most human behavior, varies widely for a multitude of reasons, many of which we may not think to or are unable to measure. Identifying groups when a traditional mixed model constructed using standard variable selection methods suggests lack of fit can be a useful step towards better understanding differences through post-hoc analyses of the groups' characteristics.

Acknowledgements Research reported in this publication was supported by the National Institutes of Health (NIH) under award numbers T32ES007271 and R01HL119058. The content is solely the responsibility of the authors and does not necessarily represent the official views of the NIH.

References

1. Young, D. R., Mohan, Y. D., Saksvig, B. I., Sidell, M., Wu, T. T., Cohen, D.: Longitudinal predictors of moderate to vigorous physical activity among adolescent girls and young women. Under review. (2017)
2. Ng, S. K., McLachlan, G. J., Wang, K., Ben-Tovim Jones, L., Ng, S. W.: A mixture model with random-effects components for clustering correlated gene-expression profiles. Bioinformatics **22**(14), 1745 (2006)
3. Zhou, C., Wakefield, J.: A Bayesian mixture model for partitioning gene expression data. Biometrics **62**(2), 515–525 (2006)
4. Richardson, S., Green, P. J.: On Bayesian analysis of mixtures with an unknown number of components (with discussion). J. Roy. Stat. Soc. B **59**(4), 731–792 (1997)
5. Dahl, D. B.: Model-based clustering for expression data via a Dirichlet process mixture model. Bayesian Inference for Gene Expression and Proteomics. 201–218 (2006)
6. Flegal, J. M., Hughes, J., Vats, D.: mcmcse: Monte Carlo standard errors for MCMC. R package version 1.2-1 (2016)

Unsupervised Classification of Categorical Time Series Through Innovative Distances

Ángel López-Oriona, José A. Vilar, and Pierpaolo D'Urso

Abstract In this paper, two novel distances for nominal time series are introduced. Both of them are based on features describing the serial dependence patterns between each pair of categories. The first dissimilarity employs the so-called association measures, whereas the second computes correlation quantities between indicator processes whose uniqueness is guaranteed from standard stationary conditions. The metrics are used to construct crisp algorithms for clustering categorical series. The approaches are able to group series generated from similar underlying stochastic processes, achieve accurate results with series coming from a broad range of models and are computationally efficient. An extensive simulation study shows that the devised clustering algorithms outperform several alternative procedures proposed in the literature. Specifically, they achieve better results than approaches based on maximum likelihood estimation, which take advantage of knowing the real underlying procedures. Both innovative dissimilarities could be useful for practitioners in the field of time series clustering.

Keywords: categorical time series, clustering, association measures, indicator processes

1 Introduction

Clustering of time series concerns the challenge of splitting a set of unlabeled time series into homogeneous groups, which is a pivotal problem in many knowledge discovery tasks [1]. Categorical time series (CTS) are a particular class of time series exhibiting a qualitative range which consists of a finite number of categories. Most of the classical statistical tools used for real-valued time series (e.g., the autocorrelation function) are not useful in the categorical case, so different types of measures than the standard ones are needed for a proper analysis of CTS. CTS

Ángel López-Oriona (✉), José A. Vilar
Research Group MODES, Research Center for Information and Communication Technologies (CITIC), University of A Coruña, Spain,
e-mail: oriona38@hotmail.com; jose.vilarf@udc.es

Pierpaolo D'Urso
Department of Social Sciences and Economics, Sapienza University of Rome, Italy,
e-mail: pierpaolo.durso@uniroma1.it

© The Author(s) 2023   233
P. Brito et al. (eds.), *Classification and Data Science in the Digital Age*,
Studies in Classification, Data Analysis, and Knowledge Organization,
https://doi.org/10.1007/978-3-031-09034-9_26

arise in an extensive assortment of fields [2, 3, 7, 8, 9]. Since only a few works have addressed the problem of CTS clustering [4, 5], the main goal of this paper is to introduce novel clustering algorithms for CTS.

2 Two Novel Feature-based Approaches for Categorical Time Series Clustering

Consider a set of s categorical time series $S = \{X_t^{(1)}, \ldots, X_t^{(s)}\}$, where the j-th element $X_t^{(j)}$ is a T_j-length partial realization from any categorical stochastic process $(X_t)_{t \in \mathbb{Z}}$ taking values on a number r of unordered qualitative categories, which are coded from 1 to r so that the range of the process can be seen as $\mathcal{V} = \{1, \ldots, r\}$. We suppose that the process $(X_t)_{t \in \mathbb{Z}}$ is bivariate stationary, i.e., the pairwise joint distribution of (X_{t-k}, X_t) is invariant in t. Our goal is to perform clustering on the elements of S in such a way that the series assumed to be generated from identical stochastic processes are placed together. To that aim, we propose two distance metrics which are based on feature extraction.

2.1 Descriptive Features for Categorical Processes

Let $\{X_t, t \in \mathbb{Z}\}$ be a bivariate stationary categorical stochastic process with range $\mathcal{V} = \{1, \ldots, r\}$. Denote by $\pi = (\pi_1, \ldots, \pi_r)$ the marginal distribution of X_t, which is, $P(X_t = j) = \pi_j > 0$, $j = 1, \ldots, r$. Fixed $l \in \mathbb{N}$, we use the notation $p_{ij}(l) = P(X_t = i, X_{t-l} = j)$, with $i, j \in \mathcal{V}$, for the lagged bivariate probability and the notation $p_{i|j}(l) = P(X_t = i | X_{t-l} = j) = p_{ij}(l)/\pi_j$ for the conditional bivariate probability.

To extract suitable features characterizing the serial dependence of a given CTS, we start by defining the concepts of perfect serial independence and dependence for a categorical process. We have perfect serial independence at lag $l \in \mathbb{N}$ if and only if $p_{ij}(l) = \pi_i \pi_j$ for any $i, j \in \mathcal{V}$. On the other hand, we have perfect serial dependence at lag $l \in \mathbb{N}$ if and only if the conditional distribution $p_{\cdot|j}(l)$ is a one-point distribution for any $j \in \mathcal{V}$. There are several association measures which describe the serial dependence structure of a categorical process at lag l. One of such measures is the so-called Cramer's v, which is defined as

$$v(l) = \sqrt{\frac{1}{r-1} \sum_{i,j=1}^{r} \frac{(p_{ij}(l) - \pi_i \pi_j)^2}{\pi_i \pi_j}}. \tag{1}$$

Cramer's v summarizes the serial dependence patterns of a categorical process for every pair (i, j) and $l \in \mathbb{N}$. However, this quantity is not appropriate for characterizing a given stochastic process, since two different processes can have the same value of $v(l)$. A better way to characterize the process X_t is by considering the matrix $V(l) = (V_{ij}(l))_{1 \leq i,j \leq r}$, where $V_{ij}(l) = \frac{(p_{ij}(l) - \pi_i \pi_j)^2}{\pi_i \pi_j}$. The elements of the matrix

$V(l)$ give information about the so-called *unsigned* dependence of the process. However, it is often useful to know whether a process tends to stay in the state it has reached or, on the contrary, the repetition of the same state after l steps is infrequent. This motivates the concept of *signed* dependence, which arises as an analogy of the autocorrelation function of a numerical process, since such quantity can take either positive or negative values. Provided that perfect serial dependence holds, we have perfect *positive* (*negative*) serial dependence if $p_{i|i}(l) = 1$ ($p_{i|i}(l) = 0$) for all $i \in \mathcal{V}$.

Since $V(l)$ does not shed light on the signed dependence structure, it would be valuable to complement the information contained in $V(l)$ by adding features describing signed dependence. In this regard, a common measure of signed serial dependence at lag l is the Cohen's κ, which takes the form

$$\kappa(l) = \frac{\sum_{j=1}^{r} (p_{jj}(l) - \pi_j^2)}{1 - \sum_{j=1}^{r} \pi_j^2}. \tag{2}$$

Proceeding as with $v(l)$, the quantity $\kappa(l)$ can be decomposed in order to obtain a complete representation of the signed dependence pattern of the process. In this way, we consider the vector $\mathcal{K}(l) = (\mathcal{K}_1(l), \dots, \mathcal{K}_r(l))$, where each $\mathcal{K}_i$ is defined as

$$\mathcal{K}_i(l) = \frac{p_{ii}(l) - \pi_i^2}{1 - \sum_{j=1}^{r} \pi_j^2}, \tag{3}$$

$i = 1, \dots, r$.

In practice, the matrix $V(l)$ and the vector $\mathcal{K}(l)$ must be estimated from a T-length realization of the process, $\{X_1, \dots X_T\}$. To this aim, we consider estimators of π_i and $p_{ij}(l)$, $\widehat{\pi}_i$ and $\widehat{p}_{ij}(l)$, respectively, defined as $\widehat{\pi}_i = \frac{N_i}{T}$ and $\widehat{p}_{ij}(l) = \frac{N_{ij}(l)}{T-l}$, where N_i is the number of variables X_t equal to i in the realization $\{X_1, \dots X_T\}$, and $N_{ij}(l)$ is the number of pairs $(X_t, X_{t-l}) = (i, j)$ in the realization $\{X_1, \dots X_T\}$. Hence, estimates of $V(l)$ and $\mathcal{K}(l)$, $\widehat{V}(l)$ and $\widehat{\mathcal{K}}(l)$, respectively, can be obtained by plugging in the estimates $\widehat{\pi}_i$ and $\widehat{p}_{ij}(l)$ in (2) and (3), respectively. This leads directly to estimates of $v(l)$ and $\kappa(l)$, denoted by $\widehat{v}(l)$ and $\widehat{\kappa}(l)$.

An alternative way of describing the dependence structure of the process $\{X_t, t \in \mathbb{Z}\}$ is to take into consideration its equivalent representation as a multivariate binary process. The so-called *binarization* of $\{X_t, t \in \mathbb{Z}\}$ is constructed as follows. Let $e_1, \dots, e_r \in \{0, 1\}^r$ be unit vectors such that e_k has all its entries equal to zero except for a one in the k-th position, $k = 1, \dots, r$. Then, the binary representation of $\{X_t, t \in \mathbb{Z}\}$ is given by the process $\{Y_t = (Y_{t,1}, \dots, Y_{t,r})^\top, t \in \mathbb{Z}\}$ such that $Y_t = e_j$ if $X_t = j$. Fixed $l \in \mathbb{N}$ and $i, j \in \mathcal{V}$, consider the correlation $\phi_{ij}(l) = Corr(Y_{t,i}, Y_{t-l,j})$, which measures linear dependence between the i-th and j-th categories with respect to the lag l. The following proposition provides some properties of the quantity $\phi_{ij}(l)$.

Proposition 1

Let $\{X_t, t \in \mathbb{Z}\}$ be a bivariate stationary categorical process with range $\mathcal{V} = \{1, \ldots, r\}$. Then the following properties hold:

1. For every $i, j \in \mathcal{V}$, the function $\phi_{ij} : \mathbb{N} \to [-1, 1]$ given by $l \to \phi_{ij}(l) = Corr(Y_{t,i}, Y_{t-l,j})$ is well-defined.
2. $\phi_{ij}(l) = 0 \Leftrightarrow p_{ij}(l) = \pi_i \pi_j$.
3. $\phi_{ij}(l) = \pm 1 \Leftrightarrow p_{ij}(l) = \pm\sqrt{\pi_i(1-\pi_i)\pi_j(1-\pi_j)} + \pi_i \pi_j$.
4. $\phi_{ij}(l) = \sqrt{\frac{\pi_j(1-\pi_i)}{\pi_i(1-\pi_j)}} \Leftrightarrow p_{i|j}(l) = 1$.

The proof of Proposition 1 is quite straightforward and it is not shown in the manuscript for the sake of brevity. According to Proposition 1, the quantity $\phi_{ij}(l)$ can be used to explain both types of dependence, signed and unsigned, within the underlying process. In fact, in the case of perfect unsigned independence at lag l, we have that $p_{ij}(l) = \pi_i \pi_j$ for all $i, j \in \mathcal{V}$ so that $\phi_{ij}(l) = 0$ for all $i, j \in \mathcal{V}$ in accordance with Property 2 of Proposition 1. Under perfect positive dependence at lag l, $p_{i|i}(l) = 1$ for all $i \in \mathcal{V}$. Then $\phi_{ii}(l) = 1$ for all $i \in \mathcal{V}$ by following Property 4 of Proposition 1. The same property allows to conclude that $\phi_{ii}(l) = -\pi_i/(1-\pi_i)$ for all $i \in \mathcal{V}$ in the case of perfect negative dependence. In sum, $\phi_{ij}(l)$ evaluates unsigned dependence when $i \neq j$ and signed dependence when $i = j$. The previous quantities can be encapsulated in a matrix $\mathbf{\Phi}(l) = (\phi_{ij}(l))_{1 \leq i,j \leq r}$, which can be directly estimated by means of $\widehat{\mathbf{\Phi}}(l) = (\widehat{\phi}_{ij}(l))_{1 \leq i,j \leq r}$, where each $\widehat{\phi}_{ij}(l)$ is computed as $\widehat{\phi}_{ij}(l) = \frac{\widehat{p}_{ij}(l) - \widehat{\pi}_i \widehat{\pi}_j}{\sqrt{\widehat{\pi}_i(1-\widehat{\pi}_i)\widehat{\pi}_j(1-\widehat{\pi}_j)}}$ (this is derived from the proof of Proposition 1).

2.2 Two Innovative Dissimilarities Between CTS

In this section we introduce two distance measures between categorical series based on the features described above. Suppose we have a pair of CTS $X_t^{(1)}$ and $X_t^{(2)}$, and consider a set of L lags, $\mathcal{L} = \{l_1, \ldots, l_L\}$. A dissimilarity based on Cramer's v and Cohen's κ, so-called d_{CC}, is defined as

$$d_{CC}(X_t^{(1)}, X_t^{(2)}) = \sum_{k=1}^{L} \left[\left\| vec\left(\widehat{\mathbf{V}}(l_k)^{(1)} - \widehat{\mathbf{V}}(l_k)^{(2)}\right) \right\|^2 \right.$$
$$\left. + \left\| \widehat{\mathcal{K}}(l_k)^{(1)} - \widehat{\mathcal{K}}(l_k)^{(2)} \right\|^2 \right] + \left\| \widehat{\boldsymbol{\pi}}^{(1)} - \widehat{\boldsymbol{\pi}}^{(2)} \right\|^2,$$

where the superscripts (1) and (2) are used to indicate that the corresponding estimations are obtained with respect to the realizations $X_t^{(1)}$ and $X_t^{(2)}$, respectively.

An alternative distance measure relying on the binarization of the processes, so-called d_B, is defined as

$$d_B(X_t^{(1)}, X_t^{(2)}) = \sum_{k=1}^{L} \left\| vec\left(\widehat{\boldsymbol{\Phi}}(l_k)^{(1)} - \widehat{\boldsymbol{\Phi}}(l_k)^{(2)}\right) \right\|^2 + \left\| \widehat{\boldsymbol{\pi}}^{(1)} - \widehat{\boldsymbol{\pi}}^{(2)} \right\|^2.$$

For a given set of categorical series, the distances d_{CC} and d_B can be used as input for traditional clustering algorithms. In this manuscript we consider the *Partition Around Medoids* (PAM) algorithm.

3 Partitioning Around Medoids Clustering of CTS

In this section we examine the performance of both metrics d_{CC} and d_B in the context of hard clustering (i.e., each series is assigned to exactly one cluster) of CTS through a simulation study.

3.1 Experimental Design

The simulated scenarios encompass a broad variety of generating processes. In particular, three setups were considered, namely clustering of (i) Markov Chains (MC), (ii) Hidden Markov Models (HMM) and (iii) New Discrete ARMA (NDARMA) processes. The generating models with respect to each class of processes are given below.

Scenario 1. Clustering of MC. Consider four three-state MC, so-called MC_1, MC_2, MC_3 and MC_4, with respective transition matrices $\boldsymbol{P}_1^1$, $\boldsymbol{P}_2^1$, $\boldsymbol{P}_3^1$ and $\boldsymbol{P}_4^1$ given by

$$\boldsymbol{P}_1^1 = Mat^3(0.1, 0.8, 0.1, 0.5, 0.4, 0.1, 0.6, 0.2, 0.2),$$
$$\boldsymbol{P}_2^1 = Mat^3(0.1, 0.8, 0.1, 0.6, 0.3, 0.1, 0.6, 0.2, 0.2),$$
$$\boldsymbol{P}_3^1 = Mat^3(0.05, 0.90, 0.05, 0.05, 0.05, 0.90, 0.90, 0.05, 0.05),$$
$$\boldsymbol{P}_4^1 = Mat^3(1/3, 1/3, 1/3, 1/3, 1/3, 1/3, 1/3, 1/3, 1/3),$$

where the operator Mat^k, $k \in \mathbb{N}$ transforms a vector into a square matrix of order k by sequentially placing the corresponding numbers by rows.

Scenario 2. Clustering of HMM. Consider the bivariate process $(X_t, Q_t)_{t \in \mathbb{Z}}$, where Q_t stands for the hidden states and X_t for the observable random variables. Process $(Q_t)_{t \in \mathbb{Z}}$ constitutes an homogeneous MC. Both $(X_t)_{t \in \mathbb{Z}}$ and $(Q_t)_{t \in \mathbb{Z}}$ are assumed to be count processes with range $\{1, \ldots, r\}$. Process $(X_t, Q_t)_{t \in \mathbb{Z}}$ is assumed to verify the three classical assumptions of a HMM. Based on previous considerations, let HMM_1, HMM_2, HMM_3 and HMM_4 be four three-state HMM with respective transition matrices $\boldsymbol{P}_1^2$, $\boldsymbol{P}_2^2$, $\boldsymbol{P}_3^2$ and $\boldsymbol{P}_4^2$ and emission matrices $\boldsymbol{E}_1^2$, $\boldsymbol{E}_2^2$, $\boldsymbol{E}_3^2$ and $\boldsymbol{E}_4^2$ given by

$$P_1^2 = Mat^3(0.05, 0.90, 0.05, 0.05, 0.05, 0.90, 0.90, 0.05, 0.05), P_2^2 = P_1^2,$$

$$P_3^2 = Mat^3(0.1, 0.7, 0.2, 0.4, 0.4, 0.2, 0.4, 0.3, 0.3),$$

$$P_4^2 = Mat^3(1/3, 1/3, 1/3, 1/3, 1/3, 1/3, 1/3, 1/3, 1/3), E_1^2 = P_1^2,$$

$$E_2^2 = Mat^3(0.1, 0.8, 0.1, 0.5, 0.4, 0.1, 0.6, 0.2, 0.2), E_3^2 = E_2^2,$$

$$E_4^2 = Mat^3(1/3, 1/3, 1/3, 1/3, 1/3, 1/3, 1/3, 1/3, 1/3).$$

Scenario 3. Clustering of NDARMA processes. Let $(X_t)_{t \in \mathbb{Z}}$ and $(\epsilon_t)_{t \in \mathbb{Z}}$, be two count processes with range $\{1, \ldots, r\}$ following the equation

$$X_t = \alpha_{t,1} X_{t-1} + \ldots + \alpha_{t,p} X_{t-p} + \beta_{t,0} \epsilon_t + \ldots + \beta_{t,q} \epsilon_{t-q},$$

where $(\epsilon_t)_{t \in \mathbb{Z}}$ is i.i.d with $P(\epsilon_t = i) = \pi_i$, independent of $(X_s)_{s < t}$, and the i.i.d multinomial random vectors

$$(\alpha_{t,1}, \ldots, \alpha_{t,p}, \beta_{t,0}, \ldots, \beta_{t,q}) \sim \text{MULT}(1; \phi_1, \ldots, \phi_p, \varphi_0, \ldots, \varphi_q),$$

are independent of $(\epsilon_t)_{t \in \mathbb{Z}}$ and $(X_s)_{s < t}$. The considered models are three three-state NDARMA(2,0) processes and one three-state NDARMA(1,0) process with marginal distribution $\pi^3 = (2/3, 1/6, 1/6)$, and corresponding probabilities in the multinomial distribution given by

$$(\phi_1, \phi_2, \varphi_0)_1^3 = (0.7, 0.2, 0.1), (\phi_1, \phi_2, \varphi_0)_2^3 = (0.1, 0.45, 0.45),$$

$$(\phi_1, \phi_2, \varphi_0)_3^3 = (0.5, 0.25, 0.25), (\phi_1, \varphi_0)_4^3 = (0.2, 0.8).$$

The simulation study was carried out as follows. For each scenario, 5 CTS of length $T \in \{200, 600\}$ were generated from each process in order to execute the clustering algorithms twice, thus allowing to analyze the impact of the series length. The resulting clustering solution produced by each considered algorithm was stored. The simulation procedure was repeated 500 times for each scenario and value of T. The computation of d_{CC} and d_B was carried out by considering $\mathcal{L} = \{1\}$ in Scenarios 1 and 2, and $\mathcal{L} = \{1, 2\}$ in Scenario 3. This way, we adapted the distances to the maximum number of significant lags existing in each setting.

3.2 Alternative Metrics and Assessment Criteria

To better analyze the performance of both metrics d_{CC} and d_B, we also obtained partitions by using alternative techniques for clustering of categorical series. The considered procedures are described below.

- *Model-based approach using maximum likelihood estimation (MLE).* The distance between two CTS is defined as the squared Euclidean distance between the corresponding vectors of fitted coefficients via MLE (d_{MLE}).
- *Model-based approach using mixtures.* [4] propose to group a set of CTS by using a mixture of first order Markov models via the EM algorithm (d_{CZ}).
- *An hybrid framework for clustering CTS.* [6] presents a dissimilarity between categorical series which evaluates both closeness between raw categorical values and proximity between dynamic patterns (d_{MV}).

Note that the approach based on the distance d_{MLE} can be seen as a strict benchmark in the evaluation task. The effectiveness of the clustering approaches was assessed by comparing the clustering solution produced by the algorithms with the true clustering partition, so-called ground truth. The latter consisted of $C = 4$ clusters in all scenarios, each group including the five CTS generated from the same process. The value $C = 4$ was provided as input parameter to the PAM algorithm in the case of d_{CC}, d_B, d_{MLE} and d_{MV}. As for the approach d_{CZ}, a number of 4 components were considered for the mixture model. Experimental and true partitions were compared by using three well-known external clustering quality indexes, the Adjusted Rand Index (ARI), the Jaccard Index (JI) and the Fowlkes-Mallows index (FMI).

3.3 Results and Discussion

Average values of the quality indexes by taking into account the 500 simulation trials are given in Tables 1, 2 and 3 for Scenarios 1, 2 and 3, respectively.

Table 1 Average results for Scenario 1.

		$T = 200$			$T = 600$	
Method	ARI	JI	FMI	ARI	JI	FMI
d_{CC}	**0.774**	**0.710**	**0.830**	**0.916**	**0.886**	**0.935**
d_B	0.729	0.661	0.792	0.861	0.878	0.893
d_{MLE}	0.704	0.633	0.772	0.841	0.792	0.876
d_{CZ}	0.712	0.648	0.786	0.915	**0.886**	0.934
d_{MV}	0.406	0.363	0.665	0.379	0.363	0.650

The results in Table 1 indicate that the dissimilarity d_{CC} is the best performing one when dealing with MC, outperforming the MLE-based metric d_{MLE}. The distance d_B is also superior to d_{MLE}. The measure d_{CZ} attains in Scenario 1 similar results than d_{CC}, specially for $T = 600$. The good performance of d_{CZ} was expected, since the assumption of first order Markov models considered by this metric is fulfilled in Scenario 1. Table 2 shows a completely different picture, indicating that the metrics d_{CC} and d_B exhibit a significantly better effectiveness than the rest of the dissimilarities. Finally, the quantities in Table 3 reveal that the model-based distance d_{MLE} attains the best results when $T = 200$, but is defeated by d_B when

Table 2 Average results for Scenario 2.

	$T = 200$			$T = 600$		
Method	ARI	JI	FMI	ARI	JI	FMI
d_{CC}	0.707	0.639	0.777	0.856	0.810	0.888
d_B	**0.760**	**0.701**	**0.812**	**0.963**	**0.949**	**0.971**
d_{MLE}	0.354	0.342	0.512	0.299	0.310	0.478
d_{CZ}	0.645	0.577	0.739	0.703	0.638	0.779
d_{MV}	0.089	0.175	0.323	0.062	0.175	0.301

Table 3 Average results for Scenario 3.

	$T = 200$			$T = 600$		
Method	ARI	JI	FMI	ARI	JI	FMI
d_{CC}	0.627	0.563	0.715	0.875	0.837	0.903
d_B	0.680	0.612	0.754	**0.925**	**0.901**	**0.941**
d_{MLE}	**0.727**	**0.656**	**0.788**	0.872	0.828	0.900
d_{CZ}	0.586	0.562	0.693	0.647	0.577	0.738
d_{MV}	0.035	0.167	0.292	-0.028	0.138	0.251

$T = 600$. The metric d_{CZ} suffers again from model misspecification. In summary, the numerical experiments carried out throughout this section show the excellent ability of both measures d_{CC} and d_B to discriminate between a broad variety of categorical processes. Specifically, these metrics either outperform or show similar behavior than distances based on estimated model coefficients, which take advantage of knowing the true underlying models.

It is worth highlighting that the methods proposed in this paper could have promising applications in some fields as the clustering of genetic data sequences.

References

1. Liao, T. W.: Clustering of time series data: A survey. Pattern Recogn. **38**, 1857-1874 (2005)
2. Churchill, G. A.: Stochastic models for heterogeneous DNA sequences. Bull. Math. Biol. **51**, 79-94 (1989)
3. Fokianos, K., Kedem, B.: Regression theory for categorical time series. Stat. Sci. **18**, 357-376 (2003)
4. Cadez, I., Heckerman, D., Meek, C., Smyth, P., White, S.: Model-based clustering and visualization of navigation patterns on a web site. Data Min. Knowl. Discov. **7**, 399-424 (2003)
5. Fruhwirth-Schnatter, S., Pamminger, C.: Model-based clustering of categorical time series. Bayesian Analysis. **5**, 345-368 (2010)
6. García Magariños, M., Vilar, J. A.: A framework for dissimilarity-based partitioning clustering of categorical time series. Data Min. Knowl. Discov. **29**, 466-502 (2015)
7. Elzinga, C. H.: Combinatorial representations of token sequences. J. Classif. **22**, 87-118 (2005)
8. Elzinga, C. H.: Sequence similarity: a nonaligning technique. Socio. Meth. Res. **32**, 3-22 (2003)
9. Elzinga, C. H.: Sequence analysis: Metric representations of categorical time series. Socio. Meth. Res. (2006)

Fuzzy Clustering by Hyperbolic Smoothing

David Masís, Esteban Segura, Javier Trejos, and Adilson Xavier

Abstract We propose a novel method for building fuzzy clusters of large data sets, using a smoothing numerical approach. The usual sum-of-squares criterion is relaxed so the search for good fuzzy partitions is made on a continuous space, rather than a combinatorial space as in classical methods [8]. The smoothing allows a conversion from a strongly non-differentiable problem into differentiable subproblems of optimization without constraints of low dimension, by using a differentiable function of infinite class. For the implementation of the algorithm, we used the statistical software R and the results obtained were compared to the traditional fuzzy C–means method, proposed by Bezdek [1].

Keywords: clustering, fuzzy sets, numerical smoothing

1 Introduction

Methods for making groups from data sets are usually based on the idea of disjoint sets, such as the classical crisp clustering. The most well known are hierarchical and k-means [8], whose resulting clusters are sets with no intersection. However, this restriction may not be natural for some applications, where the condition for

David Masís
Costa Rica Institute of Technology, Cartago, Costa Rica, e-mail: dmasis@itcr.ac.cr

Esteban Segura
CIMPA & School of Mathematics, University of Costa Rica, San José, Costa Rica,
e-mail: esteban.seguraugalde@ucr.ac.cr

Javier Trejos (✉)
CIMPA & School of Mathematics, University of Costa Rica, San José, Costa Rica,
e-mail: javier.trejos@ucr.ac.cr

Adilson E. Xavier
Universidade Federal de Rio de Janeiro, Brazil, e-mail: adilson.xavier@gmail.com

© The Author(s) 2023
P. Brito et al. (eds.), *Classification and Data Science in the Digital Age*,
Studies in Classification, Data Analysis, and Knowledge Organization,
https://doi.org/10.1007/978-3-031-09034-9_27

some objects may be to belong to two or more clusters, rather than only one. Several methods for constructing overlapping clusters have been proposed in the literature [4, 5, 8]. Since Zadeh introduced the concept of fuzzy sets [17], the principle of belonging to several clusters has been used in the sense of a degree of membership to such clusters. In this direction, Bezdek [1] introduced a fuzzy clustering method that became very popular since it solved the problem of representation of clusters with centroids and the assignment of objects to clusters, by the minimization of a well-stated numerical criterion. Several methods for fuzzy clustering have been proposed in the literature; a survey of these methods can be found in [16].

In this paper we propose a new fuzzy clustering method based on the numerical principle of hyperbolic smoothing [15]. Fuzzy C-Means method is presented in Section 2 and our proposed Hyperbolic Smoothing Fuzzy Clustering method in Section 3. Comparative results between these two methods are presented in Section 4. Finally, Section 5 is devoted to the concluding remarks.

2 Fuzzy Clustering

The most well known method for fuzzy clustering is the original Bezdek's C-means method [1] and it is based on the same principles of k-means or dynamical clusters [2], that is, iterations on two main steps: i) class representations by the optimization of a numerical criterion, and ii) assignment to the closest class representative in order to construct clusters; these iterations are made until a convergence is reached to a local minimum of the overall quality criterion.

Let us introduce the notation that will be used and the numerical criterion for optimization. Let $\mathbf{X}$ be an $n \times p$ data matrix containing p numerical observations over n objects. We look for a $K \times p$ matrix $\mathbf{G}$ that represents centroids of K clusters of the n objects and an $n \times K$ membership matrix with elements $\mu_{ik} \in [0, 1]$, such that the following criterion is minimized:

$$W(\mathbf{X}, \mathbf{U}, C) = \sum_{i=1}^{n} \sum_{k=1}^{K} (\mu_{ik})^m \|\mathbf{x}_i - \mathbf{g}_k\|^2$$

$$\text{subject to } \sum_{k=1}^{K} \mu_{ik} = 1, \text{ for all } i \in \{1, 2, \ldots, n\}$$
$$0 < \sum_{i=1}^{n} \mu_{ik} < n, \text{ for all } k \in \{1, 2, \ldots, K\},$$

(1)

where $\mathbf{x}_i$ is the i-th row of $\mathbf{X}$ and $\mathbf{g}_k$ is the k-th row of $\mathbf{G}$, representing in $\mathbb{R}^p$ the centroid of the k-th cluster.

The parameter $m \neq 1$ in (1) controls the fuzzyness of the clusters. According to the literature [16], it is usual to take $m = 2$, since greater values of m tend to give very low values of μ_{ik}, tending to the usual crisp partitions such as in k-means. We also assume that the number of clusters, K, is fixed.

Minimization of (1) represents a non linear optimization problem with constraints, which can be solved using Lagrange multipliers as presented in [1]. The solution, for each row of the centroids matrix, given a matrix $\mathbf{U}$, is:

$$\mathbf{g}_k = \sum_{i=1}^{n} (\mu_{ik})^m \mathbf{x}_i \left/ \sum_{i=1}^{n} (\mu_{ik})^m \right. . \tag{2}$$

The solution for the membership matrix, given a matrix centroids $\mathbf{G}$, is [1]:

$$\mu_{ik} = \left[\sum_{j=1}^{K} \left(\frac{||\mathbf{x}_i - \mathbf{g}_k||^2}{||\mathbf{x}_i - \mathbf{g}_j||^2} \right)^{1/(m-1)} \right]^{-1} . \tag{3}$$

The following pseudo-code shows the mains steps of Bezdek's Fuzzy C-Means method [1].

Bezdek's Fuzzy c-Means (FCM) Algorithm

1. Initialize fuzzy membership matrix $\mathbf{U} = [\mu_{ik}]_{n \times K}$
2. Compute centroids for fuzzy clusters according to (2)
3. Update membership matrix $\mathbf{U}$ according to (3)
4. If improvement in the criterion is less than a threshold, then stop; otherwise go to Step 2.

Fuzzy C-Means method starts from an initial partition that is improved in each iteration, according to (1), applying Steps 2 and 3 of the algorithm. It is clear that this procedure may lead to local optima of (1) since iterative improvement in (2) and (3) is made by a local search strategy.

3 Algorithm for Hyperbolic Smoothing Fuzzy Clustering

For the clustering problem of the n rows of data matrix $\mathbf{X}$ in K clusters, we can seek for the minimum distance between every $\mathbf{x}_i$ and its class center $\mathbf{g}_k$:

$$z_i^2 = \min_{\mathbf{g}_k \in \mathbf{G}} ||\mathbf{x}_i - \mathbf{g}_k||_2^2$$

where $|| \cdot ||_2$ is the Euclidean norm. The minimization can be stated as a sum-of-squares:

$$\min \sum_{i=1}^{n} \min_{\mathbf{g}_k \in \mathbf{G}} ||\mathbf{x}_i - \mathbf{g}_k||_2^2 = \min \sum_{i=1}^{n} z_i^2$$

leading to the following constrained problem:

$$\min \sum_{i=1}^{n} z_i^2 \text{ subject to } z_i = \min_{\mathbf{g}_k \in \mathbf{G}} ||\mathbf{x}_i - \mathbf{g}_k||_2, \text{ with } i = 1, \ldots, n.$$

This is equivalent to the following minimization problem:

$$\min \sum_{i=1}^{n} z_i^2 \text{ subject to } z_i - \|\mathbf{x}_i - \mathbf{g}_k\|_2 \leq 0, \text{ with } i = 1, \ldots, n \text{ and } k = 1, \ldots, K.$$

Considering the function: $\varphi(y) = \max(0, y)$, we obtain the problem:

$$\min \sum_{i=1}^{n} z_i^2 \text{ subject to } \sum_{k=1}^{K} \varphi(z_i - \|\mathbf{x}_i - \mathbf{g}_k\|_2) = 0 \text{ for } i = 1, \ldots, n.$$

That problem can be re-stated as the following one:

$$\min \sum_{i=1}^{n} z_i^2 \text{ subject to } \sum_{k=1}^{K} \varphi(z_i - \|\mathbf{x}_i - \mathbf{g}_k\|_2) > 0, \text{ for } i = 1, \ldots, n.$$

Given a perturbation $\epsilon > 0$ it leads to the problem:

$$\min \sum_{i=1}^{n} z_i^2 \text{ subject to } \sum_{k=1}^{K} \varphi(z_i - \|\mathbf{x}_i - \mathbf{g}_k\|_2) \geq \epsilon \text{ for } i = 1, \ldots, n.$$

It should be noted that function φ is not differentiable. Therefore, we will make a smoothing procedure in order to formulate a differentiable function and proceed with a minimization by a numerical method. For that, consider the function: $\psi(y, \tau) = \frac{y + \sqrt{y^2 + \tau^2}}{2}$, for all $y \in \mathbb{R}$, $\tau > 0$, and the function: $\theta(\mathbf{x}_i, \mathbf{g}_k, \gamma) = \sqrt{\sum_{j=1}^{P} (x_{ij} - g_{kj})^2 + \gamma^2}$, for $\gamma > 0$. Hence, the minimization problem is transformed into:

$$\min \sum_{i=1}^{n} z_i^2 \text{ subject to } \sum_{k=1}^{K} \psi(z_i - \theta(\mathbf{x}_i, \mathbf{g}_k, \gamma), \tau) \geq \epsilon, \text{ for } i = 1, \ldots, n.$$

Finally, according to the Karush–Kuhn–Tucker conditions [10, 11], all the constraints are active and the final formulation of the problem is:

$$\min \sum_{i=1}^{n} z_i^2$$

$$\text{subject to } h_i(z_i, \mathbf{G}) = \sum_{k=1}^{K} \psi(z_i - \theta(\mathbf{x}_i, \mathbf{g}_k, \gamma), \tau) - \epsilon = 0, \text{ for } i = 1, \ldots, n, \tag{4}$$

$$\epsilon, \tau, \gamma > 0.$$

Considering (4), in [15] it was stated the Hyperbolic Smoothing Clustering Method presented in the following algorithm.

Hyperbolic Smoothing Clustering Method (HSCM) Algorithm

1. Initialize cluster membership matrix $\mathbf{U} = [\mu_{ik}]_{n \times K}$
2. Choose initial values: $\mathbf{G}^0, \gamma^1, \tau^1, \epsilon^1$
3. Choose values: $0 < \rho_1 < 1, 0 < \rho_2 < 1, 0 < \rho_3 < 1$
4. Let $l = 1$
5. Repeat steps 6 and 7 until a stop condition is reached:
6. Solve problem (P): $\min f(\mathbf{G}) = \sum_{i=1}^{n} z_i^2$ with $\gamma = \gamma^l$, $\tau = \tau^l$ and $\epsilon = \epsilon^l$, $\mathbf{G}^{l-1}$

   being the initial value and $\mathbf{G}^l$ the obtained solution
7. Let $\gamma^{l+1} = \rho_1 \gamma^l$, $\tau^{l+1} = \rho_2 \tau^l$, $\epsilon^{l+1} = \rho_3 \epsilon^l$ and $l = l + 1$.

The most relevant task in the hyperbolic smoothing clustering method is finding the zeroes of the function $h_i(z_i, \mathbf{G}) = \sum_{k=1}^{K} \psi(z_i - \theta(\mathbf{x}_i, \mathbf{g}_k, \gamma), \tau) - \epsilon = 0$ for for $i = 1, \ldots, n$. In this paper, we used the Newton-Raphson method for finding these zeroes [3], particularly the BFGS procedure [12]. Convergence of the Newton-Raphson method was successful, mainly, thank to a good choice of initial solutions. In our implementation, these initial approximations were generated by calculating the minimum distance between the i-th object and the k-th centroid for a given partition. Once the zeroes z_i of the functions h_i are obtained, it is implemented the hyperbolic smoothing. The final solution for this method consists on solving a finite number of optimization subproblems corresponding to problem (P) in Step 6 of the HSCM algorithm. Each one of these subproblems was solved with the R routine *optim* [13], a useful tool for solving optimization problems in non linear programming. As far as we know there is no closed solution for solving this step. For the future, we can consider writing a program by our means, but for this paper we are using this R routine.

Since we have that: $\sum_{k=1}^{K} \psi(z_i - \theta(\mathbf{x}_i, \mathbf{g}_k, \gamma), \tau) = \epsilon$, then each entry μ_{ik} of the membership matrix is given by: $\mu_{ik} = \frac{\psi(z_i - d_k, \tau)}{\epsilon}$. It is worth to note that fuzzyness is controlled by parameter ϵ.

The following algorithm contains the main steps of the Hyperbolic Smoothing Fuzzy Clustering (HSFC) method.

Hyperbolic Smoothing Fuzzy Clustering (HSFC) Algorithm

1. Set $\epsilon > 0$
2. Choose initial values for: $\mathbf{G}^0$ (centroids matrix), γ^1, τ^1 and N (maximum number of iterations)
3. Choose values: $0 < \rho_1 < 1$, $0 < \rho_2 < 1$
4. Set $l = 1$
5. While $l \leq N$:
6. Solve the problem (P): Minimize $f(\mathbf{G}) = \sum_{i=1}^{n} z_i^2$ with $\gamma = \gamma^{(l)}$ and $\tau = \tau^{(l)}$, with an initial point $\mathbf{G}^{(l-1)}$ and $\mathbf{G}^{(l)}$ being the obtained solution
7. Set $\gamma^{(l+1)} = \rho_1 \gamma^{(l)}, \tau^{(l+1)} = \rho_2 \tau^{(l)}$, y $l = l + 1$
8. Set $\mu_{ik} = \psi(z_i - \theta(\mathbf{x}_i, \mathbf{g}_k, \gamma), \tau)/\epsilon$ for $i = 1, \ldots, n$ and $k = 1, \ldots, K$.

4 Comparative Results

Performance of the HSFC method was studied on a data table well known from the literature, the Fisher's iris [7] and 16 simulated data tables built from a semi-Monte Carlo procedure [14].

For comparing FCM and HSFC, we used the implementation of FCM in R package *fclust* [6]. This comparison was made upon the within class sum-of-squares: $W(P) = \sum_{k=1}^{K} \sum_{i=1}^{n} \mu_{ik} \|\mathbf{x}_i - \mathbf{g}_k\|^2$. Both methods were applied 50 times and the best value of W is reported. For simplicity here, for HSFC we used the following parameters: $\rho_1 = \rho_2 = \rho_3 = 0.25$, $\epsilon = 0.01$ and $\gamma = \tau = 0.001$ as initial values. In Table 1 the results for Fisher's iris are shown, in which case HSFC performs slightly better. It contains the Adjusted Rand Index (ARI) [9] between HSFC and the best FCM result among 100 runs; ARI compares fuzzy membership matrices crisped into hard partitions.

Table 1 Minimum sum-of-squares (SS) reported for the Fisher's iris data table with HSFC and FCM, K being the number of clusters, ARI comparing both methods. In bold best method.

Table	K	SS for HSFC	SS for FCM	ARI
	2	**152.348**	152.3615	1
Fisher's iris	3	**78.85567**	78.86733	0.994
	4	57.26934	57.26934	0.980

Simulated data tables were generated in a controlled experiment as in [14], with random numbers following a Gaussian distribution. Factors of the experiment were:

- The number of objects (with 2 levels, $n = 105$ and $n = 525$).
- The number of clusters (with levels $K = 3$ and $K = 7$).
- Cardinality (card) of clusters, with levels i) all with the same number of objects (coded as card(=)), and ii) one large cluster with 50% of objects and the rest with the same number (coded as card(≠)).
- Standard deviation of clusters, with levels i) all Gaussian random variables with standard deviation (SD) equal to one (coded as SD(=)), and ii) one cluster with SD=3 and the rest with SD=1 (coded as SD(≠)).

Table 2 contains codes for simulated data tables according to the codes we used.

Table 3 contains the minimum values of the sum-of-squares obtained for our HSFC and Bezdek's FCM methods; the best solution of 100 random applications for FCM in presented and one run of HSFC. It also contains the ARI values for comparing HSFC solution with that best solution of FCM. It can be seen that, generally, HSFC method tends to obtain better results than FCM, with only few exceptions. In 23 cases HSFC obtains better results, FCM is better in 5 cases, and results are in same in 17 cases. However, ARI shows that partitions tend to be very similar with both methods.

Table 2 Codes and characteristics of simulated data tables; n: number of objects, K: number of clusters, card: cardinality, DS: standard deviation.

Table	Characteristcs	Table	Characteristcs
T1	$n = 525$, $K = 3$, card(=), SD(=)	T9	$n = 525$, $K = 3$, card($\neq$), DS(=)
T2	$n = 525$, $K = 7$, card(=), SD(=)	T10	$n = 525$, $K = 7$, card($\neq$), DS(=)
T3	$n = 105$, $K = 3$, card(=), SD(=)	T11	$n = 105$, $K = 3$, card($\neq$), DS(=)
T4	$n = 105$, $K = 7$, card(=), SD(=)	T12	$n = 105$, $K = 7$, card($\neq$), DS(=)
T5	$n = 525$, $K = 3$, card(=), SD($\neq$)	T13	$n = 525$, $K = 3$, card($\neq$), DS($\neq$)
T6	$n = 525$, $K = 7$, card(=), SD($\neq$)	T14	$n = 525$, $K = 7$, card($\neq$), DS($\neq$)
T7	$n = 105$, $K = 3$, card(=), SD($\neq$)	T15	$n = 105$, $K = 3$, card($\neq$), DS($\neq$)
T8	$n = 105$, $K = 7$, card(=), SD($\neq$)	T16	$n = 105$, $K = 7$, card($\neq$), DS($\neq$)

Table 3 Minimum sum-of-squares (SS) reported for HSFC and FCM methods on the simulated data tables. Best method in bold.

Table	K	SS for HSFC	SS for FCM	ARI	Table	K	SS for HSFC	SS for FCM	ARI
	2	**7073.402**	7073.814	0.780		2	12524.31	12524.31	0.900
T1	3	3146.119	3146.119	1	T9	3	**9269.361**	9269.611	1
	4	2983.651	2983.651	1		4	6298.47	**6298.368**	1
	2	**16987.19**	16987.71	0.764		2	**5466.893**	5466.912	0.890
T2	3	11653.22	11653.22	1	T10	3	2977.58	2977.58	1
	4	**7776.855**	7777.396	1		4	**2745.721**	2746.671	1
	2	**3923.051**	3923.062	0.763		2	**2969.247**	2969.32	0.860
T3	3	2917.13	2917.13	0.754	T11	3	1912.323	1912.323	1
	4	2287.523	**2256.298**	0.993		4	1401.394	1401.394	1
	2	**1720.365**	1720.374	0.992		2	1816.056	1816.056	1
T4	3	569.3112	569.3112	1	T12	3	525.7118	525.7118	1
	4	535.5491	**535.3541**	1		4	**477.0593**	477.2696	1
	2	15595.67	15595.67	0.910		2	12804.03	12805.05	0.920
T5	3	**11724.93**	11725.28	1	T13	3	**8816.805**	8817.702	1
	4	8409.738	8409.738	0.984		4	6293.774	6293.951	1
	2	11877.96	11877.96	0.970		2	**16228.07**	16228.98	0.920
T6	3	**8299.779**	8300.718	1	T14	3	**7255.113**	7255.423	1
	4	**7212.611**	7213.725	1		4	6427.313	6427.313	1
	2	**4336.261**	4336.507	0.955		2	2616.286	2616.943	1
T7	3	3041.076	3041.076	1	T15	3	**1978.017**	1978.233	1
	4	**2395.683**	2421.333	1		4	**1526.895**	1526.953	1
	2	1767.43	1767.43	1		2	2226.923	**2226.212**	0.962
T8	3	**1380.766**	1381.019	1	T16	3	**1232.074**	1232.124	1
	4	1215.302	**1211.235**	1		4	**982.7074**	982.9721	1

5 Concluding Remarks

In hyperbolic smoothing, parameters τ, γ and ϵ tend to zero, so the constraints in the subproblems make that problem (P) tends to solve (1). Parameter ϵ controls the fuzzyness degree in clustering; the higher it is, the solution becomes more and more fuzzy; the less it is, the clustering is more and more crisp. In order to compare results and efficiency of the HSFC method, zeroes of functions h_i can be obtained with any method for solving equations in one variable or a predefined routine. According to the results we obtained so far and the implementation of the hyperbolic smoothing for fuzzy clustering, we can conclude that, generally, the HSFC method has a slightly better performance than original Bezdek's FCM on small real and simulated data tables. Further research is required for testing performance of HSFC method on very large data sets, with measures of efficiency, quality of solutions and running time. We are also considering to study further comparisons between HSFC and FCM with different indices, and writing the program for solving Step 6 in HSFC algorithm, that is the minimization of $f(G)$, by our means, instead of using the *optim* routine in R.

References

1. Bezdek, J.C.: Pattern Recognition with Fuzzy Objective Function Algorithms. Plenum Press, New York (1981)
2. H.-H. Bock: Origins and extensions of the k-means algorithm in cluster analysis. Electronic Journal for History of Probability and Statistics **4** (2008)
3. Burden, R., Faires, D.: Numerical analysis, 9th ed. Brooks/Cole, Pacific Grove (2011)
4. Diday, E.: Orders and overlapping clusters by pyramids. In J.De Leeuw et al. (eds.) Multidimensional Data Analysis, DSWO Press, Leiden (1986)
5. Dunn, J. C.: A fuzzy relative of the ISODATA process and its use in detecting compact, well separated clusters. J. Cybernetics **3**, 32–57 (1974)
6. Ferraro, M. B., Giordani, P., Serafini, A.: fclust: An R Package for Fuzzy Clustering. The R Journal **11**(1), 198-210 (2019) doi: 10.32614/RJ-2019-017
7. Fisher, R. A.: The use of multiple measurements in taxonomic problems. Annals of Eugenics **7**: 179-188 (1936)
8. Hartigan, J. A.: Clustering Algorithms. Wiley, New York, NY (1975)
9. Hubert, L., Arabie, P.: Comparing partitions. J. Classif. **2**(1), 193-218 (1985)
10. Karush, W.: Minima of Functions of Several Variables with Inequalities as Side Constraints. Master's Thesis, Dept. of Mathematics, University of Chicago, Chicago, Illinois (1939)
11. Kuhn, H., Tucker, A.: Nonlinear programming, Proc. 2nd Berkeley Symposium on Mathematical Statistics and Probability, University of California Press, Berkeley, pp. 481-492 (1951)
12. Li, D., Fukushima, M.: On the global convergence of the BFGS method for nonconvex unconstrained optimization problems. SIAM J. Optim. **11**, 1054-1064 (2001)
13. R Core Team: R: A language and environment for statistical computing. R Foundation for Statistical Computing, Vienna, Austria (2021)
14. Trejos, J., Villalobos, M. A.: Partitioning by particle swarm optimization. In: Brito, P. Bertrand, P., Cucumel G., de Carvalho, F. (eds.) Selected Contributions in Data Analysis and Classification, pp. 235-244. Springer, Berlin (2007)
15. Xavier, A.: The hyperbolic smoothing clustering method, Pattern Recognit. **43**, 731-737 (2010)
16. Yang, M. S.: A survey of fuzzy clustering. Math. Comput. Modelling **18**, 1-16 (1993)
17. Zadeh, L. A.: Fuzzy sets. Information and Control **8**(3), 338-353 (1965)

Stochastic Collapsed Variational Inference for Structured Gaussian Process Regression Networks

Rui Meng, Herbert K. H. Lee, and Kristofer Bouchard

Abstract This paper presents an efficient variational inference framework for a family of structured Gaussian process regression network (SGPRN) models. We incorporate auxiliary inducing variables in latent functions and jointly treat both the distributions of the inducing variables and hyper-parameters as variational parameters. Then we take advantage of the collapsed representation of the model and propose structured variational distributions, which enables the decomposability of a tractable variational lower bound and leads to stochastic optimization. Our inference approach is able to model data in which outputs do not share a common input set, and with a computational complexity independent of the size of the inputs and outputs to easily handle datasets with missing values. Finally, we illustrate our approach on both synthetic and real data.

Keywords: stochastic optimization, Gaussian process, variational inference, multivariate time series, time-varying correlation

1 Introduction

Multi-output regression problems arise in various fields. Often, the processes that generate such datasets are nonstationary. Modern instrumentation has resulted in increasing numbers of observations, as well as the occurrence of missing values. This motivates the development of scalable methods for forecasting in such datasets.

Multi-ouput Gaussian process models or multivariate Gaussian process models (MGP) generalise the powerful Gaussian process predictive model to vector-valued

Rui Meng (✉) · Kristofer Bouchard
Biological Systems and Engineering Division, Lawrence Berkeley National Laboratory, USA,
e-mail: rmeng@lbl.gov;kebouchard@lbl.gov

Herbert K. H. Lee
University of California, Santa Cruz, USA, e-mail: herbie@ucsc.edu

© The Author(s) 2023
P. Brito et al. (eds.), *Classification and Data Science in the Digital Age*,
Studies in Classification, Data Analysis, and Knowledge Organization,
https://doi.org/10.1007/978-3-031-09034-9_28

random fields [1]. Those models demonstrate improved prediction performance compared with independent univariate Gaussian processes (GP) because MGPs express correlations between outputs. Since the correlation information of data is encoded in the covariance function, modeling the flexible and computationally efficient cross-covariance function is of interest. In the literature of multivariate processes, many approaches are proposed to build valid cross-covariance functions including the linear model of coregionalization (LMC) [2], kernel convolution techniques [3], B-spline based coherence functions [4]. However, most of these models are designed for modelling low-dimensional stationary processes, and require Monte Carlo simulations, making inference in large datasets computationally intractable.

Modelling the complicated temporal dependencies across variables is addressed in [5, 6] by several adaptions of stochastic LMC. Such models can handle input-varying correlation across multivariate outputs. Especially for multivariate time series, [6] propose a SGPRN that captures time-varying scale, correlation, and smoothness. However, the inference in [6] is difficult to handle in applications where either the number of observations and dimension size are large or where missing data exist.

Here, we propose an efficient variational inference approach for the SGPRN by employing the inducing variable framework on all latent processes [7], taking advantage of its collapsed representation where nuisance parameters are marginalized out [8] and proposing a tractable variational bound amenable to doubly stochastic variational inference. We call our approach variational SGPRN (VSGPRN). This variational framework allows the model to handle missing data without increasing the computational complexity of inference. We numerically provide evidence of the benefits of simultaneously modeling time-varying correlation, scale and smoothness in both a synthetic experiment and a real-world problem.

The main contributions of this work are threefold:

- Learning structured Gaussian process regression networks using inducing variables on both mixing coefficients and latent functions.
- Employing doubly stochastic variational inference for structured Gaussian process regression networks by taking advantage of its collapsed representation and constructing a tractable lower bound of the loglikelihood, making it suitable for mini-batching learning.
- Demonstrating that our proposed algorithm succeeds in handling time-varying correlation on missing data under different scenarios in both synthetic data and real data.

2 Model

Assume $y(x) \in \mathbb{R}^D$ is a vector-valued function of $x \in \mathbb{R}^P$, where D is the dimension size of the outputs and P is the dimension size of the inputs. SGPRN assumes that noisy observations $y(x)$ are the linear combination of latent variables $g(x) \in \mathbb{R}^D$, corrupted by Gaussian noise $\epsilon(x)$. The coefficients $\mathbb{L}(x) \in \mathbb{R}^{D \times D}$ of the latent functions are assumed to be a stochastic lower triangular matrix with

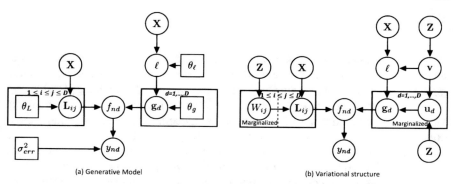

(a) Generative Model   (b) Variational structure

Fig. 1 Graphical model of VSGPRN. Left: Illustration of the generative model. Right: Illustration of the variational structure. The dashed (red) block means that we marginalize out those latent variables in the variational inference framework.

positive values on the diagonal for model identification [9, 6]. Thus, SGPRN is defined in the generative model of Figure 1 and it is $y(x) = f(x) + \epsilon(x)$, $f(x) = \mathbb{L}(x)g(x)$ with independent white noise $\epsilon(x) \overset{iid}{\sim} N(0, \sigma_{err}^2 I)$. We note that each latent function g_d in g is independently sampled from a GP with a non-stationary kernel K^g and the stochastic coefficients are modeled via a structured GP based prior as proposed in [9] with a stationary kernel K^l such that

$$g_d \overset{iid}{\sim} GP(0, K^g), d = 1, \ldots, D, \text{ and } l_{ij} \sim \begin{cases} GP(0, K^l), & i > j, \\ logGP(0, K^l), & i = j, \end{cases} \text{ where logGP}$$

denotes the log Gaussian process [10]. K^g is modelled as a Gibbs correlation function $K^g(x, x') = \sqrt{\frac{2\ell(x)\ell'(x)}{\ell(x)^2 + \ell(x')^2}} \exp\left(-\frac{\|x - x'\|^2}{\ell(x)^2 + \ell(x')^2}\right)$, $\ell \sim logGP(0, K^\ell)$, where ℓ determines the input-dependent length scale of the shared correlations in K^g for all latent functions g_d. The varying length-scale process ℓ plays an important role in modelling nonstationary time series as illustrated in [11, 6].

Let $\mathbb{X} = \{x_i\}_{i=1}^N$ be the set of observed inputs and $\mathbb{Y} = \{y_i\}_{i=1}^N$ be the set of observed outputs. Denote η as the concatenation of all coefficients and all log length-scale parameters, i.e., $\eta = (\mathbb{l}, \tilde{\ell})$ evaluated at training inputs $\mathbb{X}$. Here, $\mathbb{l}$ is a vector including the entries below the main diagonal and the entries on the diagonal in the log scale and $\tilde{\ell} = \log \ell$ is the length-scale parameters in log scale. Also, denote $\theta = (\theta_l, \theta_\ell, \sigma_{err}^2)$ as all hyper-parameters, where θ_l and θ_ℓ are the hyper-parameters in kernel K_l and K_ℓ. We note that directly inferring the posterior of the latent variables $p(\eta|\mathbb{Y}, \theta) \propto p(\mathbb{Y}|\eta, \sigma_{err}^2)p(\eta|\theta_l, \theta_\ell)$ is computationally intractable in general because the computational complexity of $p(\eta|\mathbb{Y}, \theta)$ is $O(N^3 D^3)$. To overcome this issue, we propose an efficient variational inference to significantly reduce the computational burden in the next section.

3 Inference

We introduce a shared set of inducing inputs $\mathbb{Z} = \{\mathbb{z}_m\}_{m=1}^M$ that lie in the same space as the inputs $\mathbb{X}$ and a set of shared inducing variables $\mathbb{w}_d$ for each latent function g_d evaluated at the inducing inputs $\mathbb{Z}$. Likewise, we consider inducing variables $\mathbb{u}_{ii}$ for the function $\log L_{ii}$ when $i = j$, $\mathbb{u}_{ij}$ for function L_{ij} when $i > j$, and inducing variables $\mathbb{v}$ for function $\log \ell(\mathbb{x})$ evaluated at inducing inputs $\mathbb{Z}$. We denote those collective variables as $\mathbb{l} = \{\mathbb{l}_{ij}\}_{i \geq j}$, $\mathbb{u} = \{\mathbb{u}_{ij}\}_{i \geq j}$, $\mathbb{g} = \{\mathbb{g}_d\}_{d=1}^D$, $\mathbb{w} = \{\mathbb{w}_d\}_{d=1}^D$, ℓ and $\mathbb{v}$. Then we redefine the model parameters $\eta = (\mathbb{l}, \mathbb{u}, \mathbb{g}, \mathbb{w}, \ell, \mathbb{v})$, and the prior of those model parameters is $p(\eta) = p(\mathbb{l}|\mathbb{w})p(\mathbb{w})p(\mathbb{g}|\mathbb{u}, \ell, \mathbb{v})p(\mathbb{u})p(\ell|\mathbb{v})p(\mathbb{v})$.

The core assumption of inducing point-based sparse inference is that the inducing variables are sufficient statistics for the training and testing data in the sense that the training and testing data are conditionally independent given the inducing variables. In the context of our model, this means that the posterior processes of L, g and ℓ are sufficiently determined by the posterior distribution of $\mathbb{u}$, $\mathbb{w}$ and $\mathbb{v}$. We propose a structured variational distribution and its corresponding variational lower bound. Due to the nonconjugacy of this model, instead of doing expectation in the evidence lower bound (ELBO), as is normally done in the literature, we perform the marginalization on inducing variables $\mathbb{u}$, $\mathbb{w}$ and $\mathbb{g}$, and then use the reparameterization trick to apply end-to-end training with stochastic gradient descent. We will also discuss a procedure for missing data inference and prediction.

To capture the posterior dependency between the latent functions, we propose a structured variational distribution of the model parameters η used to approximate its posterior distribution as $q(\eta) = p(\mathbb{l}|\mathbb{u})p(\mathbb{g}|\mathbb{w}, \ell, \mathbb{v})p(\ell|\mathbb{v})q(\mathbb{u}, \mathbb{w}, \mathbb{v})$. This variational structure is illustrated in Figure 1. The variational distribution of the inducing variables $q(\mathbb{u}, \mathbb{w}, \mathbb{v})$ fully characterizes the distribution of $q(\eta)$. Thus, the inference of $q(\mathbb{u}, \mathbb{w}, \mathbb{v})$ is of interest. We assume the parameters $\mathbb{u}$, $\mathbb{w}$, and $\mathbb{v}$ are Gaussian and mutually independent.

Given the definition of Gaussian process priors for the SGPRN, the conditional distributions $p(\mathbb{l}|\mathbb{u})$, $p(\mathbb{g}|\mathbb{w}, \tilde{\ell}, \mathbb{v})$, and $p(\ell|\mathbb{v})$ have closed-form expressions and all are Gaussian, except for $p(\ell|\mathbb{v})$, which is log Gaussian. The ELBO of the log likelihood of observations under our structured variational distribution $q(\eta)$ is derived using Jensen's inequality as:

$$\log p(\mathbb{Y}) \geq E_{q(\eta)} \left[\log \left(\frac{p(\mathbb{Y}|\mathbb{g}, \mathbb{l})p(\mathbb{u})p(\mathbb{w})p(\mathbb{v})}{q(\mathbb{u}, \mathbb{w}, \mathbb{v})} \right) \right] = R + A, \qquad (1)$$

where $R = \sum_{n=1}^N \sum_{d=1}^D E_{q(g_n, \mathbb{l}_n)} \log(p(y_{nd}|g_n, \mathbb{l}_n))$ is the reconstruction term and $A = \mathrm{KL}(q(\mathbb{u})||p(\mathbb{u})) + \mathrm{KL}(q(\mathbb{w})||p(\mathbb{w})) + \mathrm{KL}(q(\mathbb{v})||p(\mathbb{v}))$ is the regularization term. $\mathbb{g}_n = \{g_{dn} = (g_d)_n\}_{d=1}^D$ and $\mathbb{l}_n = \{l_{ijn} = (\mathbb{l}_{ij})_n\}_{i \geq j}$ are latent variables.

The structured decomposition trick for $q(\eta)$ has also been used by [12] to derive variational inference for the multivariate output case. The benefit of this structure is that all conditional distributions in $q(\eta)$ can be cancelled in the derivation of the lower bound in (1), which alleviates the computational burden of inference. Because of the conditional independence of the reconstruction term in (1) given $\mathbb{g}$ and $\mathbb{l}$, the lower bound decomposes across both inputs and outputs and this enables the use of stochastic optimization methods. Moreover, due to the Gaussian assumption in the prior and variational distributions of the inducing variables, all KL divergence terms in the regularization term A are analytically tractable. Next, instead of directly computing expectation, we leverage stochastic inference [13].

Stochastic inference requires sampling of $\mathbb{l}$ and $\mathbb{g}$ from the joint variational posterior $q(\eta)$. Directly sampling them would introduce much uncertainty from intermediate variables and thus make inference inefficient. To tackle this issue, we marginalize unnecessary intermediate variables $\mathbb{u}$ and $\mathbb{w}$ and obtain the marginal distributions $q(\mathbb{l}) = \prod_{i=j} \log \mathcal{N}(\mathbb{l}_{ii}|\tilde{\mu}_{ii}^l, \tilde{\Sigma}_{ii}^l) \prod_{i>j} \mathcal{N}(\mathbb{l}_{ij}|\tilde{\mu}_{ij}^l, \tilde{\Sigma}_{ij}^l)$ and $q(\mathbb{g}|\ell, \mathbb{v}) = \prod_{d=1}^{D} \mathcal{N}(\mathbb{g}_d|\tilde{\mu}_d^g, \tilde{\Sigma}_d^g)$ with a joint distribution $q(\ell, \mathbb{v}) = p(\ell|\mathbb{v})q(\mathbb{v})$, where the conditional mean and covariance matrix are easily derived. The corresponding marginal distributions $q(\mathbb{l}_n)$ and $q(\mathbb{g}_n|\ell, \mathbb{v})$ at each n are also easy to derive. Moreover, we conduct collapsed inference by marginalizing the latent variables $\mathbb{g}_n$, so then the individual expectation is

$$\mathbb{E}_{q(\mathbb{g}_n,\mathbb{l}_n)} \log(p(y_{nd}|\mathbb{g}_n,\mathbb{l}_n)) = \int (L_{nd})q(\ell_n,\mathbb{v})q(\mathbb{l}_{d\cdot n})d(\mathbb{l}_{d\cdot n},\ell_n,\mathbb{v})), \quad (2)$$

where $L_{nd} = \log \mathcal{N}(y_{nd}|\sum_{j=1}^{D} l_{djn}\tilde{\mu}_{jn}^g, \sigma_{err}^2) - \frac{1}{2\sigma_{err}^2}\sum_{j=1}^{D} l_{djn}^2\tilde{\sigma}_{jn}^{g2}$ measure the reconstruction performance for observations y_{nd}.

Directly evaluating the ELBO is still challenging due to the non-linearities introduced by our structured prior. Recent progress in black box variational inference [13] avoids this difficulty by computing noisy unbiased estimates of the gradient of ELBO, via approximating the expectations with unbiased Monte Carlo estimates and relying on either score function estimators [14] or reparameterization gradients [13] to differentiate through a sampling process. Here we leverage the reparameterization gradients for stochastic optimization for model parameters. We note that evaluating ELBO (1) involves two sources of stochasticity from Monte Carlo sampling in (2) and from data sub-sampling stochasticity [15]. The prediction procedure is based on Bayes' rule and replaces the posterior distribution by the inferred variational distribution. In the case of missing data, the only modification in (1) is in the reconstruction term, where we sum up the likelihoods of observed data instead of complete data.

4 Experiments

This section illustrates the performance of our model on multivariate time series. We first show that our approach can model the time-varying correlation and smoothness of outputs on 2D synthetic datasets in three scenarios with respect to different types of frequencies but the same missing data mechanism. Then, we compare the imputation performance on missing data with other inducing-variable based sparse multivariate Gaussian process models on a real dataset.

We conduct experiments on three synthetic time series with low frequency (LF), high frequency (HF) and varying frequency (VF) respectively. They are generated from the system of equations $y_1(t) = 5\cos(2\pi wt^s) + \epsilon_1(t)$, $y_2(t) = 5(1-t)\cos(2\pi wt^s) - 5t\cos(2\pi wt^s) + \epsilon_2(t)$, where $\{\epsilon_i(t)\}_{i=1}^2$ are independent standard white noise processes. The value of w refers to the frequency and the value of s characterizes the smoothness. The LF and HF datasets use the same $s = 1$, implying the smoothness is invariant across time. But they employ different frequencies, $w = 2$ for LF and $w = 5$ for HF (i.e., two periods and five periods in a unit time interval respectively). The VF dataset takes $s = 2$ and $w = 5$, so that the frequency of the function is gradually increasing as time increases. For all three datasets, the system shows that as time t increases from 0 to 1, the correlation between $y_1(t)$ and $y_2(t)$ gradually varies from positive to negative. Within each dataset, we randomly select 200 training data points, in which 100 time stamps are sampled on the interval $(0, 0.8)$ for the first dimension and the other 100 time stamps sampled on the interval $(0.2, 1)$ for the second dimension. For the test inputs, we randomly select 100 time stamps on the interval $(0, 1)$ for each dimension.

Table 1 Prediction measurements on three synthetic datasets and different models. LF, HF and VF refer to low-frequency, high-frequency, and time-varying datasets. Three prediction measures are root mean square error (RMSE), average length of confidence interval (ALCI), and coverage rate (CR). All three measurements are summarized by the mean and standard deviation across 10 runs with different random initializations.

Data	Model	RMSE	ALCI	CR
LF	IGPR [16]	2.25(1.33e-13)	2.18(1.88e-13)	0.835(0)
	ICM [17]	2.26(2.54e-5)	2.18(1.22e-5)	0.835(0)
	CMOGP [12]	1.43(6.12e-2)	1.36(1.98e-1)	0.651(3.00e-2)
	VGPRN [18]	1.01(0.31)	-	-
	VSGPRN	**1.00(1.43e-1)**	2.21(6.56e-2)	**0.892(1.63e-2)**
HF	IGPR [16]	1.51(6.01e-14)	3.17(1.30e-13)	0.915(2.22e-16)
	ICM [17]	1.52(1.01e-5)	3.17(1.19e-5)	0.910(0)
	CMOGP [12]	1.29(3.04e-2)	2.34(3.31e-1)	0.729(3.07e-2)
	VGPRN [18]	1.11(0.25)	-	-
	VSGPRN	**1.10(1.98e-1)**	2.74(7.94e-2)	**0.930(1.14e-2)**
VF	IGPR [16]	1.64(8.17e-14)	3.19(3.02e-13)	0.875(0)
	ICM [17]	1.66(2.37e-3)	3.16(1.49e-3)	0.880(1.50e-3)
	CMOGP [12]	2.24(3.08e-1)	2.56(9.29e-1)	0.697(1.56e-1)
	VGPRN [18]	1.04(0.67)	-	-
	VSGPRN	**1.24(1.33e-1)**	2.92(1.21e-1)	**0.887(9.80e-3)**

We quantify the model performance in terms of root mean square error (RMSE), average length of confidence interval (ALCI), and coverage rate (CR) on the test set. A smaller RMSE corresponds to better predictive performance of the model, and a smaller ALCI implies a smaller predictive uncertainty. As for CR, the better the model prediction performance is, the closer CR is to the percentile of the credible band. Those results are reported by the mean and standard deviation with 10 different random initializations of model parameters. Quantitative comparisons relating to all three datasets are in Table 1. We compare with independent Gaussian process regression (IGPR) [16], the intrinsic coregionalization model (ICM) [17], Collaborative Multi-Output Gaussian Processes (CMOGP) [12] and variational inference of Gaussian process regression networks [18] on three synthetic datasets. In both CMOGP and VSGPRN approaches, we use 20 inducing variables. We further examined model predictive performance on a real-world dataset, the PM2.5 dataset from the UCI Machine Learning Repository [19]. This dataset tracks the concentration of fine inhalable particles hourly in five cities in China, along with meteorological data, from Jan 1st, 2010 to Dec 31st, 2015. We compare our model with two sparse Gaussian process models, i.e., independent sparse Gaussian process regression (ISGPR) [20] and the sparse linear model of coregionalization (SLMC) [17]. In the dataset, we consider six important attributes and use 20% of the first 5000 standardized multivaritate for training and use the others for testing. The RMSEs on the testing data are shown in Table 2, illustrating that VSGPRN had better prediction performance compared with ISGPR and SLMC, even when using fewer inducing points.

Table 2 Empirical results for PM2.5 dataset. Each model's performance is summarized by its RMSE on the testing data. The number of equi-spaced inducing points is given in parentheses.

Data	ISGPR (100) [20]	SLMC (100) [17]	VSGPRN (50)	VSGPRN (100)	VSGPRN (200)
PM2.5	0.994	0.948	0.840	0.708	0.625

5 Conclusions

We propose a novel variational inference approach for structured Gaussian process regression networks named the variational structured Gaussian process regression network, VSGPRN. We introduce inducing variables and propose a structured variational distribution to reduce the computational burden. Moreover, we take advantage of the collapsed representation of our model and construct a tractable lower bound of the log likelihood to make it suitable for doubly stochastic inference and easy to handle missing data. In our method, the computation complexity is independent of the size of the inputs and the outputs. We illustrate the superior predictive performance for both synthetic and real data.

Our inference approach, VSGPRN can be widely used for high dimensional time series to model complicated time-varying dependence across multivariate outputs. Moreover, due to its scalability and flexibility, it can be widely applied for irregularly sampled incomplete large datatsets that widely exist in various research fields including healthcare, environmental science and geoscience.

References

1. Álvarez, M., Lawrence, N.: Computationally efficient convolved multiple output Gaussian processes. J. Mach. Learn. Res. **12**, 1459-1500 (2011)
2. Goulard, M., Voltz, M.: Linear coregionalization model: tools for estimation and choice of cross-variogram matrix. Math. Geol. **24**, 269-286 (1992)
3. Gneiting, T., Kleiber, W., Schlather, M.: Matérn cross-covariance functions for multivariate random fields. J. Am. Stat. Assoc. **105**, 1167-1177 (2010)
4. Qadir, G., Sun, Y.: Semiparametric estimation of cross-covariance functions for multivariate random fields. Biom. **77**, 547-560 (2021)
5. Gelfand, A., Schmidt, A., Banerjee, S., Sirmans, C.: Nonstationary multivariate process modeling through spatially varying coregionalization. Test. **13**, 263-312 (2004)
6. Meng, R., Soper, B., Lee, H., Liu, V., Greene, J., Ray, P.: Nonstationary multivariate Gaussian processes for electronic health records. J. Biom. Inform. **117**, 103698 (2021)
7. Titsias, M., Lawrence, N.: Bayesian Gaussian process latent variable model. Int. Conf. Artif. Intell. Stat. 844-851 (2010)
8. Teh, Y., Newman, D., Max Welling, M.: A collapsed variational Bayesian inference algorithm for latent Dirichlet allocation. In: Schölkopf, B., Platt, J., Hofmann, T. (eds.) Advances in Neural Information Processing Systems **19**, (2006)
9. Guhaniyogi, R., Finley, A., Banerjee, S., Kobe, R.: Modeling complex spatial dependencies: Low-rank spatially varying cross-covariances with application to soil nutrient data. J. Agric. Biol. Environ. Stat. **18**, 274-298 (2013)
10. Møller, J., Syversveen, A., Waagepetersen, R.: Log Gaussian Cox processes. Scand. J. Stat. **25**, 451-482 (1998)
11. Remes, S., Heinonen, M., Kaski, S.: Non-stationary spectral kernels. Adv. Neural Inf. Process. Syst. **30** (2017), https://proceedings.neurips.cc/paper/2017/file/c65d7bd70fe 3e5e3a2f3de681edc193d-Paper.pdf
12. Nguyen, T., Bonilla, E., et al.: Collaborative multi-output Gaussian processes. Uncertain. Artif. Intell. 643-652 (2014)
13. Titsias, M., Lázaro-Gredilla, M.: Doubly stochastic variational Bayes for non-conjugate inference. Int. Conf. Mach. Learn. 1971-1979 (2014)
14. Ranganath, R., Gerrish, S., Blei, D.: Black box variational inference. Int. Conf. Artif. Intell. Stat. 814-822 (2014)
15. Hoffman, M., Blei, D., Wang, C., Paisley, J.: Stochastic variational inference. J. Mach. Learn. Res. **14**, 1303-1347 (2013)
16. Rasmussen, C., Kuss, M.: Gaussian processes in reinforcement learning. Adv. Neural Inf. Process. Syst. 751-759 (2004)
17. Wackernagel, H.: Multivariate geostatistics: an introduction with applications. Springer Science & Business Media (2013)
18. Nguyen, T., Bonilla, E.: Efficient variational inference for Gaussian process regression networks. Int. Conf. Artif. Intell. Stat. 472-480 (2013)
19. Liang, X., Zou, T., Guo, B., Li, S., Zhang, H., Zhang, S., Huang, H., Chen, S.: Assessing Beijing's PM2.5 pollution: severity, weather impact, APEC and winter heating. Proc. R. Soc. A: Math. Phys. Eng. Sci. **471**, 20150257 (2015) https://royalsocietypublishing.org/doi/abs/10.1098/rspa.2015.0257
20. Snelson, E., Ghahramani, Z.: Sparse Gaussian processes using pseudo-inputs. Adv. Neural Inf. Process. Syst. 1257-1264 (2006), http://papers.nips.cc/paper/2857-sparse-g aussian-processes-using-pseudo-inputs.pdf

An Online Minorization-Maximization Algorithm

Hien Duy Nguyen, Florence Forbes, Gersende Fort, and Olivier Cappé

Abstract Modern statistical and machine learning settings often involve high data volume and data streaming, which require the development of online estimation algorithms. The online Expectation–Maximization (EM) algorithm extends the popular EM algorithm to this setting, via a stochastic approximation approach. We show that an online version of the Minorization–Maximization (MM) algorithm, which includes the online EM algorithm as a special case, can also be constructed in a similar manner. We demonstrate our approach via an application to the logistic regression problem and compare it to existing methods.

Keywords: expectation-maximization, minorization-maximization, parameter estimation, online algorithms, stochastic approximation

1 Introduction

Expectation–Maximization (EM) [6, 17] and Minorization–Maximization (MM) algorithms [15] are important classes of optimization procedures that allow for the construction of estimation routines for many data analytic models, including

Hien Duy Nguyen (✉)
School of Mathematics and Physics, University of Queensland, St. Lucia, 4067 QLD, Australia,
e-mail: h.nguyen7@uq.edu.au

Florence Forbes
Univ. Grenoble Alpes, Inria, CNRS, Grenoble INP, LJK, 38000, Grenoble, France,
e-mail: florence.forbes@inria.fr

Gersende Fort
Institut de Mathématiques de Toulouse, CNRS, Toulouse, France,
e-mail: gersende.fort@math.univ-toulouse.fr,

Olivier Cappé
ENS Paris, Universite PSL, CNRS, INRIA, France, e-mail: Olivier.Cappe@cnrs.fr

© The Author(s) 2023
P. Brito et al. (eds.), *Classification and Data Science in the Digital Age*,
Studies in Classification, Data Analysis, and Knowledge Organization,
https://doi.org/10.1007/978-3-031-09034-9_29

many finite mixture models. The benefit of such algorithms comes from the use of computationally simple surrogates in place of difficult optimization objectives.

Driven by high volume of data and streamed nature of data acquisition, there has been a rapid development of online and mini-batch algorithms that can be used to estimate models without requiring data to be accessed all at once. Online and mini-batch versions of EM algorithms can be constructed via the classic Stochastic Approximation framework (see, e.g., [2, 13]) and examples of such algorithms include those of [3, 7, 8, 10, 11, 12, 19]. Via numerical assessments, many of the algorithms above have been demonstrated to be effective in mixture model estimation problems. Online and mini-batch versions of MM algorithms on the other hand have largely been constructed following convex optimizations methods (see, e.g., [9, 14, 23]) and examples of such algorithms include those of [4, 16, 18, 22].

In this work, we provide a stochastic approximation construction of an online MM algorithm using the framework of [3]. The main advantage of our approach is that we do not make convexity assumptions and instead replace them with oracle assumptions regarding the surrogates. Compared to the online EM algorithm of [3] that this work is based upon, the Online MM algorithm extends the approach to allow for surrogate functions that do not require latent variable stochastic representations, which is especially useful for constructing estimation algorithms for mixture of experts (MoE) models (see, e.g. [20]). We demonstrate the Online MM algorithm via an application to the MoE-related logistic regression problem and compare it to competing methods.

Notation. By convention, vectors are column vectors. For a matrix A, $A^\top$ denotes its transpose. The Euclidean scalar product is denoted by $\langle a, b \rangle$. For a continuously differentiable function $\theta \mapsto h(\theta)$ (resp. twice continuously differentiable), $\nabla_\theta h$ (or simply ∇ when there is no confusion) is its gradient (resp. $\nabla^2_{\theta\theta}$ is its Hessian). We denote the vectorization operator that converts matrices to column vectors by vec.

2 The Online MM Algorithm

Consider the optimization problem

$$\arg\max_{\theta \in \mathbb{T}} \mathbb{E}\left[f\left(\theta; X\right)\right], \tag{1}$$

where $\mathbb{T}$ is a measurable open subset of $\mathbb{R}^p$, $\mathbb{X}$ is a topological space endowed with its Borel sigma-field, $f : \mathbb{T} \times \mathbb{X} \to \mathbb{R}$ is a measurable function and X is a $\mathbb{X}$-valued random variable on the probability space $(\Omega, \mathcal{F}, \mathbb{P})$. In this paper, we are interested in the setting when the expectation $\mathbb{E}\left[f\left(\theta; X\right)\right]$ has no closed form, and the optimization problem is solved by an MM-based algorithm.

Following the terminology of [15], we say that $g : \mathbb{T} \times \mathbb{X} \times \mathbb{T}, (\theta, x, \tau) \mapsto g\left(\theta, x; \tau\right)$, is a *minorizer of* f, if for any $\tau \in \mathbb{T}$ and for any $(\theta, x) \in \mathbb{T} \times \mathbb{X}$, it holds that

$$f(\theta; x) - f(\tau; x) \geq g(\theta, x; \tau) - g(\tau, x; \tau). \tag{2}$$

In our work, we consider the case when the minorizer function g has the following structure:

A1   The minorizer surrogate g is of the form:

$$g(\theta, x; \tau) = -\psi(\theta) + \langle \bar{S}(\tau; x), \phi(\theta) \rangle, \tag{3}$$

where $\psi : \mathbb{T} \to \mathbb{R}$, $\phi : \mathbb{T} \to \mathbb{R}^d$ and $\bar{S} : \mathbb{T} \times \mathbb{X} \to \mathbb{R}^d$ are measurable functions. In addition, ϕ and ψ are continuously differentiable on $\mathbb{T}$.

We also make the following assumptions:

A2   There exists a measurable open and convex set $\mathbb{S} \subseteq \mathbb{R}^d$ such that for any $s \in \mathbb{S}$, $\gamma \in [0, 1)$ and any $(\tau, x) \in \mathbb{T} \times \mathbb{X}$:

$$s + \gamma \{ \bar{S}(\tau; x) - s \} \in \mathbb{S}.$$

A3   The expectation $\mathbb{E}[\bar{S}(\theta; X)]$ exists, is in $\mathbb{S}$, and is finite whatever $\theta \in \mathbb{T}$ but it may have no closed form. Online independent oracles $\{X_n, n \geq 0\}$, with the same distribution as X, are available.

A4   For any $s \in \mathbb{S}$, there exists a unique root to $\theta \mapsto -\nabla\psi(\theta) + \nabla\phi(\theta)^\top s$, which is the unique maximum on $\mathbb{T}$ of the function $\theta \mapsto -\psi(\theta) + \langle s, \phi(\theta) \rangle$. This root is denoted by $\bar{\theta}(s)$.

Seen as a function of θ, $g(\cdot, x; \tau)$ is the sum of two functions: $-\psi$ and a linear combination of the components of $\phi = (\phi_1, \ldots, \phi_d)$. Assumption A1 implies that the minorizer surrogate is in a functional space spanned by these $(d + 1)$ functions. By (2) and A1–A3, it follows that

$$\mathbb{E}[f(\theta; X)] - \mathbb{E}[f(\tau; X)] \geq \psi(\tau) - \psi(\theta) + \langle \mathbb{E}[\bar{S}(\tau; X)], \phi(\theta) - \phi(\tau) \rangle, \tag{4}$$

thus providing a minorizer function for the objective function $\theta \mapsto \mathbb{E}[f(\theta; X)]$. By A4, the usual MM algorithm would define iteratively the sequence $\theta_{n+1} = \bar{\theta}(\mathbb{E}[\bar{S}(\theta_n; X)])$. Since the expectation may not have closed form but infinite datasets are available (see A3), we propose a novel Online MM algorithm. It defines the sequence $\{s_n, n \geq 0\}$ as follows: given positive step sizes $\{\gamma_{n+1}, n \geq 1\}$ in $(0, 1)$ and an initial value $s_0 \in \mathbb{S}$, set for $n \geq 0$:

$$s_{n+1} = s_n + \gamma_{n+1} \{ \bar{S}(\bar{\theta}(s_n); X_{n+1}) - s_n \}. \tag{5}$$

The update mechanism (5) is a Stochastic Approximation iteration, which defines an $\mathbb{S}$-valued sequence (see A2). It consists of the construction of a sequence of minorizer functions through the definition of their *parameter* s_n in the functional space spanned by $-\psi, \phi_1, \ldots, \phi_d$.

If our algorithm (5) converges, any limiting point $s_\star$ satisfies $\mathbb{E}[\bar{S}(\bar{\theta}(s_\star); X)] = s_\star$. Hence, our algorithm is designed to approximate the intractable expectation, evaluated at $\bar{\theta}(s_\star)$, where $s_\star$ satisfies a fixed point equation. The following lemma establishes the relation between the limiting points of (5) and the optimization problem (1) at hand. Namely, it implies that any limiting value $s_\star$ provides a stationary point $\theta_\star := \bar{\theta}(s_\star)$ of the objective function $\mathbb{E}\left[f\left(\theta; X\right)\right]$ (i.e., $\theta_\star$ is a root of the derivative of the objective function). The proof follows the technique of [3]. Set

$$h(s) := \mathbb{E}\left[\bar{S}\left(\bar{\theta}\left(s\right); X\right)\right] - s, \qquad \Gamma := \{s \in \mathbb{S} : h(s) = 0\}.$$

Lemma 1 *Assume that* $\theta \mapsto \mathbb{E}\left[f(\theta; X)\right]$ *is continuously differentiable on* $\mathbb{T}$ *and denote by* $\mathcal{L}$ *the set of its stationary points. If* $s_\star \in \Gamma$, *then* $\bar{\theta}(s_\star) \in \mathcal{L}$. *Conversely, if* $\theta_\star \in \mathcal{L}$, *then* $s_\star := \mathbb{E}\left[\bar{S}\left(\theta_\star; X\right)\right] \in \Gamma$.

Proof A4 implies that

$$-\nabla\psi(\bar{\theta}(s)) + \nabla\phi(\bar{\theta}(s))^\top s = 0, \qquad s \in \mathbb{S}. \tag{6}$$

Use (2) and A1, and apply the expectation w.r.t. X (under A3). This yields (4), which is available for any $\theta, \tau \in \mathbb{T}$. This inequality provides a minorizer function for $\theta \mapsto \mathbb{E}\left[f(\theta; X)\right]$: the difference is nonnegative and minimal (i.e. equal to zero) at $\theta = \tau$. Under the assumptions and A1, this yields

$$\nabla\mathbb{E}\left[f(\cdot; X)\right]|_{\theta=\tau} + \nabla\psi(\tau) - \nabla\phi(\tau)^\top\mathbb{E}\left[\bar{S}(\tau; X)\right] = 0. \tag{7}$$

Let $s_\star \in \Gamma$ and apply (7) with $\tau \leftarrow \bar{\theta}(s_\star)$. It then follows that

$$\nabla\mathbb{E}\left[f(\cdot; X)\right]|_{\theta=\bar{\theta}(s_\star)} + \nabla\psi(\bar{\theta}(s_\star)) - \nabla\phi(\bar{\theta}(s_\star))^\top s_\star = 0,$$

which implies $\bar{\theta}(s_\star) \in \mathcal{L}$ by (6). Conversely, if $\theta_\star \in \mathcal{L}$, then by (7), we have

$$\nabla\psi(\theta_\star) - \nabla\phi(\theta_\star)^\top\mathbb{E}\left[\bar{S}(\theta_\star; X)\right] = 0,$$

which, by A3 and A4, implies that $\theta_\star = \bar{\theta}\left(\mathbb{E}\left[\bar{S}(\theta_\star; X)\right]\right) = \bar{\theta}(s_\star)$. By definition of $s_\star$, this yields $s_\star = \mathbb{E}\left[\bar{S}\left(\bar{\theta}(s_\star); X\right)\right]$; i.e. $s_\star \in \Gamma$. $\qquad\square$

By applying the results of [5] regarding the asymptotic convergence of Stochastic Approximation algorithms, additional regularity assumptions on $\phi, \psi, \bar{\theta}$ imply that the algorithm (5) possesses a continuously differentiable Lyapunov function V defined on $\mathbb{S}$ and given by $V : s \mapsto \mathbb{E}\left[f(\bar{\theta}(s); X)\right]$, satisfying $\langle\nabla V(s), h(s)\rangle \leq 0$, where the inequality is strict outside the set Γ (see [3, Prop. 2]). In addition to Lemma 1, assumptions on the distribution of X and on the stability of the sequence $\{s_n, n \geq 0\}$ are provided in [5, Thm. 2 and Lem. 1], which, combined with the usual conditions on the step sizes: $\sum_n \gamma_n = +\infty$ and $\sum_n \gamma_n^2 < \infty$, yields the almost-sure convergence of the sequence $\{s_n, n \geq 0\}$ to the set Γ, and the almost-sure convergence of the sequence $\{\bar{\theta}(s_n), n \geq 0\}$ to the set $\mathcal{L}$ of the stationary points of the objective function $\theta \mapsto \mathbb{E}\left[f(\theta; X)\right]$. Due to the limited space, the exact statement of these convergence results for our Online MM framework is omitted.

3 Example Application

As an example, we consider the logistic regression problem, where we solve (1) with

$$f(\theta; x) := yw^\top\theta - \log\left\{1 + \exp\left(w^\top\theta\right)\right\}, \qquad x := (y, w),$$

where $y \in \{0, 1\}$, $w \in \mathbb{R}^p$, and $\theta \in \mathbb{T} := \mathbb{R}^p$. Here, we assume that $X = (Y, W)$ is a random variable such that $\mathbb{E}[f(\theta; X)]$ exists for each θ.

Denote by λ the standard logistic function $\lambda(\cdot) := \exp\{\cdot\}/(1+\exp\{\cdot\})$. Following [1], (2) and A1 are verified by taking

$$\psi(\theta) := 0, \qquad \phi(\theta) := \begin{bmatrix} \theta \\ \text{vec}(\theta\theta^\top) \end{bmatrix}, \qquad \bar{S}(\tau; x) = \begin{bmatrix} \bar{s}_1(\tau; x) \\ \text{vec}(\bar{S}_2(\tau; x)) \end{bmatrix}$$

where

$$\bar{s}_1(\tau; x) := \left\{y - \lambda(\tau^\top w)\right\}w + \frac{1}{4}ww^\top\tau, \qquad \bar{S}_2(\tau; x) = -\frac{1}{8}ww^\top.$$

With $\mathbb{S} := \{(s_1, \text{vec}(S_2)) : s_1 \in \mathbb{R}^p$ and $S_2 \in \mathbb{R}^{p\times p}$ is symmetric positive definite$\}$, it follows that $\bar{\theta}(s) := -(2S_2)^{-1}s_1$.

Online MM. Let $s_n = (s_{1,n}, S_{2,n}) \in \mathbb{S}$. The corresponding `Online MM` recursion is then

$$s_{1,n+1} = s_{1,n} + \gamma_{n+1}\left(Y_{n+1} - \lambda(\bar{\theta}(s_n)^\top W_{n+1})W_{n+1} + \frac{1}{4}W_{n+1}W_{n+1}^\top\bar{\theta}(s_n) - s_{1,n}\right) \quad (8)$$

$$S_{2,n+1} = S_{2,n} + \gamma_{n+1}\left(-\frac{1}{8}W_{n+1}W_{n+1}^\top - S_{2,n}\right), \quad (9)$$

where $\{(Y_{n+1}, W_{n+1}), n \geq 0\}$ are i.i.d. pairs with the same distribution as $X = (Y, W)$. Parameter estimates can then be deduced by setting $\theta_{n+1} := \bar{\theta}(s_{n+1})$.

For comparison, we also consider two Stochastic Approximation schemes directly on θ in the parameter-space: a stochastic gradient (SG) algorithm and a Stochastic Newton Raphson (SNR) algorithm.

Stochastic gradient. SG requires the gradient of $f(\theta; x)$ with respect to θ: $\nabla f(\theta; x) = \{y - \lambda(\theta^\top w)\}w$, which leads to the recursion

$$\hat{\theta}_{n+1} = \hat{\theta}_n + \gamma_{n+1}\left\{Y_{n+1} - \lambda(\hat{\theta}_n^\top W_{n+1})\right\}W_{n+1}. \quad (10)$$

Stochastic Newton-Raphson. In addition SNR requires the Hessian with respect to θ, given by $\nabla_{\theta\theta}^2 f(\theta; x) = -\lambda(\theta^\top w)\{1 - \lambda(\theta^\top w)\}ww^\top$. The SNR recursion is then

$$\hat{A}_{n+1} = \hat{A}_n + \gamma_{n+1}\left\{\nabla_{\theta\theta}^2 f(\hat{\theta}_n; X_{n+1}) - \hat{A}_n\right\} \quad (11)$$

$$G_{n+1} = -\hat{A}_{n+1}^{-1} \quad (12)$$

$$\hat{\theta}_{n+1} = \hat{\theta}_n + \gamma_{n+1}G_{n+1}\left\{Y_{n+1} - \lambda(\hat{\theta}_n^\top W_{n+1})\right\}W_{n+1}. \quad (13)$$

Equation (12) assumes that $\hat{A}_{n+1}$ is invertible. In this logistic example, we can guarantee this by choosing $\hat{A}_0$ to be invertible. Otherwise $\hat{A}_n$ is invertible after some n sufficiently large, with probability one. Again in the logistic case, observe that, from the structure of $\nabla^2_{\theta\theta} f$ and from the Woodbury matrix identity, Equations (11–12) can be replaced by

$$G_{n+1} = \frac{G_n}{1 - \gamma_{n+1}} - \frac{\gamma_{n+1}}{1 - \gamma_{n+1}} \frac{a_{n+1} G_n W_{n+1} W_{n+1}^\top G_n}{\left\{(1 - \gamma_{n+1}) + \gamma_{n+1} a_{n+1} W_{n+1}^\top G_n W_{n+1}\right\}}.$$

where $a_{n+1} := \lambda(\hat{\theta}_n^\top W_{n+1}) \left\{1 - \lambda(\hat{\theta}_n^\top W_{n+1})\right\}$,

It appears that the Online MM recursion in the s-space defined by (8) and (9) is equivalent to the SNR recursion above (i.e., (11)–(13)) when the Hessian $\nabla^2_{\theta\theta} f(\theta; x)$ is replaced by the lower bound $-\frac{1}{4} ww^\top$. This observation holds whenever g is quadratic in $(\theta - \tau)$.

Polyak averaging. In practice, for Online MM, SG, and SNR recursions, it is common to consider Polyak averaging [21], starting from some iteration n_0, chosen such as to avoid the initial highly volatile estimates. Set $\hat{\theta}_{n_0}^A := 0$, and for $n \geq n_0$,

$$\hat{\theta}_{n+1}^A = \hat{\theta}_n^A + \alpha_{n-n_0+1}(\hat{\theta}_n - \hat{\theta}_n^A), \qquad (14)$$

where α_n is usually set to $\alpha_n := n^{-1}$.

Numerical illustration. We now demonstrate the performance of the Online MM algorithm for logistic regression – defined by (5) and the derivations above. To do so, a sequence $\{X_i = (Y_i, W_i), i \in \{1, \ldots, n_{max}\}\}$ of $n_{max} = 10^5$ i.i.d. replicates of $X = (Y, W)$ is simulated: $W = (1, U)$, where $U \sim N(0, 1)$ and $[Y|W = w] \sim$ Ber $\left(\lambda\left(\theta_0^\top w\right)\right)$, where $\theta_0 = (3, -3)$. Online MM is run using the learning rate $\gamma_n = n^{-0.6}$, as suggested in [3]. The algorithm is initialized with $\hat{\theta}_0 = (0, 0)$ and $s_0 = \sum_{i=1}^2 \bar{S}\left(\hat{\theta}_0; X_i\right)/2$.

For comparison, we also show, on Figure 1, the SG, SNR estimates and their Polyak averaged values in θ-space. As is usually recommended with Stochastic Approximation, the first few volatile estimations are discarded. Similarly, for Polyak averaging, we set $n_0 = 10^3$. As expected, we observe that the Online MM and the SNR recursions are very close but with the SNR showing more variability. Their comparison after Polyak averaging shows very close trajectories while the SG trajectory is clearly different and shows more bias. Final estimates [Polyak averaged estimates] of θ_0 from the SG, SNR, and Online MM algorithms are respectively: $(2.67, -2.66)$ $[(2.51, -2.48)]$, $(3.03, -3.03)$ $[(2.99, -3.03)]$, and $(3.01, -3.03)$ $[(2.98, -3.02)]$, which we can compare to the batch maximum likelihood estimate $(3.00, -3.05)$ (obtained via the glm function in R). Notice the remarkable closeness between the online MM and batch estimates.

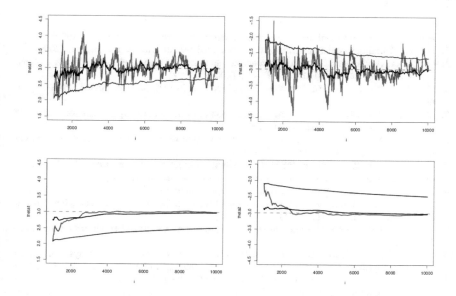

Fig. 1 Logistic regression example: the first row shows Online MM (black), SG (blue), and SNR (red) recursions. The second row shows the respective Polyak averaging recursions. The estimates of the first θ (first column) and the second (second column) components of θ are plotted started from $n = 10^3$ for readability.

4 Final Remarks

Remark 1 For a parametric statistical model indexed by θ, let $f(\theta; x)$ be the log-density of a random variable X with stochastic representation $f(\theta; x) = \log \int_{\mathbb{Y}} p_\theta(x, y) \mu(\mathrm{d}y)$, where $p_\theta(x, y)$ is the joint density of (X, Y) with respect to the positive measure μ for some latent variable $Y \in \mathbb{Y}$. Then, via [15, Sec. 4.2], we recover the Online EM algorithm by using the minorizer function g:

$$g(\theta, x; \tau) := \int_{\mathbb{Y}} \log p_\theta(x, y) \ p_\tau(x, y) \exp(-f(\tau; x)) \mu(\mathrm{d}y).$$

Remark 2 Via the minorization approach of [1] (as used in Section 3) and the mixture representation from [19], we can construct an Online MM algorithm for MoE models, analogous to the MM algorithm of [20]. We shall provide exposition on such an algorithm in future work.

Acknowledgements Part of the work by G. Fort is funded by the *Fondation Simone et Cino Del Duca, Institut de France*. H. Nguyen is funded by ARC Grant DP180101192. The work is supported by Inria project LANDER.

References

1. Bohning, D.: Multinomial logistic regression algorithm. Ann. Inst. Stat. Math. (1992)
2. Borkar, V.S.: Stochastic approximation: A dynamical systems viewpoint. Springer (2009)
3. Cappé, O., Moulines, E.: On-line expectation-maximization algorithm for latent data models. J. Roy. Stat. Soc. B Stat. Meth. **71**, 593–613 (2009)
4. Cui, Y., Pang, J.: Modern nonconvex nondifferentiable optimization. SIAM, Philadelphia (2022)
5. Delyon, B., Lavielle, M., Moulines, E.: Convergence of a stochastic approximation version of the EM algorithm. Ann. Stat. **27**, 94–128 (1999)
6. Dempster, A. P., Laird, N. M., Rubin, D. B.: Maximum likelihood from incomplete data via the EM algorithm. J. Roy. Stat. Soc. B Stat. Meth. **39**, 1–38 (1977)
7. Fort, G., Gach, P., Moulines, E.: Fast incremental expectation maximization for finite-sum optimization: nonasymptotic convergence. Stat. Comput. **31**, 1–24 (2021)
8. Fort, G., Moulines, E., Wai, H. T.: A stochastic path-integrated differential estimator expectation maximization algorithm. In: Proceedings of the 34th Conference on Neural Information Processing Systems (NeurIPS) (2020)
9. Hazan, E.: Introduction to online convex optimization. Foundations and Trends in Optimization. **2** (2015)
10. Karimi, B., Miasojedow, B., Moulines, E., Wai, H. T.: Non-asymptotic analysis of biased stochastic approximation scheme. Proceedings of Machine Learning Research. **99**, 1–31 (2019)
11. Karimi, B., Wai, H. T., Moulines, R., Lavielle, M.: On the global convergence of (fast) incremental expectation maximization methods. In: Proceedings of the 33rd Conference on Neural Information Processing Systems (NeurIPS) (2019)
12. Kuhn, E., Matias, C., Rebafka, T.: Properties of the stochastic approximation EM alpgorithm with mini-batch sampling. Stat. Comput. **30**, 1725–1739 (2020)
13. Kushner, H. J., Yin, G. G.: Stochastic Approximation and Recursive Algorithms and Applications. Springer, New York (2003)
14. Lan, G.: First-order and Stochastic Optimization Methods for Machine Learning. Springer, Cham (2020)
15. Lange, K.: MM Optimization Algorithms. SIAM, Philadelphia (2016)
16. Mairal, J.: Stochastic majorization-minimization algorithm for large-scale optimization. In: Advances in Neural Information Processing Systems, pp. 2283–2291 (2013)
17. McLachlan, G. J., Krishnan, T.: The EM Algorithm And Extensions. Wiley, New York (2008)
18. Mokhtari, A., Koppel, A.: High-dimensional nonconvex stochastic optimization by doubly stochastic successive convex approximation. IEEE Trans. Signal Process. **68**, 6287–6302 (2020)
19. Nguyen, H.D., Forbes, F., McLachlan, G. J.: Mini-batch learning of exponential family finite mixture models. Stat. Comput. **30**, 731–748 (2020)
20. Nguyen, H. D., McLachlan, G. J.: Laplace mixture of linear experts. Comput. Stat. Data Anal. **93**, 177–191 (2016)
21. Polyak, B. T., Juditsky, A. B.: Acceleration of stochastic approximation by averaging. SIAM J. Contr. Optim. **30**, 838–855 (1992)
22. Razaviyayn, M., Sanjabi, M., Luo, Z.: A stochastic successive minimization method for non-smooth nonconvex optimization with applications to transceiver design in wireless communication networks. Math. Program. Series B. 515–545 (2016)
23. Shalev-Shwartz, S.: Online learning and online convex optimization. Foundations and Trends in Machine Learning. **4**, 107–194 (2011)

Detecting Differences in Italian Regional Health Services During Two Covid-19 Waves

Lucio Palazzo and Riccardo Ievoli

Abstract During the first two waves of Covid-19 pandemic, territorial healthcare systems have been severely stressed in many countries. The availability (and complexity) of data requires proper comparisons for understanding differences in performance of health services. We apply a three-steps approach to compare the performance of Italian healthcare system at territorial level (NUTS 2 regions), considering daily time series regarding both intensive care units and ordinary hospitalizations of Covid-19 patients. Changes between the two waves at a regional level emerge from the main results, allowing to map the pressure on territorial health services.

Keywords: regional healthcare, time series, multidimensional scaling, cluster analysis, trimmed k-means

1 Introduction

During the Covid-19 pandemic, the evaluation of similarities and differences between territorial health services [23] is relevant for decision makers and should guide the governance of countries [15] through the so-called "waves". This type of analysis becomes even more crucial in countries where the National healthcare system is regionally-based, which is the case of Italy (or Spain) among others. Italy is one of the countries in Europe which has been mostly affected by the pandemic, and the pressure on Regional Health Services (RHS) has been producing dramatic effects also in the economic [2] and the social [3] spheres. Regional Covid-19-related health

Lucio Palazzo (✉)
Department of Political Sciences, University of Naples Federico II, via Leopoldo Rodinò 22 - 80138 Napoli, Italy, e-mail: lucio.palazzo@unina.it

Riccardo Ievoli
Department of Chemical, Pharmaceutical and Agricultural Sciences, University of Ferrara, via Luigi Borsari 46 - 44121 Ferrara, Italy, e-mail: riccardo.ievoli@unife.it

© The Author(s) 2023
P. Brito et al. (eds.), *Classification and Data Science in the Digital Age*,
Studies in Classification, Data Analysis, and Knowledge Organization,
https://doi.org/10.1007/978-3-031-09034-9_30

273

indicators are extremely relevant for monitoring the pandemic territorial widespread [21], and to impose (or relax) restrictions in accordance with the level of health risk.

The aim of this work is to exploit the potential of Multidimensional Scaling (MDS) to detect the main imbalances occurred in the RHS, observing the hospital admission dynamics of patients with Covid-19 disease. Both daily time series regarding patients treated in Intensive Care (IC) units and individuals hospitalized in other hospital wards are used to evaluate and compare the reaction to healthcare pressure in 21 geographical areas (NUTS 2 Italian regions), considering the first two waves [4] of pandemic. Indeed, territorial imbalances in terms of RHS' performance [24] should be firstly driven by the geographical propagation flows of the virus (first wave). Then, different reactions to pandemic shock may be provided by RHSs, and changes of imbalances can be observed in the second wave.

Our proposal consists of three subsequent steps. Firstly, a matrix of distances between regional time series through a dissimilarity metric [29] is obtained. Therefore, we apply a (weighted) MDS [19, 22] to map similarity patterns in a reduced space, adding also a weighting scheme considering the number of neighbouring regions. Finally, we perform a cluster analysis to identify groups according to RHS performance in the two waves.

The paper is organized as follows: Section 2 describes the methodological approach used to compare and cluster time series, while Section 3 introduces data and descriptive analysis. Results regarding RHSs are depicted and discussed in Section 4, while Section 5 concludes with some remarks and possible advances.

2 Time Series Clustering

Given a matrix $T \times n$, where T represents the days and n the number of regions, our methodological approach consists of three subsequently steps:

Step 1. Compute a dissimilarity matrix D based on a given measure;

Step 2. Apply a weighted multidimensional scaling (wMDS) procedure, storing the coordinates of the first two components;

Step 3. Perform cluster analysis on the MDS reduced space to identify groups between the n regions.

In the first step, a dissimilarity measure is computed for each pair of regional time series. The objective is to obtain a dissimilarity matrix D (with elements $d_{i,j}$) for estimating synthetic measures of the differences between regions. There are different alternatives to compare time series, some comprehensive overviews are in [29, 13].

A reasonable choice is the the Fourier dissimilarity $d_F(\mathbf{x}, \mathbf{y})$, which applies the n-point Discrete Fourier Transform [1] on two time series, allowing to compare the similarity between two time sequences after converting them into a combination of structural elements, such as trend and/or cycle.

In the second step, we implement a multidimensional scaling [31]. Due to its flexibility, MDS has been introduced also in time series analysis [25] and recently applied to different topics [30, 9, 16].

Since our aim is to take into account the degree of proximity between regions, we also employ a weighted multidimensional scaling technique (wMDS) [17, 14]. The $\mathcal{L}_2$ norm is multiplied by a set of weights $\omega = (\omega_1, \ldots, \omega_n)$ such that high weights have a stronger influence on the result than low weights.

The reduced space generated by MDS can be used as starting point for subsequent analyses. Then, a cluster algorithm can be performed on the coordinates (of the reduced space) of MDS [18]. Different procedures should be suitable to perform a cluster analysis on the wMDS coordinates map. For an overview of modern clustering techniques in time series, see e.g. [26].

In our case, both the geographical spread of the pandemic and population density can determine remarkable differences in terms of hospitalization rates [12]. To mitigate the risk of regional outliers in the data, generating potential *spurious* clusters, we employ the trimmed k-means algorithm [8, 11]. A relevant topic in cluster analysis is related to the choice of the k number of groups. Our strategy is purely data-driven and it is based on the minimization of the within-cluster variance.

3 Data and Descriptive Statistics

Daily regional time series reporting a) the number of patients treated in IC units and b) the number of patients admitted in the other hospital wards are retrieved through the official website of Italian Civil Protection[1]. All patients were positive for the Covid-19 test (nasal and oropharyngeal swab). To take into account the different sizes in terms of inhabitants, both a) and b) are normalized according to the population of each territorial unit (estimated at 2020/01/01). The rates of patients treated in IC units and hospitalized (HO) patients in other hospital wards, are then multiplied by 100,000.

The whole dataset contains two identified waves[2] of Covid-19, as follows:

Wave 1 (W1): $T = 109$ days from February 24 to June 11, 2020
Wave 2 (W2): $T = 109$ days from September 14 to December 31, 2020

The date/trend may also depend on external factors, such as the implementation of restrictive measures introduced by the Italian Government [27, 6], which influenced the observed differences between W1 and W2. We have to remark that a full national lockdown was held between March 9th and May 18th 2020.

Figure 1 shows the time series for HO and IC (rows), according to the two waves of Covid-19 (columns). The anomaly of the small Italian region (Valle D'Aosta) emerges both in the first (in particular concerning IC) and second waves (also for

---

[1] Source: www.dati-covid.italia.it
[2] Refer to [7] for further details.

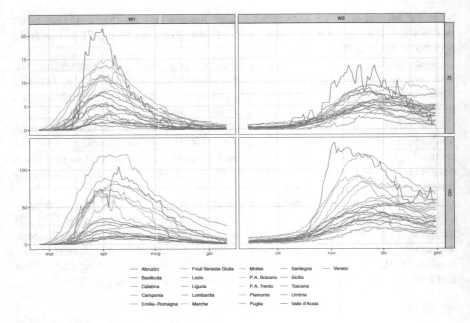

Fig. 1 Time series distributions of Italian regions.

HO), while Lombardia, which is the largest and most populous region, dominates other territories especially when considering HO of W1. The upper panel of Figure 1 helps to understand differences between the two waves in terms of admission to intensive cares: while regions with high, medium and low IC rate can be directly identified through the eyeball of the series during W1, in W2 more homogeneity is observed. Furthermore, with the exception of Valle D'Aosta, the IC rate remains always less than 10 for all considered observations.

For what concerns HO rate, (lower panels of Figure 1), Lombardia reaches values greater than 100 in W1 (especially in April), while during W2 this threshold had exceeding by Valle D'Aosta and Piemonte (both in November). Again, if W1 opposes regions with high and (moderately) low HO rates, in W2 the following situation arises: a) Valle D'Aosta and Piemonte reach values over 100, b) four regions (Liguria, Lazio, P.A. Trento and P.A. Bolzano) present values over 75, and c) the majority of territories share similar trends with peaks always lower than 75.

4 Grouping Regions by Clustering and Discussion

In order to confirm and deepen the descriptive results of Section 3, we perform a cluster analysis following the scheme proposed in Section 2. We compute wMDS equipped with the Fourier distance[3], using a set of weights ω proportional to the number of neighbourhoods for each region, ensuring a spatial feature into the model.

Figure 2 displays the main results of wMDS, distinguishing between four levels of critical issues experienced by the RHS. Outlying performances are coloured in **Violet**. A first cluster (in **Red**) includes "critical" regions while a group depicted in Orange contains territories with high pressure in their RHS. Regions involved in the Green cluster experimented a moderate pressure on RHS, while colour **Blue** indicates territories suffering from a low pressure. These clusters may also be interpreted as a ranking of the health service risk.

As regards the HO during W1, leaving apart the two outliers (Lombardia and P.A. Bolzano) the "red" cluster is composed by three Northern regions (Piemonte, Valle d'Aosta and Emilia-Romagna). The group of high pressure is composed by Liguria, Marche and P.A. Trento, while the green cluster involves Lazio, Abruzzo and Toscana (from the centre of Italy) and Veneto. The last group includes nine regions, 7 of which are located in the southern Italy. In W2 the clustering procedure Piemonte and Valle d'Aosta are identified as outliers, while the high-pressure group is composed by two autonomous provinces (Trendo and Bolzano), Lombardia and Liguria. The "orange" group is constituted by regions located in the North-East (Friuli-Venezia Giulia, Emilia-Romagna and Veneto), along with Abruzzo and Lazio. Southern regions are allocated in the "green" coloured group (together with Umbria, Toscana and Marche), while Molise, Calabria and Basilicata remain in the low-pressure cluster.

Regarding IC rates, during W1 Lombardia and Valle d'Aosta are considered as outliers while the "red" cluster is composed by four northern Italian regions (Emilia-Romagna, P.A. of Trento, Piemonte and Liguria), and Marche (located in the centre). The "orange" cluster contains Toscana, Veneto and P.A. Bolzano, while the moderate-pressure cluster involves three areas of centre Italy (Lazio and Umbria), among with the Friuli-Venezia Giulia (from the north-east) and Abruzzo. The last cluster includes only regions from the south. According to the bottom right panel of Figure 2, apart from Valle D'Aosta, the procedure identifies Calabria as an outlier. The "red" group acquires two observations from the Centre of Italy such as Toscana and Umbria, while the majority of regions are classified in the moderately pressured group. Only three Southern Italian areas are allocated in the last group (in green).

If the geography of the disease appears fundamental in W1, especially regarding adjoining territories of Lombardia, in W2 this effect is less evident. Thus, regions improving (e.g. Emilia-Romagna) or worsening (such as Lazio and Abruzzo) their clustering "ranking" can be easily observed. As mentioned, the differences of restrictive measures imposed by the Government in the two periods may have a role on these results.

---

[3] We remark that other distance measures have been applied. Moreover, a) the Fourier one shows better performance in terms of goodness of fit; b) the results are not sensitive with respect to the choice of distance.

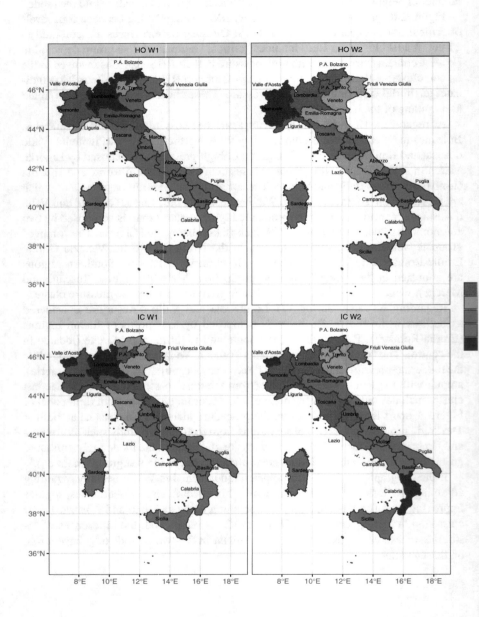

Fig. 2 Map of the identified regional clusters.

5 Concluding Remarks

The Covid-19 pandemic has put a strain on the Italian healthcare system. The reactions of RHS play a relevant role to mitigate the health crisis at territorial level and to guarantee an equitable access to healthcare.

This work helps to understand similarities and divergences between the Italian regions in relation to the health pressure of the first two waves of the virus. Considering crucial measures such as HO and IC rates, the comparison between two waves allows to understand differences in the reactions to pandemic shocks of RHS. Although the northern Italy represented the epicentre of the Covid-19 spread in the first wave, some regions (e.g. Veneto and Friuli-Venezia Giulia) seem to have succeeded in avoiding hospitals overcrowding, while Southern regions (and Islands) definitively suffered from less pressure. Furthermore, in the second wave, the difference appears slightly smoothed and the cluster sizes seem more homogeneous. Moreover, there are some exceptions, such as the Emilia-Romagna, which seems to have been less affected by the second wave, compared to the other regions. The detection of clusters represents a starting point for the improvement of health governance and can be used to monitor potential imbalances in future unfortunate waves.

Further analysis may employ other dedicated indicators coming, for instance, from the Italian National Institute of Statistics[4], or using different proposals for combining wMDS with dissimilarity measures and clustering [28]. Following a different methodological approach, the recent method proposed in [10] should be applied on those data to include more complex spatial relationships between territories.

References

1. Agrawal, R., Faloutsos, C., Swami, A.: Efficient similarity search in sequence databases. In: International Conference on Foundations of Data Organization and Algorithms, pp. 69-84. Springer, Berlin (1993)
2. Ascani, A., Faggian, A., Montresor, S.: The geography of COVID-19 and the structure of local economies: The case of Italy. Journal of Regional Science, **61**(2), 407-441 (2021)
3. Beria, P., Lunkar, V.: Presence and mobility of the population during the first wave of Covid-19 outbreak and lockdown in Italy. Sustainable Cities and Society, **65**, 102616 (2021)
4. Bontempi, E.; The Europe second wave of COVID-19 infection and the Italy "strange" situation. Environmental Research, **193**, 110476 (2021)
5. Capolongo, S., Gola, M., Brambilla, A., Morganti, A., Mosca, E. I., Barach, P.: COVID-19 and Healthcare facilities: A decalogue of design strategies for resilient hospitals. Acta Bio Medica: Atenei Parmensis, **91**(9-S), 50 (2020)
6. Chirico, F., Sacco, A., Nucera, G., Magnavita, N.: Coronavirus disease 2019: the second wave in Italy. Journal of Health Research (2021).
7. Cicchetti, A., Damiani, G., Specchia, M. L., Basile, M., Di Bidino, R., Di Brino, E., Tattoli, A.: Analisi dei modelli organizzativi di risposta al Covid-19. ALTEMS (2020). link: `https://altems.unicatt.it/altems-report47.pdf`
8. Cuesta-Albertos, J. A., Gordaliza, A., Matrán, C.: Trimmed k-means: An attempt to robustify quantizers. The Annals of Statistics, **25**(2), 553-576 (1997).

---

4 see for example the BES indicators of the domain "Health" and "Quality of services"
`https://www.istat.it/it/files//2021/03/BES\_2020.pdf`

9. Di Iorio, F., Triacca, U.: Distance between VARMA models and its application to spatial differences analysis in the relationship GDP-unemployment growth rate in Europe. In: International Work-Conference on Time Series Analysis, pp. 203-215. Springer, Cham (2017)
10. D'Urso, P., De Giovanni, L., Disegna, M., Massari, R.: Fuzzy clustering with spatial-temporal information. Spatial Statistics, **30**, 71-102 (2019)
11. Garcia-Escudero, L. A., Gordaliza, A.: Robustness properties of k-means and trimmed k-means. Journal of the American Statistical Association, **94**(447), 956–969 (1999) doi: 10.2307/2670010
12. Giuliani, D., Dickson, M. M., Espa, G., Santi, F.: Modelling and predicting the spatio-temporal spread of COVID-19 in Italy. BMC infectious diseases, **20**(1), 1-10 (2020)
13. Górecki, T., Piasecki, P.: A comprehensive comparison of distance measures for time series classification. In: Steland, A., Rafajłowicz, E., Okhrin, O. (Eds.) Workshop on Stochastic Models, Statistics and their Application, pp. 409-428. Springer, Nature (2019)
14. Greenacre, M.: Weighted metric multidimensional scaling. In: New developments in Classification and Data Analysis, pp. 141-149. Springer, Berlin, Heidelberg (2005)
15. Han, E., Tan, M. M. J., Turk, E., Sridhar, D., Leung, G. M., Shibuya, K., Legido-Quigley, H.: Lessons learnt from easing COVID-19 restrictions: an analysis of countries and regions in Asia Pacific and Europe. The Lancet, **396**(10261), 1525–1534 (2020)
16. He, J., Shang, P., Xiong, H.: Multidimensional scaling analysis of financial time series based on modified cross-sample entropy methods. Physica A: Statistical Mechanics and its Applications, **500**, 210-221 (2018)
17. Kent, J. T., Bibby, J., Mardia, K. V.: Multivariate Analysis. Amsterdam: Academic Press (1979)
18. Kruskal, J.: The relationship between multidimensional scaling and clustering. In: Classification and Clustering, pp. 17-44. Academic Press (1977)
19. Kruskal, J. B.: Multidimensional Scaling (No. 11). Sage (1978)
20. Mardia, K. V.: Some properties of classical multi-dimensional scaling. Communications in Statistics-Theory and Methods, **7**(13), 1233-1241 (1978)
21. Marziano, V., Guzzetta, G., Rondinone, B. M., Boccuni, F., Riccardo, F., Bella, A., Merler, S.: Retrospective analysis of the Italian exit strategy from COVID-19 lockdown. Proceedings of the National Academy of Sciences, **118**(4) (2021)
22. Mead, A.: Review of the development of multidimensional scaling methods. Journal of the Royal Statistical Society: Series D (The Statistician), **41**(1), 27-39 (1992)
23. Pecoraro, F., Luzi, D., Clemente, F.: Analysis of the different approaches adopted in the Italian regions to care for patients affected by COVID-19. International Journal of Environmental Research and Public Health, **18**(3), 848 (2021)
24. Pecoraro, F., Clemente, F., Luzi, D.: The efficiency in the ordinary hospital bed management in Italy: An in-depth analysis of intensive care unit in the areas affected by COVID-19 before the outbreak. PLoS One, **15**(9), e0239249 (2020)
25. Piccolo, D.: Una rappresentazione multidimensionale per modelli statistici dinamici. In: Atti della XXXII Riunione Scientifica della SIS, **2**, pp. 149-160 (1984)
26. Saxena, A., Prasad, M., Gupta, A., Bharill, N., Patel, O. P., Tiwari, A., Lin, C. T.: A review of clustering techniques and developments. Neurocomputing, **267**, 664-681 (2017)
27. Sebastiani, G., Massa, M., Riboli, E.: Covid-19 epidemic in Italy: evolution, projections and impact of government measures. European Journal of Epidemiology, **35**(4), 341-345 (2020)
28. Shang, D., Shang, P., Liu, L.: Multidimensional scaling method for complex time series feature classification based on generalized complexity-invariant distance. Nonlinear Dynamics, **95**(4), 2875-2892 (2019)
29. Studer, M., Ritschard, G.: What matters in differences between life trajectories: A comparative review of sequence dissimilarity measures. Journal of the Royal Statistical Society: Series A (Statistics in Society), **179**(2), 481-511 (2016)
30. Tenreiro Machado, J. A., Lopes, A. M., Galhano, A. M.: Multidimensional scaling visualization using parametric similarity indices. Entropy, **17**(4), 1775-1794 (2015)
31. Torgerson, W. S.: Multidimensional scaling: I. Theory and method. Psychometrika, **17**(4), 401-419 (1952)

Political and Religion Attitudes in Greece: Behavioral Discourses

Georgia Panagiotidou and Theodore Chadjipadelis

Abstract The research presented in this paper attempts to explore the relationship between religious and political attitudes. More specifically we investigate how religious behavior, in terms of belief intensity and practice frequency, is related to specific patterns of political behavior such as ideology, understanding democracy and his set of moral values. The analysis is based on the use of multivariable methods and more specifically Hierarchical Cluster Analysis and Multiple Correspondence Analysis in two steps. The findings are based on a survey implemented in 2019 on a sample of 506 respondents in the wider area of Thessaloniki, Greece. The aim of the research is to highlight the role of people's religious practice intensity in shaping their political views by displaying the profiles resulting from the analysis and linking individual religious and political characteristics as measured with various variables. The final output of the analysis is a map where all variable categories are visualized, bringing forward models of political behavior as associated together with other factors such as religion, moral values and democratic attitudes.

Keywords: political behavior, religion, democracy, multivariate methods, data analysis

1 Introduction

In this research we present the analysis results of a survey, which was implemented in April 2019 to 506 respondents in Thessaloniki, focusing on their religious profile as well as their political attitudes, their moral profile and the way they comprehend democracy. The aim of the analysis is to investigate and highlight the role of religious practice in shaping political behavior. In the political behavior analysis field, religion

Georgia Panagiotidou (✉)
Aristotle University of Thessaloniki, Greece, e-mail: gvpanag@polsci.auth.gr

Theodore Chadjipadelis
Aristotle University of Thessaloniki, Greece, e-mail: chadji@polsci.auth.gr

© The Author(s) 2023
P. Brito et al. (eds.), *Classification and Data Science in the Digital Age*,
Studies in Classification, Data Analysis, and Knowledge Organization,
https://doi.org/10.1007/978-3-031-09034-9_31

and more specifically church practice has emerged as one of the main pillars that form the political attitudes of voters. Religious habits seem to have a decisive influence on electoral choices, as derives from Lazarsfeld's research at Columbia University in 1944 [3], followed by the work of Butler and Stokes in 1969 [1] and the research of Michelat and Simon in France [6]. More specifically in the comparative study of Rose in 1974 [9], it turns out that the more religious voters appear to be more conservative by choosing to place themselves on the right side of the ideological "left-right" axis, while the non-religious voters opt for the left political parties. The research and analysis of Michelat and Simon [6] brings to the surface two opposing cultural models: on the one hand we have the deeply religious voters, who belong to the middle and upper classes, residing in the cities or in the countryside, while on the other hand we have the non-religious left voters with working class characteristics. The first framework is articulated around religion and those who belong to it identifying themselves as religious people, is inspired by a conservative value system, put before the value of the individual, the family, the ancestral heritage and tradition. The second cultural context is articulated around class rivalries and socio-economic realities; those who belong to this context identify themselves as "us workers towards others". They believe in the values of collective action, vote for left-wing parties, participate actively in unions and defend the interests of the working class. To measure the influence of religious practice on political behavior, applied research uses measurement scales about the intensity of religious beliefs and the frequency of church service practice as an indicator of the level of one's religious integration.

To measure religious intensity level, variables are used such as how often they go to the service, how much do they believe in the existence of God, of afterlife, in the dogmas of the church and so on. Since the 90's there is a rapid decline in the frequency with which the population attends church service or self-identifies strongly in terms of religiousness. Nevertheless, the strong correlation between electoral preference and religious practice remains strong [5]. The most significant change for non-religious people is that the left is losing its universal influence as many of these voters expand also to the center. Strongly religious people continue to support the right more and, in some cases, strengthen the far right. In this paper, apart from attempting to explore and verify the existing literature over the effect of religion on political behavior, focusing on the Greek case, the approach exploits methods used to achieve the visualization of all existing relationships between different sets of variables. To link together numerous variables and their categories to construct a model of religious and political behavior, multiple applications of Hierarchical Cluster analysis (HCA) are being made followed by Multiple Correspondence Analysis (MCA) for the emerging clusters. In this way, a semantic map is constructed [7], which visualizes discourses of political and religious behavior and the inner antagonisms between the behavioral profiles.

2 Methodology

For the implementation of the research a poll was conducted on a random sample of 506 people in the greater area of Thessaloniki in Greece, during April 2019. A questionnaire was used as a research tool which was distributed with an on-site approach of the random respondents. The questionnaire consisted of three sections: a) the first section included seven questions for demographic data of the respondent such as gender, age, educational level, marital status, household income, occupation and social class to which the respondent considers belonging; b) the second part contained seven questions, ordinal variables, related to the religious practice and beliefs of the respondent: i) how often does one go to church? ii) how often does one pray? iii) how close does one feel to God, Virgin Mary (or to another seven religious concepts) during church service? iv) how strongly does one have seven different feelings during church service? v) does one believe or not in the saints, miracles, prophecies (and another six religious concepts)? Two more questions investigating their profile in terms of what is taught in the Christian dogma were included vi) one asking if one can progress only by being an ethical person and vii) another one asking if they agree on the pain/righteousness scheme, that is if one suffers in this life will be rewarded later or in the afterlife; c) questions concerning the political profile of the respondent are developed in the third part of the questionnaire: i) one's self-positioning on the ideological left-right axis, ii) a set of nine ordinal variables requiring one's agreement or disagreement level on sentences that reflect the dimensions of liberalism-authoritarianism and left-right iii) this last section also includes two different sets of pictures, used as symbolic representation for the "democratic self" and the "moral self" [4]. The first set of twelve pictures represent various conceptualizations of democracy, and one is asked to select three pictures that represent democracy. The second set of pictures represent moral values in life, and one is asked to choose three pictures that represent one's set of personal values. Variables are ordinal, using a five-point Likert scale, apart from the question regarding whether one believes or not in prophecies magic etc. and the two last questions with the pictures, where we are using a binary scale of yes-no or zero-one where zero is for a non selected picture and one is for a selected picture.

Data analysis was implemented with the use of M.A.D software (Méthodes d'Analyse des Données), developed by Professor Dimitris Karapistolis (more about M.A.D software at www.pylimad.gr). Firstly, Hierarchical Cluster Analysis (HCA) using chi-quare distance and Ward's linkage, assigns subjects into distinct groups based on their response patterns. This first step produces a cluster membership variable, assigning each subject into a group. In addition to this, the behavior typology of each group is examined, seeing the connection of each variable level to each cluster using two proportion z test (significance level set at 0.05) between respondents belonging to cluster i and those who do not belong in cluster i for a variable level. The number of clusters is determined by using the empirical criterion of the change in the ratio of between-cluster inertia to total inertia, when moving from a partition with r clusters to a partition with $r - 1$ clusters [8]. In the second step of the analysis, the cluster membership variable is analyzed together with the existing variables using

MCA on the Burt table [2]. All associations among the variable categories are given on a set of orthogonal axes, with the least possible loss of the original information of the original Burt table. Next, we apply HCA for the coordinates of variable categories on the total number of dimensions of the reduced space resulting from the MCA. In this way we cluster the variable, as previously we clustered the subjects. By clustering the variable response categories, we detect the various discourses of behavior, where each cluster of categories stands as a behavioral profile linked with a set of responses and characteristics. To produce the final output, the semantic map, we created a table including the output variables of the questionnaire, including demographics and variables for political behavior. Using the same two-step procedure using HCA and MCA for this final table, the semantic map is constructed, positioning the variable categories on a bi plot created by the two first dimensions of MCA.

3 Results

In the first step of the analysis, we apply HCA for each set of variables in each question. In the question: "How close do you feel during the service 1-To God, 2-To the Virgin, 3-To Christ, 4-To some Saint, Angel, 5-To the other churchgoers, 6-To Paradise, 7-To Hell, 8-To the divine service, 9-To his preaching priest", we get four clusters (Figure 1).

Cluster	Responses related to the cluster	%
e19837	"not at all" in everything	7,9%
e19882	"enough" in 1,2,3 / "little" or "not at all" in 5,6,9	55,1%
e19883	"a little" in 1,2,3 / "not at all" in 4,5,6,8,9	19,5%
e19884	"absolutely" in everything and "enough" in 5,6,9	17,5%

Fig. 1 Four clusters on how close the respondents feel during church service.

For the question: "How strongly you feel after the end of the service 1-The Grace of God in me, 2-Power of the soul, 3-Forgiveness for those who have hurt me, 4-Forgiveness for my sins, 5-Peace, 6-Relief it is over", we get six clusters (Figure 2).

Cluster	Responses related to the cluster	%
e21902	in everything "absolutely"	9,0%
e21904	"absolutely" peace, strength of soul / "not at all" forgiveness, relief	23,4%
e21905	in all "absolutely" / "not at all" relief	11,8%
e21906	"quite" relief / in all others "a little"	16,8%
e21907	in everything "not at all"	5,9%
e21908	in all "enough"	33,0%

Fig. 2 Six clusters on how the respondents feel at the end of church service.

Five clusters (Figure 3) for the question: "Do you believe in 1-Bad (magic influence) 2-Magic? 3- Destiny? 4-Miracles? 5-Prophecies of the Saints? 6- Do you have pictures of holy figures in your house? 7-in your workplace? 8-Do you have a family Saint?".

Cluster	Responses related to the cluster	%
e22877	yes to miracles and images	23,8%
e22872	yes to miracles, prophecies and pictures	12,0%
e22874	not at all	8,4%
e22875	yes in bad influence, magic, miracles, prophecies and pictures	17,4%
e22879	yes to all	37,8%

Fig. 3 Five clusters on the beliefs of the respondents on various aspects of the Christian faith.

Six clusters are detected (Figure 4) for the question: "How do you feel when you come face to face with a religious image 1-Peace, 2-Awe, 3-The presence of God, 4-Emotion, 5-The need to pray, 6-Contact with the person in the picture".

Cluster	Responses related to the cluster	%
e23856	in everything "not at all"	5,1%
e23887	in all other "moderately" (a little in awe, emotion / enough in prayer)	16,9%
e23890	"not at all" in prayer and person in the picture / in everything else "a little"	9,8%
e23892	in everything "absolutely"	15,3%
e23893	"not at all" in awe / in everything else "a little"	12,4%
e23894	in everything "enough"	40,4%

Fig. 4 Six clusters on how the respondents feel when facing a religious image.

We proceed with the clustering of the replies on political views and we get seven clusters of political profiles (Figure 5).

Cluster	Responses related to the cluster	%
e29881	"strongly agrees" with drachma, individualism, anti-immigrant, anti-EU, welfare state, not leader	7,8%
e29885	"agrees" with welfare state agrees, "disagrees" with all the rest	8,2%
e29886	"agrees" with strong leader, tax cuts	27,6%
e29887	"disagrees" with the right to violence, "agrees" with all the rest	8,9%
e29889	"agrees" with drachma, individualism, anti-immigrants, welfare state, not leader (difference with 881, here simply "agrees" and not interested in EU)	14,0%
e29890	"agrees" with drachma, "disagrees" with all the rest	11,4%
e29891	"agrees" with tax cuts, drachma, anti-immigrant, anti-EU, individualism, strong leader	22,0%

Fig. 5 Seven clusters according to the political views- profile of the respondents.

For the symbolic representation of the democratic self, when choosing three pictures that represent democracy for the respondent, we find eight clusters (Figure 6), and eight clusters for the symbolic representation of the moral self for the respondents, as show in Figure 7.

Cluster	Responses related to the cluster	%
e31892	direct democracy, money, revolution, riot	5,4%
e31893	parliament, money	2,4%
e31914	direct democracy	11,6%
e31916	parliament, council, church	10,9%
e31918	protest, revolution	10,7%
e31920	e-gov	14,2%
e31921	protest, council, revolution	13,3%
e31924	protest, ancient Greece, parliament, volunteering, church	31,5%

Fig. 6 Eight clusters on how the respondents understand democracy.

Cluster	Responses related to the cluster	%
e30970	Christ, intimacy, volunteering, family	24,9%
e30953	fun, intimacy, meditation, win, rebellion	2,2%
e30958	Christ, family, army	13,7%
e30960	meditation, win	7,6%
e30961	fun, career, intimacy, money	7,4%
e30972	career, win, fun, career	17,2%
e30966	career, peace, family	9,4%
e30968	Christ, peace, family	17,6%

Fig. 7 Eight clusters on the different sets of moral values of the respondents.

In the second step of the analysis, we jointly process the cluster membership variables. MCA produces the coefficients of each variable category which are now positioned in a two-dimensional map as seen in Figure 9. HCA is then applied again to the coefficients of the items, which bring forward three main clusters, modeling political and religious behavior. In Figure 8, Cluster 77 is connected to centre and moderate religious behaviour, cluster 78 reflects the voters of the right, with strong religious habits and beliefs, individualistic attitudes and more authoritarian and nationalistic political views, whereas cluster 79 represents the leftists, non-religious voters, closer to revolutionary political views and collective goods. Examining the antagonisms on the behavioral map (Figure 9), the first horizontal axis which explains 22.8% of the total inertia, is created by the antithesis between right political ideology - strong religious behavior and left political ideology-no religious behavior (cluster 78 opposite to cluster 79). The second axis (vertical) accounts for 7% of the inertia, and is explained as the opposition between the center (moderate religious behavior) against the left and right (cluster 77 opposite to both clusters 78 and 79).

Variables	77	78	79
Ethical person	Enough, a little	Absolutely	Not at all
Pain / Righteousness:	A little / moderately	Enough / Very / Absolutely	Not at all
Ideology	Centre	Right	Left
Praying		I pray often	I pray sometimes / I never pray
Church service		I go to church often	I rarely go to church
Political attitudes	{pro-drachma, individualism, anti-immigrant, anti-EU, welfare state, not leader [strongly agrees] { {strong leader, tax cuts (agrees)} {tax cuts, pro- drachma, against immigrants, against EU, person first, strong leader}	{in all others agrees / disagrees on the right to violence} {agrees with drachma, individualism, against immigrants, welfare state, not leader (difference with 881, here simply agrees and there is no EU)} {better with drachma, everything else disagrees}	{welfare state agrees, all the rest disagrees}
Democratic self	{parliament, money} {direct democracy} {e-gov}	{parliament, council, church} {protest, ancient Greece, parliament, volunteering, church}	{direct democracy, money, revolution, riot} {protest, revolution} {protest, council, revolution}
Moral self	{Meditation, win} {Fun, Career, intimacy, Money} {Career, peace, family} {Christ, Peace, family}	{Christ, Family, Army} {Christ, intimacy, volunteering, family}	{fun, intimacy, meditation, win, rebellion} {Career, win, Fun, Career}

Fig. 8 Three main behavioral discourses linking all variable categories together.

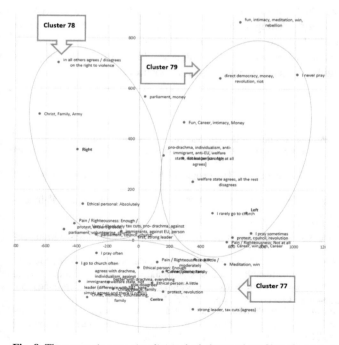

Fig. 9 The semantic map visualizing the behavioral profiles of voters, and the inner antagonisms.

4 Discussion

The analysis uncovers the strong existing relationship between religious habits and political views, for the Greek case. The semantic map indicates two main antagonistic cultural discourses, including both religious, political and moral characteristics: The first discourse (cluster 77) is described as moderately religious practice and beliefs, connected to the ideological center. These voters have political attitudes that belong to the space between the center-left and the center-right. They understand democracy as a connection to money, direct democracy and electronic democracy. Their moral set of values is naturalistic and individualistic. The next behavioral discourse (cluster 78) describes the voters of right ideology, with strong religious beliefs andfrequent religious practice. They appear as very ethical and believe in the concept of pain and righteousness. Regarding their political attitudes these more religious voters are against violence, have more authoritarian and nationalistic positions. They view democracy as parliamentary, representative, ancient Greece but also as church, while their moral set of values appear clearly naturalistic, Christian and nationalistic.

Cluster 79 reflects the exact opposite discourse compared to 78. These voters belong to the left ideology and are non-religious. They do not adopt the ideas of the ethical person, or the scheme of pain and righteousness as mentioned in the Christian dogma. In terms of political attitudes, they are pro-welfare state. These non-religious and left voters understand democracy as direct with the need for revolution, protest and riot and support collective goods. Interpreting further the antagonisms as visualized on the semantic map, the main competition exists between the "right political ideology - strong religious behavior individualism" discourse and the "left political ideology-no religious behavior collectivism" discourse. A secondary opposition is found between the "center ideology- moderate religious behavior" discourse against the left and right extreme positions.

References

1. Butler, D., Stokes, D.: Political Change in Britain. Macmillan, London (1969)
2. Greenacre, M.: Correspondence Analysis in Practice. Chapman and Hall/CRC Press, Boca Raton (2007)
3. Lazarsfeld, P. F., Berelson, B., Gaudet, H.: The People's Choice. Columbia University Press (1944)
4. Marangudakis, M., Chadjipadelis, T.: The Greek Crisis and its Cultural Origins. Palgrave-Macmillan, New York (2019)
5. Mayer, N.: Les Modèles Explicatifs du Vote. L'Harmatan, Paris (1997)
6. Michelat, G., Simon, M.: Classe, Religion et Comportement Politique. PFNSP-Editions Sociales, Paris (1977)
7. Panagiotidou, G., Chadjipadelis, T.: First-time voters in Greece: views and attitudes of youth on Europe and democracy. In T. Chadjipadelis, B. Lausen, A. Markos, T. R. Lee, A. Montanari and R. Nugent (Eds.), Data Analysis and Rationality in a Complex World, Studies in Classification, Data Analysis and Knowledge Organization, pp. 415-429. Springer (2020)
8. Papadimitriou, G., Florou, G.: Contribution of the Euclidean and chi-square metrics to determining the most ideal clustering in ascending hierarchy (in Greek). In Annals in Honor of Professor I. Liakis, 546-581. University of Macedonia, Thessaloniki (1996)
9. Rose, R.: Electoral Behavior: a Comparative Handbook. Free Press, New York (1974)

Supervised Classification via Neural Networks for Replicated Point Patterns

Kateřina Pawlasová, Iva Karafiátová, and Jiří Dvořák

Abstract A spatial point pattern is a collection of points observed in a bounded region of $\mathbb{R}^d$, $d \geq 2$. Individual points represent, e.g., observed locations of cell nuclei in a tissue ($d = 2$) or centers of undesirable air bubbles in industrial materials ($d = 3$). The main goal of this paper is to show the possibility of solving the supervised classification task for point patterns via neural networks with general input space. To predict the class membership for a newly observed pattern, we compute an empirical estimate of a selected functional characteristic (e. g., the pair correlation function). Then, we consider this estimated function to be a functional variable that enters the input layer of the network. A short simulation example illustrates the performance of the proposed classifier in the situation where the observed patterns are generated from two models with different spatial interactions. In addition, the proposed classifier is compared with convolutional neural networks (with point patterns represented by binary images) and kernel regression. Kernel regression classifiers for point patterns have been studied in our previous work, and we consider them a benchmark in this setting.

Keywords: spatial point patterns, pair correlation function, supervised classification, neural networks, functional data

Kateřina Pawlasová (✉)
Charles University, Faculty of Mathematics and Physics, Ke Karlovu 3, 121 16 Praha 2, Czech Republic, e-mail: pawlasova@karlin.mff.cuni.cz

Iva Karafiátová
Charles University, Faculty of Mathematics and Physics, Ke Karlovu 3, 121 16 Praha 2, Czech Republic, e-mail: karafiatova@karlin.mff.cuni.cz

Jiří Dvořák
Charles University, Faculty of Mathematics and Physics, Ke Karlovu 3, 121 16 Praha 2, Czech Republic, e-mail: dvorak@karlin.mff.cuni.cz

© The Author(s) 2023
P. Brito et al. (eds.), *Classification and Data Science in the Digital Age*,
Studies in Classification, Data Analysis, and Knowledge Organization,
https://doi.org/10.1007/978-3-031-09034-9_32

1 Introduction

Spatial point processes have recently received increasing attention in a broad range of scientific disciplines, including biology, statistical physics, or material science [9]. They are used to model the locations of objects or events randomly occurring in $\mathbb{R}^d$, $d \geq 2$. We distinguish between the stochastic model (point process) and its realization observed in a bounded observation window (point pattern).

Typically, analyzing spatial point pattern data means working with just one pattern, which comes from a specific physical measurement. In this paper, we take another perspective: we suppose that a collection of patterns, which are independent realizations of some underlying stochastic models, is to be analyzed simultaneously. These independent realizations are then referred to as replicated point patterns. Recently, this type of data has become more frequent, encouraging the adaptation of methods such as supervised classification to the point pattern setting.

Since we are talking about supervised classification, our task is to predict the label variable (indicating class membership) for a newly observed point pattern, using the knowledge about a sample collection of patterns with known labels (training data). In the literature, this problem has been studied to a limited extent. Properties of a classifier constructed specifically for the situation where the observed patterns were generated by inhomogeneous Poisson point processes with different intensity functions are discussed in [5]. However, this method is based on the special properties of the Poisson point process, and its use is thus limited to a small class of models. On the other hand, no assumptions about the underlying stochastic models are made in [12], where the task for replicated point patterns is transformed, with the help of multidimensional scaling [16], to the classification task in $\mathbb{R}^2$. In [10, 11], the kernel regression classifier for functional data [4] is adapted for replicated point patterns. Instead of classifying the patterns themselves, a selected functional characteristic (e.g. the pair correlation function) is estimated for each pattern. These estimated values are considered functional observations, and the classification if performed in the context of functional data. The idea of linking point patterns to functional data also appears in [12] -- the dissimilarity matrix needed for the multidimensional scaling is based on the same type of dissimilarity measure that is used for the kernel regression classifier in [10, 11]. Finally, [17] briefly discusses the model-based supervised classification. Unsupervised classification is explored in [2].

In this paper, our goal is to discuss the use of classifiers based on artificial neural networks in the context of replicated point patterns. We pay special attention to the procedure described in [14], where both functional and scalar observations enter the input layer. Hence, similarly as in [10, 11], each pattern can be represented by estimated values of a selected functional characteristic and the classification is performed in the context of functional data. The resulting decision about class membership is based on the spatial properties of the observed patterns that can be described by the selected characteristic. Therefore, with a thoughtfully chosen characteristic, this method has great potential within a wide range of possible classification scenarios. Moreover, it can be used without assuming stationarity of the underlying point processes, and it can be easily extended to more complicated settings (e.g., point patterns in non-Euclidean spaces or realizations of random sets).

We present a short simulation experiment that illustrates the behaviour of the neural network described in [14]. Binary classification is performed on realizations of two different point process models – the Thomas process (model for attractive interactions among pairs of points) and the Poisson point process (benchmark model for no interactions among points). This approach is then compared to the classification based on convolutional neural networks (CNNs) [8], where each pattern enters the network as a binary image. Finally, both methods based on artificial neural networks are compared to the kernel regression classifier studied in [10, 11] which can be considered a benchmark in the context of replicated point patterns.

This paper is organized as follows. Section 2 provides a brief theoretical background on spatial point processes and their functional characteristics, including the definition of the pair correlation function, which plays a crucial role in the sequel. Section 3 summarizes the methodology introduced in [14] about neural network models with general input space. Section 4 is devoted to a short simulation example.

2 Point Processes and Point Patterns

This section presents the necessary definitions from the point process theory. Our exposition closely follows the book [13]. For detailed explanation of the theoretical foundations, see, e.g., [7]. Throughout the paper, a simple point process X is defined as a random locally finite subset of $\mathbb{R}^d$, $d \geq 2$, where each point $x \in X$ corresponds to a specific object or event occurring at the location $x \in \mathbb{R}^d$. In applications, X can be used as a mathematical tool to model random locations of cell nuclei in a tissue (with $d = 2$) or centers of undesirable air bubbles in industrial materials ($d = 3$). We distinguish between the mathematical model X, which is called a point process, and its observed realization X, which is called a point pattern. Examples of four different point patterns are given in Figure 1.

Before proper definition of the pair correlation function, a functional characteristic that plays a key role in the sequel, we need to define some moment properties of X. The *intensity function* $\lambda(\cdot)$ is a non-negative measurable function on $\mathbb{R}^d$ such that $\lambda(x)\,dx$ corresponds to the probability of observing a point of X in a neighborhood of x with an infinitesimally small area dx. If X is stationary (its distribution is translation invariant in $\mathbb{R}^d$), then $\lambda(\cdot) = \lambda$ is a constant function and the constant λ is called the *intensity* of X. In this case, λ is interpreted as the expected number of points of X that occur in a set with unit d-dimensional volume. Similarly, the *second-order product density* $\lambda^{(2)}(\cdot, \cdot)$ is a non-negative measurable function on $\mathbb{R}^d \times \mathbb{R}^d$ such that $\lambda^{(2)}(x, y)\,dx\,dy$ corresponds to the probability of observing two points of X that occur jointly at the neighborhoods of x and y with infinitesimally small areas dx and dy.

Assuming the existence of λ and $\lambda^{(2)}$, the *pair correlation function* $g(x, y)$ is defined as $\lambda^{(2)}(x, y)/(\lambda(x)\lambda(y))$, for $\lambda(x)\lambda(y) > 0$. If $\lambda(x) = 0$ or $\lambda(y) = 0$, we set

$g(x, y) = 0$. We write $g(x, y) = g(x - y)$ when g is translation invariant and $g(x, y) = g(\|x - y\|)$ when g is also isotropic (invariant under rotations around the origin). For the Poisson point process, a model for complete spatial randomness, $\lambda^{(2)}(x, y) = \lambda(x)\lambda(y)$ and $g \equiv 1$. Thus, $g(x, y)$ quantifies how likely it is to observe two points in X jointly occurring in infinitesimally small neighbourhoods of x and y, relative to the "no interactions" benchmark.

A large variety of characteristics (both functional and numerical) have been developed to capture various hypotheses about the stochastic models that generated the observed point patterns at hand. We have focused on the pair correlation function g mainly because of its widespread use in practical applications and ease of interpretation. Other popular characteristics are based on g, e.g., its cumulative counterpart, traditionally called the K-function. Others are based on inter-point distances, such as the nearest-neighbor distance distribution function G and the spherical contact distribution function F. A comprehensive summary of commonly used characteristics, including the list of possible empirical estimators, is presented in [9, 13]. Estimators of g, K, G, and F are implemented in the R package spatstat [3].

3 Neural Networks with General Input Space

This section prepares the theoretical background for the supervised classification of replicated point patterns via artificial neural networks. The recent approach of [14, 15] is the cornerstone of our proposed classifier, and hence we focus on its description in the following paragraphs. On the other hand, the approach based on CNNs is more established in the literature. We use it primarily for comparison and thus we refer the reader to [8] for a detailed description.

Following the setup in [14], let us assume that we want to build a neural network such that it takes $K \in \mathbb{N}$ functional variables and $J \in \mathbb{N}$ scalar variables as input. In detail, suppose that we have $f_k : \tau_k \longrightarrow \mathbb{R}$, $k = 1, 2, \ldots, K$ (τ_k are possibly different intervals in $\mathbb{R}$), and $z_j^{(1)} \in \mathbb{R}$, $j = 1, 2, \ldots, J$. Furthermore, suppose that the first layer of the network contains $n_1 \in \mathbb{N}$ neurons. We then want the i-th neuron of the first layer to transfer the value

$$z_i^{(2)} = g\left(\sum_{k=1}^{K} \int_{\tau_k} \beta_{ik}(t) f_k(t) \, dt + \sum_{j=1}^{J} w_{ij}^{(1)} z_j^{(1)} + b_i^{(1)}\right), \quad i = 1, 2, \ldots, n_1,$$

where $b_i^{(1)} \in \mathbb{R}$ is the bias and $g : \mathbb{R} \longrightarrow \mathbb{R}$ is the activation function. Two types of weights appear in the formula: the functional weights $\{\beta_{ik} : \tau_k \longrightarrow \mathbb{R}\}$, and the scalar weights $\{w_{ij}^{(1)}, b_i^{(1)}\}$. The optimal value of all these weights should be found during the training of the network. To overcome the difficulty of finding the optimal weight functions β_{ik}, we can express β_{ik} as a linear combination of $\phi_1, \ldots, \phi_{m_k}$, where $\phi_1, \ldots, \phi_{m_k}$ are the basis functions (from the Fourier or B-spline basis) and m_k is chosen by the user. The sum $\sum_{k=1}^{K} \int_{\tau_k} \beta(t)_{ik} f_k(t) \, dt$ can

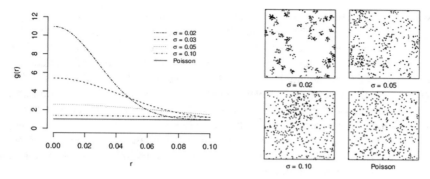

Fig. 1 Theoretical values of the pair correlation function g for the Poisson point process and the Thomas process with different values of the model parameter σ. For these models, g is translation invariant and isotropic. A single realization of the Poisson point process and the Thomas process with parameter σ set to 0.1, 0.05 and 0.02 respectively, is illustrated in the right part of the figure.

be expressed as $\sum_{k=1}^{K} \sum_{l=1}^{m_k} c_{ilk} \int_{\tau_k} \phi_l(t) f_k(t) \, dt$, where the integrals $\int_{\tau_k} \phi_l(t) f_k(t) \, dt$ can be calculated a priori and the coefficients of the linear combination of the basis functions $\{c_{ilk}\}$ act as scalar weights of the first layer and are learned by the network. The scalar values $z_i^{(2)}$, $i = 1, \ldots, n_1$, then propagate through the next fully connected layers as usual. An in-depth analysis of the computational point of view is provided in [14]. In the software R, neural networks with general input space are covered by the package FuncNN [15] built over the packages keras [6] and tensorflow [1]. The last two packages are used to handle CNNs.

4 Simulation Example

This section presents a simple simulation experiment in which we illustrate the performance of the classification rule based on the neural network with general input space. Binary classification is considered, where the group membership indicates whether a point pattern was generated by a stationary Poisson point process or a stationary Thomas process, the latter exhibiting attractive interactions among pairs of points [13]. The sample realizations can be seen in Figure 1.

We consider the Thomas process to be a model with one parameter σ. Small values of σ indicates strong, attractive short-range interactions between points, while larger values of σ result in looser clusters of points. Attractive interactions between the points of a Thomas process result in the values of the pair correlation function being greater than the constant 1, which corresponds to the Poisson case. The effect of σ on the shape of the theoretical pair correlation function of the Thomas process (which is translation invariant and isotropic) is illustrated in Figure 1.

Since the model parameter σ affects the strength and range of attractive interactions between points of the Thomas process, the complexity of the binary classification task described above increases with increasing values of σ [10, 11]. Therefore, this experiment focuses on the situation where σ is set to 0.1, and all realizations are observed on the unit square $[0, 1]^2$. We fix the intensity of the two models to 400 (in spatial statistics, patterns with several hundreds of points are standard nowadays). In this framework, we expect the classification task to be challenging enough to observe differences in the performance of the considered classifiers. On the other hand, it is still reasonable to distinguish (w.r.t. the chosen observation window) the realizations of the model with attractive interactions from the realizations corresponding to the complete spatial randomness.

Two different collections of labelled point patterns are considered as training sets. The first, referred to as *Training data 1*, is composed of 1 000 patterns per group. The second, called *Training data 2*, is then composed of 100 patterns per group. The test and validation sets have the same size and composition as the *Training data 2*. Table 1 presents the accuracy of three classification rules (described below) with respect to the test set. For the first two rules, the accuracy is in fact averaged over five runs corresponding to different settings of initial weights in the underlying neural network. Concerning the network architecture, we fix the ReLU function to be the activation function for all layers, except the output one. The output layer consists of one neuron with sigmoid activation function. The loss function is the binary cross-entropy. A detailed description of the individual layers is given below.

Rule 1 is based on the neural network with general input space. We set K and J from Sect. 3 to be 1 and 0, respectively, and $\tau_1 = (0, 0.25)$. The value 0.25 is related to the observation window of the point patterns at hand being $[0, 1]^2$. Then, f_1 is the vector of the estimated values of the pair correlation function g (estimated by the function `pcf.ppp` from the package `spatstat` [3] with default settings but the option `divisor` set to d), considered as a functional observation. Furthermore, we set $m_1 = 29$, and consider the Fourier basis. The data preparation (estimation of g, computation of integrals from Sect. 3) takes 740 s of elapsed time (w.r.t. the *Training data 1*, on a standard personal computer). To tune the hyperparameters of the final neural network (number of hidden layers, number of neurons per hidden layers, dropout, etc.), we performed a rough grid search (models with various combinations of the hyperparameters were trained on *Training data 1* and we used the loss function and the accuracy computed on the validation set to compare the performances). The resulting network consists of one hidden layer with 128 neurons followed by a dropout layer with a rate of 0.3. We use the Adam optimizer, and the learning rate is decaying exponentially, with initial value 0.001 and decay parameter 0.05. In total, the network has 3 969 trainable parameters. To train the network, we perform 50 epochs with an average elapsed time of 200 ms per epoch (w.r.t. *Training data 1*).

Rule 2 uses CNNs. Similarly to the previous case, our decision about the network architecture is based on a rough grid search. The final network has two convolutional layers, each of them with 8 filters, a squared kernel matrix with 36 (first layer) or 16 rows (second layer), and a following average pooling layer with the pool size fixed at 2×2. We add a dropout layer after the pooling, with a rate of 0.3 (after the first

Table 1 Accuracy for the three presented classification rules w.r.t. the testing set. For *Rule 1* and *Rule 2*, the accuracy is averaged over five runs corresponding to five different choices of initial weights in the underlying neural networks. In addition, the standard deviation computed from the five accuracy values is reported. Values close to 1 indicate a nearly perfect classification.

	Rule 1	Rule 2	Rule 3
Training data 1	**0.947** ±0.003	0.934 ±0.032	0.935
Training data 2	0.895 ±0.010	0.512 ±0.028	**0.925**

pooling) and 0.2 (after the second pooling). The batch size is set to 32. We use the Adam optimizer, and the learning rate is decaying exponentially, with initial value 0.001 and decay parameter 0.1. The total number of trainable parameters is equal to 32 785 and we perform 50 epochs with the average elapsed time per epoch (w.r.t. *Training data 1*) equal to 930 s. Data preparation (converting point patterns to binary images) takes less than 10 s of the elapsed time (w.r.t. *Training data 1*).

Rule 3 is the kernel regression classifier studied in [10, 11]. We use the Epanechnikov kernel together with an automatic procedure for the selection of the smoothing parameter. The underlying dissimilarity measure for point patterns is constructed as the integrated squared difference of the corresponding estimates of the pair correlation function g; for more details, see [10]. The elapsed time needed to compute the upper triangle of the dissimilarity matrix (containing dissimilarities between every pair of patterns from *Training data 1*) is equal to 390 s. To predict the class membership for the testing set (w.r.t. *Training data 1*), 206 s elapsed. During the classification procedure, no random initialization of any weights is needed. Thus, there is no reason to average the accuracy in Table 1 over multiple runs.

For *Training data 1*, Table 1 shows that the highest accuracy was achieved for the neural network with general input space. The standard deviation of the five different accuracy values is significantly higher for CNN which has almost ten times more trainable parameters than the network with general input space. For *Training data 2*, the kernel regression method achieved the highest accuracy. In this situation, the performance of the classifier is stable even in the case of small training data. For the first two rules, the neural network models chosen with the help of the grid search (where the networks were trained w.r.t. the bigger training set) are now trained w.r.t. the smaller training set. The resulting accuracy is still around 0.90 for the network with general input space, but it drops to 0.5 (random assignment of labels) for CNN. The size of *Training data 2* seems to be too small to successfully optimize the large amount of trainable parameters of the convolutional network.

To conclude, our simulation example suggests that the classifier based on CNN (using information about the precise configuration of points) is in the presented situation outperformed by the classifiers based on the estimated values of the pair correlation function (using information about the interactions between pairs of points). The high number of trainable parameters of the CNN makes its use rather demanding with respect to computational time. The approach based on neural networks with general input space proved to be competitive with or even outperform the current benchmark method (kernel regression classifier), especially for large datasets. Also, it has the lowest demands regarding computational time. In the case of a small dataset, the low number of hyperparameters speaks in favor of kernel regression. Finally, in the simple classification scenario that we have presented, the choice of the pair correlation function was adequate. In practical applications, a problem-specific characteristic should be constructed to achieve satisfactory performance.

Acknowledgements The work of Kateřina Pawlasová and Iva Karafiátová has been supported from the Grant schemes at Charles University, project no. CZ.02.2.69/0.0/0.0/19 073/0016935. The work of Jiří Dvořák has been supported by the Czech Grant Agency, project no. 19-04412S.

References

1. Allaire, J. J., Eddelbuettel, D., Golding, N., Tang, Y.: tensorflow: R Interface to TensorFlow (2016) Available at GitHub. https://github.com/rstudio/tensorflow.Cited10Jan 2022
2. Ayala, G., Epifanio, I., Simo, A., Zapater, V.: Clustering of spatial point patterns. Comput. Stat. Data. Anal. **50**, 1016–1032 (2006)
3. Baddeley, A., Rubak, E., Turner, R.: Spatial Point Patterns: Methodology and Applications with R. Chapman & Hall/CRC Press, Boca Raton (2015)
4. Ferraty, F., Vieu, P.: Nonparametric Functional Data Analysis. Theory and Practice. Springer-Verlag, New York (2006)
5. Cholaquidis, A., Forzani, L., Llop, P., Moreno, L.: On the classification problem for Poisson point processes. J. Multivar. Anal. **153**, 1–15 (2017)
6. Chollet, F., Allaire, J. J. and others: R Interface to Keras (2017) Available via GitHub. https://github.com/rstudio/keras.Cited10Jan2022
7. Daley, D., Vere-Jones, D.: An Introduction to the Theory of Point Processes. Vol II., 2nd edn. Springer-Verlag, New York (2008)
8. Goodfellow, I., Bengio, Y., Courville, A.: Deep Learning. MIT Press, Cambridge (2016)
9. Illian, J., Penttinen, A., Stoyan, H., Stoyan, D.: Statistical Analysis and Modelling of Spatial Point Patterns. Wiley, Chichester (2004)
10. Koňasová, K., Dvořák, J.: Techniques from functional data analysis adaptable for spatial point patterns (2021) In: Proceedings of the 22nd European Young Statisticians Meeting. https://www.eysm2021.panteion.gr/publications.html.Cited10Jan2022
11. Koňasová, K., Dvořák, J.: Supervised nonparametric classification in the context of replicated point patterns. Submitted (2021)
12. Mateu, J., Schoenberg, F. P., Diez, D. M., González, J. A., Lu, W.: On measures of dissimilarity between point patterns: classification based on prototypes and multidimensional scaling. Biom. J. **57**, 340–358 (2015)
13. Møller, J., Waagepetersen, R.: Statistical Inference and Simulation for Spatial Point Processes. Chapman & Hall/CRC, Boca Raton (2004)
14. Thind, B., Multani, K., Cao, J.: Deep Learning with Functional Inputs (2020) Available via arxiv. https://arxiv.org/pdf/2006.09590.pdf.Cited10Jan2022
15. Thind, B., Wu, S., Groenewald, R., Cao, J.: FuncNN: An R Package to Fit Deep Neural Networks Using Generalized Input Spaces (2020) Available via arxiv. https://arxiv.org/pdf/2009.09111.pdf.Cited10Jan2022
16. Torgerson, W.: Multidimensional Scaling: I. Theory and Method. Psychometrika. **17**, 401–419 (1952)
17. Vo, B. N., Dam, N., Phung, D., Tran, Q. N., Vo, B. T.: Model-based learning for point pattern data. Pattern Recognit. **84**, 136–151 (2018)

Parsimonious Mixtures of Seemingly Unrelated Contaminated Normal Regression Models

Gabriele Perrone and Gabriele Soffritti

Abstract In recent years, the research into linear multivariate regression based on finite mixture models has been intense. With such an approach, it is possible to perform regression analysis for a multivariate response by taking account of the possible presence of several unknown latent homogeneous groups, each of which is characterised by a different linear regression model. For a continuous multivariate response, mixtures of normal regression models are usually employed. However, in real data, it is not unusual to observe mildly atypical observations that can negatively affect the estimation of the regression parameters under a normal distribution in each mixture component. Furthermore, in some fields of research, a multivariate regression model with a different vector of covariates for each response should be specified, based on some prior information to be conveyed in the analysis. To take account of all these aspects, mixtures of contaminated seemingly unrelated normal regression models have been recently developed. A further extension of such an approach is presented here so as to ensure parsimony, which is obtained by imposing constraints on the group-covariance matrices of the responses. A description of the resulting parsimonious mixtures of seemingly unrelated contaminated regression models is provided together with the results of a numerical study based on the analysis of a real dataset, which illustrates their practical usefulness.

Keywords: contaminated normal distribution, ECM algorithm, mixture of regression models, model-based cluster analysis, seemingly unrelated regression

Gabriele Perrone (✉)
Department of Statistical Sciences, University of Bologna, via delle Belle Arti 41, 40126 Bologna, Italy, e-mail: gabriele.perrone4@unibo.it

Gabriele Soffritti
Department of Statistical Sciences, University of Bologna, via delle Belle Arti 41, 40126 Bologna, Italy, e-mail: gabriele.soffritti@unibo.it

P. Brito et al. (eds.), *Classification and Data Science in the Digital Age*,
Studies in Classification, Data Analysis, and Knowledge Organization,
https://doi.org/10.1007/978-3-031-09034-9_33

1 Introduction

Seemingly unrelated (SU) regression equations are usually employed in a multivariate regression analysis whenever the dependence of a vector $\mathbf{Y} = (Y_1, \ldots, Y_M)'$ of M continuous variables on a vector $\mathbf{X} = (X_1, \ldots, X_P)'$ of P regressors has to be modelled by allowing the error terms in the different equations to be correlated and, thus, the regression parameters of the M equations have to be jointly estimated [14]. With such an approach, the researcher is also enabled to convey prior information on the phenomenon under study into the specification of the regression equations by defining a different vector of regressors for each dependent variable. This latter feature is particularly useful in any situation in which different regressors are expected to be relevant in the prediction of different responses, such as in [3, 6, 16]. This approach has been recently embedded into the framework of Gaussian mixture models, leading to multivariate SU normal regression mixtures [7]. In these models, the effect of the regressors on the dependent variables changes with some unknown latent sub-populations composing the population that has generated the sample of observations to be analysed. Thus, when the sample is characterised by unobserved heterogeneity, model-based cluster analysis is simultaneously carried out.

Another source of complexity which could affect the data and make the prediction of $\mathbf{Y}$ a difficult task to perform is represented by mildly atypical observations [13]. Robust methods of parameter estimation insensitive to the presence of such observations in a sample characterised by unobserved heterogeneity have been introduced in [9], where the conditional distribution $\mathbf{Y}|\mathbf{X} = \mathbf{x}$ is modelled through a mixture of K multivariate contaminated normal models, where K is the number of the latent sub-populations. A limitation associated with these latter models is that the same vector of regressors has to be specified for the prediction of all the dependent variables. To overcome this limitation while preserving all the features mentioned above, a more flexible approach which employs mixtures of multivariate SU contaminated normal regression models has been recently introduced in [11]. These latter models are able to capture the linear effects of the regressors on the dependent variables from sample observations coming from heterogeneous populations. The researcher is also enabled to specify a different vector of regressors for each dependent variable. Finally, a robust estimation of the regression parameters and the detection of mild outliers in the data are ensured.

In the presence of many responses and many latent sub-populations, analyses based on these latter models can become unfeasible in practical applications because of a large number of model parameters. In order to keep this number as low as possible, an approach due to [4], based on the spectral decompositions of the K covariance matrices of $\mathbf{Y}|\mathbf{X} = \mathbf{x}$, is exploited here so as to obtain fourteen different covariance structures. The resulting parsimonious mixtures of SU contaminated regression models are described in Section 2. The usefulness of these new models is illustrated through a study aiming at determining the effect of prices and promotional activities on sales of canned tuna in the US market. A summary of the obtained results is provided in Section 3.

2 Parsimonious SU Contaminated Normal Regression Mixtures

In a system of M SU regression equations for modelling the linear dependence of $\mathbf{Y}$ on $\mathbf{X}$, let $\mathbf{X}_m = (X_{m_1}, X_{m_2}, \ldots, X_{m_{P_m}})'$ be the P_m-dimensional sub-vector of $\mathbf{X}$ composed of the P_m regressors expected to be relevant for the explanation of Y_m, for $m = 1, \ldots, M$. Furthermore, let $\mathbf{X}_m^* = (1, \mathbf{X}_m')'$. The mixture of K SU normal regression models described in [7] can be defined as follows:

$$
\mathbf{Y} = \begin{cases} \tilde{\mathbf{X}}^{*\prime} \boldsymbol{\beta}_1^* + \boldsymbol{\epsilon}, & \boldsymbol{\epsilon} \sim N_M(\mathbf{0}_M, \boldsymbol{\Sigma}_1) \text{ with probability } \pi_1, \\ \cdots \\ \tilde{\mathbf{X}}^{*\prime} \boldsymbol{\beta}_K^* + \boldsymbol{\epsilon}, & \boldsymbol{\epsilon} \sim N_M(\mathbf{0}_M, \boldsymbol{\Sigma}_K) \text{ with probability } \pi_K, \end{cases} \tag{1}
$$

where π_k is the prior probability of the kth latent sub-population, with $\pi_k > 0$ for $k = 1, \ldots, K$; $\sum_{k=1}^K \pi_k = 1$; $\tilde{\mathbf{X}}^*$ is the following $(P^* + M) \times M$ partitioned matrix:

$$
\tilde{\mathbf{X}}^* = \begin{bmatrix} \mathbf{X}_1^* & \mathbf{0}_{P_1+1} & \cdots & \mathbf{0}_{P_1+1} \\ \mathbf{0}_{P_2+1} & \mathbf{X}_2^* & \cdots & \mathbf{0}_{P_2+1} \\ \vdots & \vdots & & \vdots \\ \mathbf{0}_{P_M+1} & \mathbf{0}_{P_M+1} & \cdots & \mathbf{X}_M^* \end{bmatrix},
$$

with $\mathbf{0}_{P_m+1}$ denoting the $(P_m + 1)$-dimensional null vector; $P^* = \sum_{m=1}^M P_m$; $\boldsymbol{\beta}_k^* = (\boldsymbol{\beta}_{k1}^{*\prime}, \ldots, \boldsymbol{\beta}_{km}^{*\prime}, \ldots, \boldsymbol{\beta}_{kM}^{*\prime})'$ is the $(P^* + M)$-dimensional vector containing all the linear effects on the M responses in the kth latent sub-population, with $\boldsymbol{\beta}_{km}^* = (\beta_{0k,m}, \boldsymbol{\beta}_{km}')'$, for $m = 1, \ldots, M$; $\boldsymbol{\epsilon} = (\epsilon_1, \ldots, \epsilon_M)'$ is the vector of the errors, which are supposed to be independent and identically distributed; $N_M(\mathbf{0}_M, \boldsymbol{\Sigma}_k)$ denotes the M-dimensional normal distribution with mean vector $\mathbf{0}_M$ and positive-definite covariance matrix $\boldsymbol{\Sigma}_k$. From now on, this mixture regression model is denoted as MSUN. When $\mathbf{X}_m = \mathbf{X}\ \forall m$ (the P regressors are employed in all the M equations), model (1) reduces to the mixtures of K normal (MN) regression models (see [8]).

When the data are contaminated by the presence of mild outliers, departures from the normal distribution could be observed within any of the K latent sub-populations. A model able to manage this situation has been recently introduced in [11]. It has been obtained from equation (1) by replacing the normal distribution with the contaminated normal distribution. Under this latter distribution, the probability density function (p.d.f.) of $\boldsymbol{\epsilon}$ within the kth sub-population is equal to $h(\boldsymbol{\epsilon}; \boldsymbol{\vartheta}_k) = \alpha_k \phi_M(\boldsymbol{\epsilon}; \mathbf{0}_M, \boldsymbol{\Sigma}_k) + (1 - \alpha_k)\phi_M(\boldsymbol{\epsilon}; \mathbf{0}_M, \eta_k \boldsymbol{\Sigma}_k)$, where $\phi_M(\cdot; \boldsymbol{\mu}, \boldsymbol{\Sigma})$ denotes the p.d.f. of the distribution $N_M(\mathbf{0}_M, \boldsymbol{\Sigma}_k)$, $\alpha_k \in (0.5, 1)$ and $\eta_k > 1$ are the proportion of typical observations within the kth sub-population and a parameter that inflates the elements of $\boldsymbol{\Sigma}_k$, respectively, and $\boldsymbol{\vartheta}_k = (\alpha_k, \eta_k, \boldsymbol{\Sigma}_k)$. As a consequence, a mixture of K SU contaminated normal (MSUCN) regression models is given by:

$$
\mathbf{Y} = \begin{cases} \tilde{\mathbf{X}}^{*\prime} \boldsymbol{\beta}_1^* + \boldsymbol{\epsilon}, & \boldsymbol{\epsilon} \sim CN_M(\alpha_1, \eta_1, \mathbf{0}_M, \boldsymbol{\Sigma}_1) \text{ with probability } \pi_1, \\ \cdots \\ \tilde{\mathbf{X}}^{*\prime} \boldsymbol{\beta}_K^* + \boldsymbol{\epsilon}, & \boldsymbol{\epsilon} \sim CN_M(\alpha_K, \eta_K, \mathbf{0}_M, \boldsymbol{\Sigma}_K) \text{ with probability } \pi_K, \end{cases} \tag{2}
$$

where $CN_M(\alpha_k, \eta_k, \mathbf{0}_M, \Sigma_k)$ denotes the M-dimensional contaminated normal distribution described by the p.d.f. $h(\epsilon; \boldsymbol{\vartheta}_k)$. The parameter vector of model (2) is $\boldsymbol{\psi} = (\boldsymbol{\psi}_1, \ldots, \boldsymbol{\psi}_k, \ldots, \boldsymbol{\psi}_K)$, where $\boldsymbol{\psi}_k = (\pi_k, \boldsymbol{\theta}_k)$, $\boldsymbol{\theta}_k = (\boldsymbol{\beta}_k^*, \boldsymbol{\vartheta}_k)$. The number of free elements of $\boldsymbol{\psi}$ is $n_\psi = 3K - 1 + K(P^* + M) + n_\sigma$, where n_σ denotes the total number of free variances and covariances, with $n_\sigma = Kn_\Sigma$ and $n_\Sigma = \frac{M(M+1)}{2}$. When $\mathbf{X}_m = \mathbf{X} \ \forall m$, model (2) coincides with the mixture of K contaminated normal (MCN) regression models described in [9]. For $\alpha_k \to 1$ or $\eta_k \to 1 \ \forall k$, model (2) reduces to model (1). Conditions ensuring identifiability of models (2) are provided in [11]. The ML estimation of $\boldsymbol{\psi}$ in equation (2) can be carried out by means of a sample $S = \{(\mathbf{x}_1, \mathbf{y}_1), \ldots, (\mathbf{x}_I, \mathbf{y}_I)\}$ of I independent observations drawn from model (2) and an expectation-conditional maximisation (ECM) algorithm [10]. Details about this algorithm, including strategies for the initialisation of $\boldsymbol{\psi}$ and convergence criteria, are illustrated in [11]. In practical applications, the value of K is generally unknown and has to be properly chosen. This task can be carried out by resorting to model selection criteria, such as the Bayesian information criterion [15]: $BIC = 2\ell(\hat{\boldsymbol{\psi}}) - n_\psi \ln I$, where $\hat{\boldsymbol{\psi}}$ is the maximum likelihood estimator of $\boldsymbol{\psi}$. Another commonly used information criterion is the integrated completed likelihood [2], which admits two slightly different formulations: $ICL_1 = BIC + 2\sum_{i=1}^{I}\sum_{k=1}^{K} \text{MAP}(\hat{z}_{ik}) \ln \hat{z}_{ik}$ and $ICL_2 = BIC + 2\sum_{i=1}^{I}\sum_{k=1}^{K} \hat{z}_{ik} \ln \hat{z}_{ik}$, where $\hat{z}_{ik}$ is the estimated posterior probability that the ith sample observation come from the kth sub-population (for further details see [11]), $\text{MAP}(\hat{z}_{ik}) = 1$ if $\max_h\{\hat{z}_{ih}\}$ occurs when $h = k$ ($\text{MAP}(\hat{z}_{ik}) = 0$ otherwise). Whenever the specification of the subvectors $\mathbf{X}_m$, $m = 1, \ldots, M$, to be considered in the M equations of the multivariate regression model is questionable, such criteria can also be employed to perform subset selection.

As the number of free parameters n_ψ increasases quadratically with M, analyses based on model (2) can become unfeasible in real applications. A way to manage this problem can be based on the introduction of suitable constraints on the elements of Σ_k, $k = 1, \ldots, K$, based on the following eigen-decomposition [4]: $\Sigma_k = \lambda_k \mathbf{D}_k \mathbf{A}_k \mathbf{D}_k'$, where $\lambda_k = |\Sigma_k|^{1/M}$, $\mathbf{A}_k$ is a diagonal matrix with entries (sorted in decreasing order) proportional to the eigenvalues of Σ_k (with the constraint $|\mathbf{A}_k| = 1$) and $\mathbf{D}_k$ is a $M \times M$ orthogonal matrix of the eigenvectors of Σ_k (ordered according to the eigenvalues). This decomposition allows to obtain variances and covariances in Σ_k from λ_k, $\mathbf{A}_k$ and $\mathbf{D}_k$. From a geometrical point of view, λ_k determines the volume, $\mathbf{A}_k$ the shape and $\mathbf{D}_k$ the orientation of the kth cluster of sample observations detected by the fitted model. By constraining λ_k, $\mathbf{A}_k$ and $\mathbf{D}_k$ to be equal or variable across the K clusters, a class of fourteen mixtures of K SUCN regression models is obtained (see Table 1). With variable volumes, shapes and orientations (VVV in Table 1), the resulting model coincides with (2). When $K > 1$, the other covariance structures allow to obtain thirteen different parsimonious mixtures of K SUCN regression models (i.e.: with a reduced n_σ). When $K = 1$, the possible covariance structures for Σ_1 are: diagonal with different entries, diagonal with the same entries and fully unconstrained. The ML estimation of $\boldsymbol{\psi}$ under model (2) with any of these parameterisations can be carried out through an ECM algorithm in which the CM-step update for Σ_k can be computed either in closed form or using iterative procedures, depending on the parameterisation to be employed (see [4]).

Table 1 Features of the parameterisations for the covariance matrices Σ_k, $k = 1, \ldots, K$ $(K > 1)$.

Acronym	Covariance structure	Volume	Shape	Orientation	CM step	n_σ
EEE	$\lambda\mathbf{DAD}'$	Equal	Equal	Equal	Closed	n_Σ
VVV	$\lambda_k\mathbf{D}_k\mathbf{A}_k\mathbf{D}'_k$	Variable	Variable	Variable	Closed	$K n_\Sigma$
EII	$\lambda\mathbf{I}$	Equal	Spherical	–	Closed	1
VII	$\lambda_k\mathbf{I}$	Variable	Spherical	–	Closed	K
EEI	$\lambda\mathbf{A}$	Equal	Equal	Axis-aligned	Closed	M
VEI	$\lambda_k\mathbf{A}$	Variable	Equal	Axis-aligned	Iterative	$M + K - 1$
EVI	$\lambda\mathbf{A}_k$	Equal	Variable	Axis-aligned	Closed	$MK - (K-1)$
VVI	$\lambda_k\mathbf{A}_k$	Variable	Variable	Axis-aligned	Closed	MK
EEV	$\lambda\mathbf{D}_k\mathbf{AD}'_k$	Equal	Equal	Variable	Iterative	$K n_\Sigma - (K-1)M$
VEV	$\lambda_k\mathbf{D}_k\mathbf{AD}'_k$	Variable	Equal	Variable	Iterative	$K n_\Sigma - (K-1)(M-1)$
EVE	$\lambda\mathbf{DA}_k\mathbf{D}'$	Equal	Variable	Equal	Iterative	$n_\Sigma - (K-1)(M-1)$
VVE	$\lambda_k\mathbf{DA}_k\mathbf{D}'$	Variable	Variable	Equal	Iterative	$n_\Sigma - (K-1)M$
VEE	$\lambda_k\mathbf{DAD}'$	Variable	Equal	Equal	Iterative	$n_\Sigma - (K-1)$
EVV	$\lambda\mathbf{D}_k\mathbf{A}_k\mathbf{D}'_k$	Equal	Variable	Variable	Iterative	$K n_\Sigma - (K-1)$

3 Analysis of U.S. Canned Tuna Sales

The models illustrated in Section 2 have been fitted to a dataset [5] containing the volume of sales (Move), a measures of the display activity (Nsale) and the log price (Lprice) for seven of the top 10 U.S. brands in the canned tuna product category in the $I = 338$ weeks between September 1989 and May 1997. The goal of the analysis is to study the dependence of canned tuna sales on prices and promotional activites for two products: Star Kist 6 oz. (SK) and Bumble Bee Solid 6.12 oz. (BBS). To this end, the following vectors have been considered: $\mathbf{Y}' = (Y_1 = \text{Lmove SK}, Y_2 = \text{Lmove BBS})$, $\mathbf{X}' = (X_1 = \text{Nsale SK}, X_2 = \text{Lprice SK}, X_3 = \text{Nsale BBS}, X_4 = \text{Lprice BBS})$, where Lmove denotes the logarithm of Move. The analysis has been carried out using all the parameterisations of the MSUN, MN, MCSUN and MCN models for each $K \in \{1, 2, 3, 4, 5, 6\}$. Furthermore, MSUN and MCSUN models have been fitted by considering all possible subvectors of $\mathbf{X}$ as vectors $\mathbf{X}_m$, $m = 1, 2$, for each K. In this way, best subset selections for Lmove SK and Lmove BBS have been included in the analysis both with and without contamination. The overall number of fitted models is 37376, including the fully unconstrained models (i.e., with the VVV parameterisation) previously employed in [11] to perform the same analysis.

Table 2 reports some information about the nine models which best fit the analysed dataset according to the three model selection criteria over the six examined values of K within each model class. An analysis based on a single linear regression model $(K = 1)$, both with and without contamination, appears to be inadequate according to all criteria. All the examined criteria indicate that the overall best model for studying the effect of prices and promotional activities on sales of SK and BBS tuna is a parsimonious mixture of two SU contaminated Gaussian linear regression models with the EVE parameterisation for the covariance matrices in which the log unit sales of SK tuna are regressed on the log prices and the promotional activites of the same brand, while the regressors selected for the BBS log unit sales are the log prices of both brands and the promotional activites of BBS. Thus, the analysis suggests that two sources of complexity affect the analysed dataset: unobserved heterogeneity over time ($K = 2$ clusters of weeks have been detected) and the presence of mildly atypical observations. Since the two estimated proportions of typical observations are quite similar (see the values of $\hat{\alpha}_k$ in Table 3), contamination seems to characterise the two clusters of weeks detected by the model almost in the same way. As far as the strength of the contaminating effects on the conditional variances and covariances of $\mathbf{Y}|\mathbf{X} = \mathbf{x}$ is concerned, it appears to be stronger in the first cluster, where the estimated inflation parameter is larger ($\hat{\eta}_1 = 15.70$). By focusing the attention on the other estimates, it appears that also some of the estimated regression coefficients, variances and covariances are affected by heterogeneity over time. Sales of SK tuna results to be negatively affected by prices and positively affected by promotional activites of the same brand within both clusters detected by the model, but with effects which are sligthly stronger in the first cluster of weeks. A similar behavior is detected for the estimated regression equation for Lmove BBS, which also highlights that Lmove BBS are positively affected by the log prices of SK tuna, especially in the first cluster of weeks. Furthermore, typical weeks in the first cluster show values of Lmove SK which are more homogeneous than those of Lmove BBC; the opposite holds true for the typical weeks belonging to the second cluster. Also the correlation between log sales of SK and BBS products results to be affected by heterogeneity over time: while in the largest cluster of weeks this correlation has been estimated to be slightly positive (0.200), the first cluster is characterised by a mild estimated negative correlation (−0.151). An interesting feature of this latter cluster is that 17 out of the 20 weeks which have been assigned to this cluster are consecutive from week no. 58 to week no. 74, which correspond to the period from mid-October 1990 to mid-February 1991 characterised by a worldwide boycott campaign encouraging consumers not to buy Bumble Bee tuna because Bumble Bee was found to be buying yellow-fin tuna caught by dolphin-unsafe techniques [1]. Such events could represent one of the sources of the unobserved heterogeneity detected by the model. According to the overall best model, some weeks have beed detected to be mild outliers. In the first cluster, this has happened for week no. 60 (immediately after Halloween 1990) and week no. 73 (two weeks immediately before Presidents day 1999). The analysis of the estimated sample residuals $\mathbf{y}_i - \hat{\boldsymbol{\mu}}_1(\mathbf{x}_i; \hat{\boldsymbol{\beta}}_1^*)$ for the 20 weeks belonging to the first cluster (see the scatterplot on the left side of Figure 1) clearly show that weeks 60 and 73 noticeably deviates from the other weeks. Among the 318 weeks of the second cluster, 32 have resulted to be mild outliers, most of which are associated with holidays and special events that took place between September 1989 and mid-October 1990 or between mid-February and May 1997 (see the scatterplot on the right side of Figure 1). These results are almost equal to those obtained using the best overall fully unconstrained fitted model in the analysis presented in [11]. However, the EVE parameterisation for the MSUCN model has allowed to obtain a better trade-off among the fit, the model complexity and the uncertainty of the estimated partition of the weeks; furthermore, it has led to a slightly lower number of mild outliers in the second cluster of weeks.

Table 2 Maximised log-likelihood $\ell(\hat{\psi})$ and values of BIC, ICL_1 and ICL_2 for nine models selected from the classes MSUCN, MCN, MSUN and MN in the analysis of tuna sales.

Model class	K	Acronym	$\mathbf{X}_1$	$\mathbf{X}_2$	$\ell(\hat{\psi})$	n_ψ	BIC	ICL_1	ICL_2
MSUCN	2	EVE	X_1, X_2	X_2, X_3, X_4	−242.9	23	−619.8	−625.7	−635.8
MCN	2	EVI	$\mathbf{X}$	$\mathbf{X}$	−239.6	28	−642.2	−648.9	−663.2
MCN	2	EEV	$\mathbf{X}$	$\mathbf{X}$	−240.8	29	−650.6	−650.8	−652.0
MCN	3	EVI	X_1, X_2, X_4	X_1, X_2, X_4	−214.2	36	−638.0	−703.1	−788.6
MSUN	2	VEV	X_1, X_2	X_3, X_4	−279.3	18	−663.4	−673.1	−692.1
MSUN	3	EEV	X_2, X_3	X_2, X_3, X_4	−259.8	28	−682.7	−684.7	−688.0
MSUN	5	VVV	X_2, X_3	X_1, X_4	−167.4	49	−620.0	−701.1	−780.3
MN	3	EEV	X_2, X_3, X_4	X_2, X_3, X_4	−258.7	31	−697.9	−699.6	−702.1
MN	4	VVE	X_2, X_4	X_2, X_4	−216.6	36	−642.9	−725.3	−832.9

Table 3 Parameter estimates of the overall best model for the analysis of tuna sales.

$\hat{\psi}$	$k = 1$	$k = 2$
$\hat{\pi}_k$	0.062	0.938
$\hat{\alpha}_k$	0.810	0.844
$\hat{\eta}_k$	15.70	6.94
$\hat{\beta}_{k1}^{\prime *}$	$(8.87, 0.56, -4.70)$	$(8.64, 0.27, -3.09)$
$\hat{\beta}_{k2}^{\prime *}$	$(15.04, 3.92, 2.83, -17.76)$	$(9.98, 0.25, 0.12, -3.83)$
$\hat{\Sigma}_k$	$\begin{pmatrix} 0.034 & -0.009 \\ -0.009 & 0.105 \end{pmatrix}$	$\begin{pmatrix} 0.121 & 0.012 \\ 0.012 & 0.030 \end{pmatrix}$

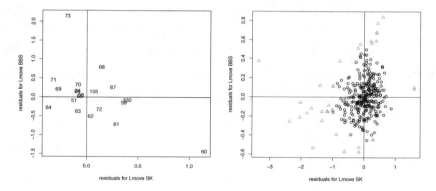

Fig. 1 Scatterplots of the estimated residuals for the weeks assigned to the first (left) and second (right) clusters detected by the overall best model. Points of the first scatterplot are labelled with the number of the corresponding weeks. Black circle and red triangle in the second scatterplot correspond to typical and outlying weeks, respectively.

4 Conclusions

The parsimonious mixtures of seemingly unrelated linear regression models for contaminated data introduced here can account for heterogeneous regression data both in the presence of mild outliers and multivariate correlated dependent variables, each of which is regressed on a different vector of covariates. Models from this class allow for simultaneous robust clustering and detection of mild outliers in multivariate regression analysis. They encompass several other types of Gaussian mixture-based linear regression models previously proposed in the literature, such as the ones illustrated in [7, 8, 9], providing a robust and flexible tool for modelling data in practical applications where different regressors are considered to be relevant for the prediction of different dependent variables. Previous research (see [9, 11]) demonstrated that BIC and ICL could be effectively employed to select a proper value for K in the presence of mildly contaminated data. Thanks to an imposition of an eigen-decomposed structure on the K variance-covariance matrices of $\mathbf{Y}|\mathbf{X} = \mathbf{x}$, the presented models are characterised by a reduced number of variance-covariance parameters to be included in the analysis, thus improving flexibility, usefulness and effectiveness of an approach to multivariate linear regression analysis based on finite Gaussian mixture models in real data applications.

References

1. Baird, I. G., Quastel, N.: Dolphin-safe tuna from California to Thailand: localisms in environmental certification of global commodity networks. Ann. Assoc. Am. Geogr. **101**, 337–355 (2011)
2. Biernacki, C., Celeux, G., Govaert, G.: Assessing a mixture model for clustering with the integrated completed likelihood. IEEE Trans. Pattern Anal. Mach. Intell. **22**, 719–725 (2000)
3. Cadavez, V. A. P., Hennningsen, A.: The use of seemingly unrelated regression (SUR) to predict the carcass composition of lambs. Meat. Sci. **92**, 548–553 (2012)
4. Celeux, G., Govaert, G.: Gaussian parsimonious clustering models. Pattern Recognit. **28**, 781–793 (1995)
5. Chevalier, J. A., Kashyap, A. K., Rossi, P. E.: Why don't prices rise during periods of peak demand? Evidence from scanner data. Am. Econ. Rev. **93**, 15–37 (2003)
6. Disegna, M., Osti, L.: Tourists' expenditure behaviour: the influence of satisfaction and the dependence of spending categories. Tour. Econ. **22**, 5–30 (2016)
7. Galimberti, G., Soffritti, G.: Seemingly unrelated clusterwise linear regression. Adv. Data Anal. Classif. **14**, 235–260 (2020)
8. Jones, P. N., McLachlan, G. J.: Fitting finite mixture models in a regression context. Aust. New Zeal. J. Stat. **34**, 233–240 (1992)
9. Mazza, A., Punzo, A.: Mixtures of multivariate contaminated normal regression models. Stat. Pap. **169**, 787–822 (2020)
10. Meng, X. L., Rubin, D. B.: Maximum likelihood estimation via the ECM algorithm: A general framework. Biometrika. **80**, 267–278 (1993)
11. Perrone, G., Soffritti, G.: Seemingly unrelated clusterwise linear regression for contaminated data. Under review (2021)
12. R Core Team R: a language and environment for statistical computing. R Foundation for Statistical Computing, Vienna, Austria (2022) http://www.R-project.org
13. Ritter, G.: Robust cluster analysis and variable selection. Chapman & Hall, Boca Raton (2015)
14. Srivastava, V. K., Giles, D. E. A.: Seemingly unrelated regression equations models. Marcel Dekker, New York (1987)
15. Schwarz, G.: Estimating the dimension of a model. Ann. Stat. **6**, 461–464 (1978)
16. White, E. N., Hewings, G. J. D.: Space-time employment modelling: some results using seemingly unrelated regression estimators. J. Reg. Sci. **22**, 283–302 (1982)

Penalized Model-based Functional Clustering: a Regularization Approach via Shrinkage Methods

Nicola Pronello, Rosaria Ignaccolo, Luigi Ippoliti, and Sara Fontanella

Abstract With the advance of modern technology, and with data being recorded continuously, functional data analysis has gained a lot of popularity in recent years. Working in a mixture model-based framework, we develop a flexible functional clustering technique achieving dimensionality reduction schemes through a L_1 penalization. The proposed procedure results in an integrated modelling approach where shrinkage techniques are applied to enable sparse solutions in both the means and the covariance matrices of the mixture components, while preserving the underlying clustering structure. This leads to an entirely data-driven methodology suitable for simultaneous dimensionality reduction and clustering. Preliminary experimental results, both from simulation and real data, show that the proposed methodology is worth considering within the framework of functional clustering.

Keywords: functional data analysis, L_1-penalty, silhouette width, graphical LASSO, mixture model

Nicola Pronello (✉)
Department of Neurosciences, Imaging and Clinical Sciences, University of Chieti-Pescara, Chieti, Italy, e-mail: `nicola.pronello@unich.it`

Rosaria Ignaccolo
Department of Economics and Statistics "Cognetti de Martiis", University of Torino, Torino, Italy, e-mail: `rosaria.ignaccolo@unito.it`

Luigi Ippoliti
Department of Economics, University of Chieti-Pescara, Pescara, Italy, e-mail: `luigi.ippoliti@unich.it`

Sara Fontanella
National Heart and Lung Institute, Imperial College London, London, United Kingdom, e-mail: `s.fontanella@imperial.ac.uk`

© The Author(s) 2023
P. Brito et al. (eds.), *Classification and Data Science in the Digital Age*,
Studies in Classification, Data Analysis, and Knowledge Organization,
https://doi.org/10.1007/978-3-031-09034-9_34

1 Introduction

In recent decades, technological innovations have produced data that are increasingly complex, high dimensional, and structured. A large amount of these data can be characterized as functions defined on some continuous domain and their statistical analysis has attracted the interest of many researchers. This surge of interests is explained by the ubiquitous examples of functional data that can be found in different application fields (see for example [2], and references therein for specific examples). With functions as the basic units of observation, the analysis of functional data poses significant theoretical and practical challenges to statisticians. Despite these difficulties, methodology for clustering functional data has advanced rapidly during the past years; recent surveys of functional data clustering are presented in [7] and [2]. Popular approaches have extended classical clustering concepts for vector-valued multivariate data to functional data.

In this paper, we consider a finite mixture as a flexible model for clustering. In particular, applying a functional model-based clustering algorithm with an L_1-penalty function on a set of projection coefficients, we extend the results of [8] and [9] for vector-valued multivariate data to a functional data framework. This approach appears particularly appealing in all cases in which the functions are spatially heterogeneous, meaning that some parts of the function can be smoother than in other parts, or that there may be distant parts of the function that are correlated with each other. Furthermore, the introduction of a shrinkage penalty allows to look for directions in the feature space (that is now the space of expansion/projection coefficients) that are the most useful in separating the underlying groups without first applying dimensionality reduction techniques.

In Section 2 we present at first the methodology along with some details on model estimation (subsection 2.2). Secondly, in Section 3, we perform a validation study with simulated and real data for which the classes are known a-priori.

2 Shrinkage Method for Model-based Clustering for Functional Data

Here we consider the problem of clustering a set of n observed curves into K homogeneous groups (or clusters). To this end, we propose a flexible model based on a finite mixture of Gaussian distributions, with a L_1 penalized likelihood, which we name *Penalized model-based Functional Clustering* (PFC-L_1).

2.1 Model Definition

We consider a set of n observed curves, $x_1, \ldots, x_n$, that are independent realizations of a continuous stochastic process $X = \{X(t)\}_{t \in [0,T]}$ taking values in $L_2[0, T]$. In practice, such curves/trajectories are available only at a discrete set of the domain points $\{t_{is} : i = 1, \ldots, n, \ s = 1, \ldots, m_i\}$ and the n curves need to be reconstructed. To this goal, it is common to assume that the curves belong to a finite dimensional space spanned by a basis of functions, so that given a basis of functions $\mathbf{\Phi} = \{\psi_1, \ldots, \psi_p\}$ each curve $x_i(t)$ admits the following decomposition:

$$x_i(t) = \sum_{j=1}^{p} \beta_{j,i} \psi_j(t), \qquad i = 1, \ldots, n; \tag{2.1}$$

that is the stochastic process X admits a corresponding truncated basis expansion

$$X(t) = \sum_{j=1}^{p} \beta_j(X)\psi_j(t),$$

where $\boldsymbol{\beta} = \{\beta_1(X), \ldots, \beta_p(X)\}$ is a random vector in $\mathbb{R}^p$. By considering observations with a sampling error, such that

$$x_i^{obs}(t) = x_i(t) + \epsilon_i, \qquad i = 1, \ldots, n, \tag{2.2}$$

with $\epsilon_i \sim \mathcal{N}(0, \sigma_\epsilon^2)$, the realizations of the random coefficients $\beta_{j,i}$ for $j = 1, \ldots, p$ describing each curve can be obtained via least squares as $\hat{\boldsymbol{\beta}}_i = (\mathbf{\Theta}_i' \mathbf{\Theta}_i)^{-1} \mathbf{\Theta}_i' \mathbf{X}_i^{obs}$ where $\mathbf{\Theta}_i = (\psi_j(t_{is})), 1 \le j \le p, 1 \le s \le m_i$ contains the basis functions evaluated at the fixed domain points and $\mathbf{X}_i^{obs} = (x_i^{obs}(t_{i1}), \ldots, x_i^{obs}(t_{im_i}))'$ is the vector of observed values of the i-th curve.

With the goal of dividing into K homogeneous groups the observed curves $x_1, \ldots, x_n$, let us assume that it exists an unobservable grouping variable $\mathbf{Z} = (Z_1, \ldots, Z_K) \in [0, 1]^K$ indicating the cluster membership: $z_{i,k} = 1$ if x_i belongs to cluster k, 0 otherwise (and $z_{i,k}$ is indeed what we want to predict for each curve).

In adopting a model-based clustering approach, we denote with π_k the (a-priori) probabilities of belonging to a group:

$$\pi_k = \mathbb{P}(Z_k = 1), \qquad k = 1, \ldots, K,$$

such that $\sum_{k=1}^{K} \pi_k = 1$ and $\pi_k > 0$ for each k, and we assume that, conditionally on Z, the random vector $\boldsymbol{\beta}$ follows a multivariate Gaussian distribution, that is for each cluster

$$\boldsymbol{\beta}|(Z_k = 1) = \boldsymbol{\beta}_k \sim \mathcal{N}(\boldsymbol{\mu}_k, \boldsymbol{\Sigma}_k)$$

where $\boldsymbol{\mu}_k = (\mu_{1,k}, \ldots, \mu_{p,k})^T$ and $\boldsymbol{\Sigma}_k$ are respectively the mean vector and the covariance matrix of the k-th group. Then the marginal distribution of $\boldsymbol{\beta} = \{\beta_1, \ldots, \beta_p\}$ can be written as a finite mixture with mixing proportions π_k as

$$p(\boldsymbol{\beta}) = \sum_{k=1}^{K} \pi_k f(\boldsymbol{\beta}_k; \boldsymbol{\mu}_k, \boldsymbol{\Sigma}_k),$$

where f is the multivariate Gaussian density function. The log-likelihood function can then be written as

$$l(\theta;\beta) = \sum_{i=1}^{n} log \sum_{k=1}^{K} \pi_k f(\beta_i; \mu_k, \Sigma_k),$$

where $\theta = \{\pi_1, \ldots, \pi_K; \mu_1, \ldots, \mu_K; \Sigma_1, \ldots, \Sigma_K\}$ is the vector of parameters to be estimated and $\beta_i = (\beta_{1,i}, \ldots, \beta_{p,i})^T$ is the vector of projection coefficients of the i-th curve.

In this modeling framework, we consider a very general situation without introducing any kind of constraints neither for cluster means nor for covariance matrices, that can be different in each cluster. This flexibility, however, leads to overparameterization and, as an alternative to any kind of constraints, we consider a penalty that allows regularized parameters' estimation.

To define a suitable penalty term, we follow the penalized approach introduced by Zhou et al. [8] in the high-dimensional setting, and so we consider a penalty composed by two terms: the first one on the mean vector of each cluster μ_k, and the second one on the inverse of the covariance matrix in each group $\mathbf{W}_k = \Sigma_k^{-1}$, otherwise said "precision" matrix, with elements $W_{k;j,l}$. The proposed penalized log-likelihood function, given the projection coefficients β_i, is

$$l_P(\theta;\beta) = \sum_{i=1}^{n} log \sum_{k=1}^{K} \pi_k f(\beta_i; \mu_k, \Sigma_k) - \lambda_1 \sum_{k=1}^{K} ||\mu_k||_1 - \lambda_2 \sum_{k=1}^{K} \sum_{j,l}^{P} |W_{k;j,l}|,$$

where $||\mu_k||_1 = \sum_{j=1}^{P} |\mu_{k,j}|$, $\lambda_1 > 0$ and $\lambda_2 > 0$ are penalty parameters to be suitably chosen.

The penalty term on the cluster mean vectors allow for component selection in the functional data framework (whereas it would be variable selection in the multivariate case), considering that when the j-th component in the basis expansion is not useful in separating groups it has a common mean across groups, that is $\mu_{1,j} = \ldots = \mu_{K,j} = 0$. Then to realize component selection the considered term is $\sum_{k=1}^{K} ||\mu_k||_1$.

The second part of the penalty, namely $\sum_{k=1}^{K} \sum_{j,l}^{P} |W_{k;j,l}|$, imposes a shrinkage on the elements of the precision matrices, thus avoiding possible singularity problems and facilitating the estimation of large and sparse covariance matrices.

2.2 Model Estimation via E-M Algorithm

Since the membership of each observation to a cluster is unobservable, data related to the grouping variable $\mathbf{Z}$ is inevitably missing and the maximum penalized log-likelihood estimator can be obtained by means of the E-M algorithm [4], that iterates over two steps: expectation (E) of the complete data (penalized) log-likelihood by considering the unknown parameters equal to those obtained at the previous iteration (with initialization values), and maximization (M) of a lower bound of the obtained expected value with respect to the unknown parameters.

In particular, at the d-th iteration, given a current estimate $\theta^{(d)}$, the lower bound after the E-step assumes the following form:

$$Q_P(\theta;\theta^{(d)})=\sum_{k=1}^{K}\sum_{i=1}^{n}\tau_{k,i}^{(d)}[\log \pi_k+\log f\,(\beta_i;\mu_k,\Sigma_k)]-\lambda_1\sum_{k=1}^{K}||\mu_k||_1-\lambda_2\sum_{k=1}^{K}\sum_{j,l}^{P}|W_{k;j,l}|,$$

where $\tau_{k,i} = \mathbb{P}(Z_k = 1|X = x_i)$ is the posterior probability of observation i to belong to group k. The M-step maximizes the function Q_P in order to update the estimate of θ.

As suggested by [9], it is possible to maximize each of the K term using a "graphical lasso" (GLASSO) algorithm (first proposed by [5]), thanks to the close connection between fitting Gaussian mixture models and Gaussian graphical models. Indeed, in GLASSO the objective function looks like $\log \det(\mathbf{W}) - \mathrm{tr}(\mathbf{SW}) - \lambda \sum_{j,l}^{P}|W_{j,l}|$ so that the algorithm implemented in the R package "glasso" can be used with $\mathbf{W} = \mathbf{W}_k$, $S = \tilde{\mathbf{S}}_k$ and $\lambda = \frac{2\lambda_2}{\sum_{i=1}^{n}\tau_{k,i}^{(d)}}$ for each k to obtain the elements $\widehat{W}_{k;j,l}^{(d+1)}$ of the precision matrices.

2.3 Model Selection via Silhouette Profile

A fundamental, and probably unsolved, problem in cluster analysis is determining the "true" number of groups in a dataset. To this purpose, for simplicity, here we approach the problem choosing the number of groups as cluster validation problem and use the *average silhouette width* index as a model selection heuristic. The silhouette value for curve i is given by

$$s(i) = \frac{b(i) - a(i)}{\max\{a(i), b(i)\}}$$

where $a(i)$ is the average distance of curve i to all other curves h assigned to the same cluster (if i is the only observation in its cluster, then $s(i) = 0$), and $b(i)$ is the minimum average distance of curve i to observations h which are assigned to a different cluster. This definition ensures that $s(i)$ takes values in $[-1, 1]$, where values close to one indicate "better"clustering solutions. Conditional on K and a pair of values (λ_1, λ_2), we thus assess the overall cluster solution using the total average of silhouette values

$$S(K, \lambda_1, \lambda_2) = \frac{1}{n}\sum_{i=1}^{n} s(i).$$

In particular, by doing a grid search for the triple $(K, \lambda_1, \lambda_2)$, the best cluster solution is obtained by looking for the largest value of the *average silhouette width* (ASW) index. Note that, to evaluate $s(i)$, $i = 1, \ldots, n$, and then the objective function $S(K, \lambda_1, \lambda_2)$, we need to compute a distance between pairs of curves X_i and X_h. One possibility is to compute the euclidean distance

$$d_E^2(i, h) = \int \|X_i(t) - X_h(t)\|^2 dt.$$

3 Experimental Results

3.1 Simulation

We present here a simulated scenario in order to investigate the effectiveness of the L_1 regularization in removing noise while preserving dominant local features, accommodating for spatial heterogeneity of the curves.

The statistical analysis is illustrated for data simulated by means of a finite mixture of multivariate Gaussian distributions. In particular, based on equation (2.1) and (2.2), the curves are simulated using a combination of $p = 25$ Fourier basis functions defined over a one-dimensional regular grid with 100 observations. We consider a mixture of four ($K = 4$) multivariate Gaussian distributions with isotropic covariance matrices, i.e.

$$\beta_k \sim \mathcal{N}(\mu_k; \mathcal{I}_k) \text{ where } \epsilon_i \sim \mathcal{N}(0; 0.5), \quad k = 1, \ldots, 4.$$

With the exclusion of 3 entries per group, the means μ_k are all zero mean vectors. Under this scenario, the simulated curves (25 per group) and the non-zero group expansion coefficients are represented in Figure 1. For this simple simulation setting, estimation results suggest that, using euclidean distance to computed the ASW, the grid search procedure is always able to correctly select the cluster-relevant basis functions. This is confirmed by Figure 2 which shows both the distribution (over 100 replications) of the selected basis functions and the data projected on these bases that clearly highlight the identification of 4 clusters. Under this scenario, the quality of the estimated clusters thus appears very good as the analysis of the misclassification rate suggests an 100% of accuracy in all the replicated datasets.

Similar results hold for more complex simulation designs, where we consider different structure of the covariance matrices in the data generating process.

3.2 Performance on Real Data Sets

We evaluate the PFC-L_1 model on a well-known benchmark data set, namely the electrocardiogram (ECG) data set (data can be found at the UCR Time Series Classification Archive [3]).

The ECG data set comprises a set of 200 electrocardiograms from 2 groups of patients, myocardial infarction and healthy, sampled at 96 time instants in time.

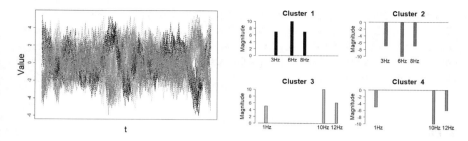

Fig. 1 Left: 25 simulated curves for each group. Right: Vector of expansion coefficients for each group, with only three non-zero coefficients corresponding to basis functions with specific periodicities (Hertz values).

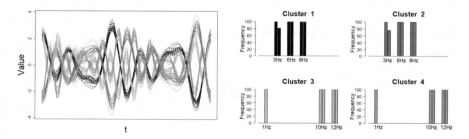

Fig. 2 Left: Data projected on cluster specific functional subspace generated by the selected basis functions. Right: Distribution (over 100 replications) of the selected basis functions shown for pairs of sine and cosine basis functions, according to the Hertz values.

This data set were previously used to compare the performance of several functional clustering models in [1]. The results in Table 5 of [1] show that the FunFEM models, compared to other state of the art methodologies, achieved the best performances in terms of accuracy. Hence, here, we limit the comparison to the results obtained with the PFC-L_1 and the FunFEM models. Although FunFEM models relay on a mixture of Gaussian distributions describing the likelihood of the data similarly to our proposal, they differ on facing the intrinsic high dimension of the problem by estimating a latent discriminant subspace in parallel with the steps of an EM algorithm.

For all the data, we reconstruct the functional form from the sampled curves choosing arbitrarily 20 cubic spline basis of functions. We tested the PFC-L_1 models considering five different values for the number of clusters, $K = \{2, 3, 4, 5, 6\}$, and six values for $\lambda_1 = \{0.5, 1, 5, 10, 15, 20\}$.

Considering that the GLASSO penalty parameter λ depends linearly from λ_2, the choice of λ_2 has to provide suitable values for λ. A practical approach is to choose values avoiding convergence problems with GLASSO. Here λ_2 was set to $\{5, 7.5, 10, 12, 15, 20\}$ for the ECG data. Both PFC-L_1 and FunFEM algorithms were initialized using a K-means procedure.

The clustering accuracies, computed with respect to the known labels, are 69% for FunFEM DFM$_{[\alpha_{kj}\beta_k]}$ (choosing among 12 different model parameterizations with BIC index), and 75% for PFC-L$_1$ [$\lambda_1 = 0.5$, $\lambda_2 = 5$] (values of tuning parameters chose by ASW index) . Thus PFC-L_1 achieves good performance, with an increase in the accuracy about 9%.

4 Discussion

In this paper we tried to investigate the potential of shrinkage methods for clustering functional data. Our numerical examples show the advantages of performing clustering with features selection, such as uncover interesting structures underlying the data while preserving good clustering accuracy. To the best of our knowledge, this is the first proposal that considers a penalty for both means and covariances of mixture components in functional model-based clustering. In the model selection section we defined an heuristic criterion to choose among different model parameterizations based on average silhouette index. It may be interesting to evaluate different distances (i.e. not euclidean) to compute this index in future research. Moreover, we will consider more complex simulation designs to investigate the robustness of the proposal and extend the comparison with the state of the art methodologies on more benchmark datasets.

References

1. Bouveyron, C., Come, E., Jacques, E.: The discriminative functional mixture model for a comparative analysis of bike sharing systems. Ann. Appl. Stat. **9**, 1726–1760 (2015)
2. Chamroukhi, F., Nguyen, H.: Model-based clustering and classification of functional data. Wiley Interdiscip. Rev.: Data Min. and Knowl. Discov. **9**, e1298, 1–36 (2019)
3. Dau, H. A., Keogh, E., Kamgar, K., Yeh, C.-C. M., Zhu, Y., Gharghabi, S., Ratanamahatana, C. A., Yanping, Hu, B., Begum, N., Bagnall, A., Mueen, A., Batista, G., Hexagon-ML: The UCR Time Series Classification Archive (October 2018)
   `https://www.cs.ucr.edu/$\sim$eamonn/time\_series\_data\_2018/`
4. Dempster, A., Laird, N., Rubin, D.: Maximum likelihood from incomplete data via the EM algorithm. J. R. Stat. Soc. Ser. B (Methodol.). **39**, 1–38 (1977)
5. Friedman, J., Hastie, T., Tibshirani, R.: Sparse inverse covariance estimation with the graphical lasso. Biostat. **9**, 432–41 (2008)
6. Friedman, J., Hastie, T., Tibshirani, R.: glasso: Graphical Lasso: Estimation of Gaussian Graphical Models, R package version 1.11 (2019).
   `https://CRAN.R-project.org/package=glasso`
7. Jacques, J., Preda, C.: Functional data clustering: A survey. Adv. Data Anal. Classif. **8**, 231–255 (2013)
8. Pan, W., Shen, X.: Penalized model-based clustering with application to variable selection. J. Mach. Learn. Res. **8**, 1145–1164 (2007)
9. Zhou, H., Pan, W., Shen, X.: Penalized model-based clustering with unconstrained covariance matrices. Electron. J. Stat. **3**, 1473–1496 (2009)

Emotion Classification Based on Single Electrode Brain Data: Applications for Assistive Technology

Duarte Rodrigues, Luis Paulo Reis, and Brígida Mónica Faria

Abstract This research case focused on the development of an emotion classification system aimed to be integrated in projects committed to improve assistive technologies. An experimental protocol was designed to acquire an electroencephalogram (EEG) signal that translated a certain emotional state. To trigger this stimulus, a set of clips were retrieved from an extensive database of pre-labeled videos. Then, the signals were properly processed, in order to extract valuable features and patterns to train the machine and deep learning models. There were suggested 3 hypotheses for classification: recognition of 6 core emotions; distinguishing between 2 different emotions and recognising if the individual was being directly stimulated or merely processing the emotion. Results showed that the first classification task was a challenging one, because of sample size limitation. Nevertheless, good results were achieved in the second and third case scenarios (70% and 97% accuracy scores, respectively) through the application of a recurrent neural network.

Keywords: emotions, brain-computer interface, EEG, supervised learning, machine and deep learning

Duarte Rodrigues
Faculty of Engineering of University of Porto (FEUP), Rua Dr. Roberto Frias, s/n 4200-465 Porto, Portugal, e-mail: up201705420@fe.up.pt

Luis Paulo Reis
Faculty of Engineering of University of Porto (FEUP) and Artificial Intelligence and Computer Science Laboratory (LIACC), Rua Dr. Roberto Frias, s/n 4200-465 Porto, Portugal, e-mail: lpreis@fe.up.pt

Brígida Mónica Faria (✉)
School of Health, Polytechnic of Porto (ESS-P.PORTO) and Artificial Intelligence and Computer Science (LIACC), Rua Dr. Roberto Frias, s/n 4200-465 Porto, Portugal, e-mail: monica.faria@ess.ipp.pt

© The Author(s) 2023
P. Brito et al. (eds.), *Classification and Data Science in the Digital Age*, Studies in Classification, Data Analysis, and Knowledge Organization, https://doi.org/10.1007/978-3-031-09034-9_35

1 Introduction

Emotions are a part of our lives, as humans we know how to identify the tiniest of microexpressions to unveil what someone is feeling, but also how to use them to express our hearts. From the youngest of ages we see and interact with others and build a database of patterns of, for example, what joy is and how different it is from fear or sadness. Computers, on the other hand, do not have any idea of what an emotion is or how to recognize it. Or do they?

The Artificial Intelligence and Computer Science Laboratory (LIACC) established 2 projects where emotion recognition can be of the utmost importance. The first project, the "IntellWheels 2.0" [1], intends to develop an interactive and intelligent electric wheelchair. This innovative equipment will have a diverse set of features, such as an adaptive control system (through eye gaze, a brain-computer interface, hand orientation, among others) and a personalized multi-modal interface which will allow communication to multiple devices both from the patients and the caregivers. In this case, having information about the mood of the patient is very beneficial, because the interface can give updates to the nursing staff of the emotional condition of the patient. The second project, the "Sleep at the Wheel" [2], focuses on the research of an interface that can sense and predict a driver's drowsiness state, being able to detect if he fell asleep while driving and, consequently, support an alarm system to provide safer routing and driving. Here the state of mind of the driver is a very important aspect, as different emotions, like anger or fear, can provoke dangerous situations or unpredictable scenarios, making the driver less attentive to his surroundings.

In this work, emotions will be sensed through a brain-computer interface (BCI). These are commercial devices that allow to acquire a surface electroencephalogram (EEG). This signal is used to measure the electrical activity of the brain, that fluctuates according to the firing of the neurons in the brain, being quantified in micro-volts. In this research, the BCI used was the "NeuroSky MindWave2" which possesses one single electrode on the forehead, from which it collects a signal from the activity of the frontal lobe. This brain area is responsible for the higher executive functions, including emotional regulation, planning, reasoning and problem solving [3].

The study of emotion recognition started with psychologist Paul Ekman that defined, based on a cross cultural study, six core emotions - Fear, Anger, Happiness, Sadness, Surprise and Disgust [4]. Later, psychologist Robert Plutchik established a model called "Wheel of Emotions", a diagram where every emotion can be derived from the core 6.

It is also important to have a way to measure what someone is feeling or what emotion they are experiencing. An easy way to do this is through the "Discrete Emotion Questionnaire", a psychological validated questionnaire to verify the intensity of a certain emotion. This assessment presents the 6 core emotions to the subjects asking them to rate the intensity they felt, from 1 to 7 [5].

As a first approach in this area, the current work aims to be able to identify the core emotions using EEG signals collected with the BCI.

2 Experimental Methodology

In order to correctly identify the core emotions, the first step is to trigger them in an efficient way for the brain data collected to be as informative as possible.To do so, the emotions were prompted via a set of video clips, that lasted 5-7 seconds. These videos were selected from a certified database, where the videos were labeled according to the intensity and kind of emotion it caused in the subjects [6]. For each of the 6 core emotions, the 4 videos classified with the biggest intensity were selected to be presented to the participants of this research work.

For each of the 24 video clips (4 videos per each of the 6 emotions), 3 EEG samples are collected. The first is before the display of the video, where a fixation cross is presented, in order to collect the idle/blank state of the user, where he is asked to relax. The second sample is the EEG during the video (active visual stimulus); and the third sample is after the video finishes where the volunteer is processing the emotion triggered (higher level thinking), while getting back to the initial relaxed state, where the fixation cross is presented again. To confirm that the volunteers experience the same emotion defined in the pre-determined label, they are a prompted to answer the "Discrete Emotion Questionnaire", after the 3 EEG samples are collected.

Regarding the physiological signal processing, this step is important because the raw EEG signal that comes directly from the BCI has a low signal-to-noise ratio, as well as many surrounding artifacts that contaminate the readings, especially eye blinks and facial movements triggered by the various emotions. These interfering signals caused by the latter, denominated electromyograms (EMG), are characterized by high frequencies (50-150 Hz) that make the underlying signal very noisy. Every time a person blinks, the EEG signal shows a very high peak with a very low frequency (<1Hz). To remove these muscle artifacts, a 5th order utterworth bandpass filter (this type of filter was chosen because it has the flattest frequency response, which leads to less signal distortion) with cut-off frequencies in 1 Hz and 50 Hz [7].The attenuation of very low frequencies is important to remove the eye blinks artifacts. Considering the top cut-off frequency, it is very convenient to use 50 Hz since it mitigates the effects of the power line noise and the EMG artifacts. Like this, no important brain data is lost. At this step, the EEG was segmented in the brain waves of interest, i.e., the alpha and beta brain waves. The best way to perform this is to apply bandpass filters (same filter type as before) in the corresponding bandwidths, 8-13Hz and 13-32 Hz, to have alpha and beta bands, respectively.

The EEG signals, at this stage possess the "emotional data" exposed allowing to extract the features. To do so, multiple mathematical equations were applied to obtain relevant information from the signals. Feature extraction methods depend on the domain, as will be seen ahead [8]. Most strategies to extract features from the EEG are formulas applied in the time domain, such as, the common statistical equations, the Hjorth statistical parameters, the mean and zero crossings (number of times the signal crosses these 2 thresholds) [8]. Besides these, there were applied more advanced feature extraction methods, based on fractal dimensions and entropy analysis (methods to assess the complexity, or irregularity, of a time-series) [9].

Regarding frequency domain approaches, these features can only be calculated in the filtered EEG and not in the brain waves, as their spectrum is very narrow. In terms of the pure frequency band, the only feature computed was the Power Spectral Density (PSD), based on the Welch method. These domains can be combined creating the time-frequency domain, leading to more sophisticated methods, like the Hilbert – Huang Transform, where the original signal is decomposed in intrinsic mode functions (IMF) [10].

The resulting number of features is too high to compute machine learning models, because the correlation between most of the features is very low, which means that between different classes the information is virtually the same. This would introduce uncertainty in the weights for each class in the models, thus the number of features needs to be reduced. To do this the "Min Redundancy Max Relevance" (MRMR) method was applied, with the objective of finding the optimal number of features to have a higher inter-class variability, in order to find distinct patterns between emotions [11]. The features were used raw, normalized or standardized to train the models.

In this study, all the models implemented are based on supervised learning and fully depend on the data that is inputted. Concerning emotion classification there is not a specific machine learning approach that is optimal, thus 9 different types of models were implemented to verify which has the best performance. These models are designed to be able to adapt to various kinds of input data, through the definition of hyper-parameters. Hence, to tune them to the best possible configuration, it was performed a GridSearchCV. This method exhaustively searches over a given list of possible parameters applying cross validation between them. In the end, the model with the best performance is chosen to be trained with the resulting feature matrix.

A deep learning model was also implemented, based on recurrent neural network (RNN), a very common architecture in classification problems using EEG. A particularity of this network is that it has a GRU, i.e., a layer that helps to mitigate the problem of vanishing gradients (common issue on artificial neural networks), giving long term memory to the model [12].

3 Evaluation and Discussion of Results

In this experiment, 12 subjects volunteered to participate. Each EEG recording is labeled according to the emotion registered in the original database, as well as if it was before video, during or after the video. The answers of the "Discrete Emotion Questionnaire" were used to validate if the emotion triggered by the video was as expected and, if so, the data was used. With this dataset structure, 3 hypotheses were tested and their results are discussed ahead.

An important aspect to have in consideration is that the EEG collected while the subject is relaxing, i.e., while the fixation cross presented before the video, does not have relevant cognitive information regarding emotions. Therefore, these segments were not considered to train any of the models.

3.1 Core Emotions Classification

This first hypothesis describes the main goal of the project where a model was developed to classify 6 emotions.

First, the feature extraction was computed. At this step, the optimal number of features to get selected was tested, iterating from 5 to 50, 5 at a time. The best number found was 30, which gave the best accuracies, with a balanced computation time and power. This value was chosen for the 3 feature matrixes (raw, normalized and standardized). The dataset was then divided into training and testing with an 80% ratio and fully independent of one another. Each model was then trained and assessed, by computing the accuracy in the test dataset. Table 1 presents the results for each model.

Table 1 Results of the 6 Core Emotions Classification.

Classification Models	Raw Features	Normalized Features	Standardized features
		Accuracy (%)	
Gaussian Naïve Bayes Classifier	12.07	12.93	10.34
Support Vector Classifier	12.07	12.93	16.38
Decision Tree Classifier	18.96	18.10	18.10
Random Forest Classifier	24.13	18.10	20.69
K Nearest Neighbors	21.55	18.96	16.38
Logistic Regression	**25.00**	14.66	18.10
Linear Discriminant Analysis	24.13	14.65	18.96
Linear Support Vector Classifier	18.10	13.79	19.82
Multi-Layer Perceptron	20.69	13.79	12.93
Recurrent Neutral Network	13.79	20.69	23.27

When comparing the various models, the average accuracy is around 16-18%, logically due to the number of classes in the problem (100%/6 = 16,6%). Despite this, the best result reached was 25% accuracy, with the features in their raw state, since the magnitude information was not lost, so patterns in different emotions could be more easily identified due to the high discrepancy in the values. These results are not discouraging since the main objective of the study is very ambitious, as we are trying to create a model to define universally what an emotion is. There is no work more subjective or abstract, and the only way to achieve this universal standardization would be with a sample population as wide and diverse as possible with different beliefs, nationalities, age groups, etc. Although this is an initial study, it shows that it is possible to register and identify differences in the electrical changes of the prefrontal cortex and, with that information, categorize what someone is feeling.

3.2 One vs One – Dual Emotion Classification

As the results in the previous hypothesis could not precisely identify an emotion when compared to the other 5, the problem was narrowed down and a new hypothesis was tested, to continue the proposed research. In this experiment, the model was trained to discern between only 2 emotions, decided *a priori*. For demonstration purposes, a concrete example can be seen in Table 2 where it compares "fear" vs "surprise".

Table 2 Results of "Fear vs Surprise" Classification.

Classification Models	Raw Features	Normalized Features	Standardized features
		Accuracy (%)	
Gaussian Naïve Bayes Classifier	48.27	55.17	53.44
Support Vector Classifier	51.72	51.72	53.44
Decision Tree Classifier	56.89	50.00	44.83
Random Forest Classifier	48.27	50.00	60.34
K Nearest Neighbors	46.55	44.82	50.00
Logistic Regression	50.00	53.45	53.45
Linear Discriminant Analysis	50.00	48.28	53.44
Linear Support Vector Classifier	50.00	51.72	55.17
Multi-Layer Perceptron	50.00	50.00	58.62
Recurrent Neutral Network	**69.23**	51.23	56.21

In this case, most of the machine learning algorithms have accuracies in the order of the 50-53%. This results are not ideal, as they are no better than a random choice between the two classes, however this can be justified by the low population sample, which is not high enough to bring to the surface concrete patterns on the features. Regarding the deep learning approach, the RNN has an advantage in this case, giving a final accuracy of 69%. This result shows that this model is reliable, and in the majority of the cases the 2 emotions can be distinguished. In this particular case, the facial expressions and their muscle activity, can induce big artifacts in the EEG. Someone who feels surprised has the tendency to raise their eyebrows and open the mouth. These movements can lead to a difference in the EEG and, consequently, in the patterns of the features, making the distinction between surprise and fear more noticeable. The same thinking applies to other emotions that trigger facial movement, like laugh, frowning, among others.

3.3 Stimulus vs No Stimulus Classification

Besides the good results presented in the last premise, one last hypothesis was assessed, regarding the difference between experiencing the emotion while watching the video (direct stimulus), and after, when the fixation cross is presented, while the volunteer is simply thinking and cognitively processing the emotion.

Table 3 summarizes the results of the various models.

Table 3 Results of Stimulus vs No Stimulus classification.

Classification Models	Raw Features	Normalized Features	Standardized features
		Accuracy (%)	
Gaussian Naïve Bayes Classifier	61.20	58.62	85.34
Support Vector Classifier	58.62	58.62	91.37
Decision Tree Classifier	39.65	58.62	89.65
Random Forest Classifier	39.65	58.62	91.37
K Nearest Neighbors	37.93	58.62	89.65
Logistic Regression	34.48	58.62	87.06
Linear Discriminant Analysis	29.31	37.06	80.17
Linear Support Vector Classifier	34.48	58.62	87.06
Multi-Layer Perceptron	31.03	58.62	88.79
Recurrent Neutral Network	**96.55**	61.20	88.79

As it can be seen, for this experiment, most models did fairly well using the standardized feature, being all accuracies higher than 80%. However, when testing the deep learning approach, this architecture revealed to fit almost perfectly to the testing data, with an accuracy higher than 96%. This hypothesis is the proof of concept that the characteristics of the signal collected during the stimulus itself are very different from the ones from a signal obtained when the person is simply thinking and cognitively processing the emotion (this change would be obvious if the EEG was collected from the occipital lobe, which is responsible for the visual perception, but is remarkable when spotted in the prefrontal cortex).

4 Conclusions

In conclusion, as a first approach, the results achieved are very satisfactory and reveal a high potential to be greatly efficient in the proposed applications both in "IntellWheels2.0" and "Sleep at the Wheel projects". Nevertheless by collecting more data the models will get more generalized resulting in more realistic patterns and, consequently, increasing the prediction's accuracies.

Comparing to the literature, using simple visual stimuli to distinguish six emotions, in a relaxed state, is a novel tactic. Most studies, complement the stimulus with forced facial expression, introducing different characteristics to the signal, leading to better results. Other studies use BCIs with more electrodes (channels), covering a wider cranial surface and, consequently, getting more EEG and information, which leads to more robust results.

As future work, the preprocessing of the data could be polished, improving the removal of artifacts and enhancing the underlying information of the EEG's. To obtain better results, it could also be used a transfer learning approach, by pre-training the models with another emotion related EEG databases.

Acknowledgements This work was financially supported by Base Funding - UIDB/00027/2020 of the Artificial Intelligence and Computer Science Laboratory – LIACC - funded by national funds through the FCT/MCTES (PIDDAC), Sono ao Volante 2.0 - Information system for predicting sleeping while driving and detecting disorders or chronic sleep deprivation (NORTE-01-0247-FEDER-039720), and Intellwheels 2.0 - IntellWheels2.0 – Intelligent Wheelchair with Flexible Multimodal Interface and Realistic Simulator (POCI-01-0247-FEDER-39898), supported by Norte Portugal Regional Operational Programme (NORTE 2020), under the PORTUGAL 2020 Partnership Agreement.

References

1. IntellWheels2.0 – Intelligent Wheelchair with Flexible Multimodal Interface and Realistic Simulator. Optimizer, Lda, FEUP, UA, Rehapoint, GroundControl. Available at `http://www.intellwheels.com/en/client/skins/geral.php?id=25` Cited 24 May 2021

2. Sono ao Volante 2.0 - Information system for predicting sleeping while driving and detecting disorders or chronic sleep deprivation. Optimizer, Lda, FEUP, IS, IPCA. Available at `http://sonoaovolante.com/en/client/skins/geral.php?id=25` Cited 24 May 2021

3. Lobes of the Brain. UQ-Queensland Brain Institute (2018). Available at `https://qbi.uq.edu.au/brain/brain-anatomy/lobes-brain.` Cited26May2021

4. Eckman, P.: Facial Expressions of Emotion: New Findings, New Questions In: Psychological Science, 34-38. Sage Journals (1992)

5. Harmon-Jones, C., Bastian, B., Harmon-Jones, E.: The Discrete Emotions Questionnaire: A New Tool for Measuring State Self-Reported Emotions. In: PLoS One **11**(8), e0159915 (2016) doi: 10.1371/journal.pone.0159915.

6. Cowen, A., Keltner, D.: Self-report captures 27 distinct categories of emotion bridged by continuous gradients In: Proceedings of the National Academy of Sciences of the United States of America **14**(38), E7900-E7909 (2017) doi: 10.1073/pnas.1702247114.

7. López-Gil, J.-M., Virgili-Gomá, J., Gil, R., Guilera, T., Batalla, I., Soler-González, J., García, R.: Method for Improving EEG Based Emotion Recognition by Combining It with Synchronized Biometric and Eye Tracking Technologies in a Non-invasive and Low Cost Way. In: Frontiers in Computational Neuroscience **10**, 85 (2016) doi: 10.3389/fncom.2016.00119

8. Jenke, R., Peer, A., Buss, M.: Feature Extraction and Selection for Emotion Recognition from EEG. In: IEEE Transactions on Affective Computing, **5**(3), 327-339, (2014) doi: 10.1109/TAFFC.2014.2339834

9. Richman, J. S., Moorman, J. R.: Physiological time-series analysis approximate entropy and sample entropy. In: American Journal of Physiology-Heart and Circulatory Physiology (2000) doi: 10.1152/ajpheart.2000.278.6.H2039

10. Junsheng, C., Dejie, Y., Yu, Y.: Research on the intrinsic mode function (IMF) criterion in EMD method In: Mechanical Systems and Signal Processing, **20**(4), 817-824. (2006) doi: 10.1016/j.ymssp.2005.09.011

11. Ding, C., Peng, H.: Minimum redundancy feature selection from microarray gene expression data. In: Bioinformatics Computation Biol., **3**(2) 185–205, (2003) doi: 10.1142/S0219720005001004

12. Zain, M. A.: Predicting Emotions Using EEG Data with Recurrent Neural Networks. Geek Culture (2021) Available at `https://medium.com/geekculture/predicting-emotions-using-eeg-data-with-recurrent-neural-networks-8acf384896f5` Cited 19 May 2021

The Death Process in Italy Before and During the Covid-19 Pandemic: a Functional Compositional Approach

Riccardo Scimone, Alessandra Menafoglio, Laura M. Sangalli, and Piercesare Secchi

Abstract In this talk, based on [1], we propose a spatio-temporal analysis of daily death counts in Italy, collected by ISTAT (Italian Statistical Institute), in Italian provinces and municipalities. While in [1] the focus was on the elderly class (70+ years old), we here focus on the middle class (50-69 years old), carrying out analogous analyses and comparative observations. We analyse historical provincial data starting from 2011 up to 2020, year in which the impacts of the Covid-19 pandemic on the overall death process are assessed and analysed. The cornerstone of our analysis pipeline is a novel functional compositional representation for the death counts during each calendar year: specifically, we work with mortality densities over the calendar year, embedding them in the Bayes space B^2 of probability density functions. This Hilbert space embedding allows for the formulation of functional linear models, which are used to split each yearly realization of the mortality density process in a predictable and an unpredictable component, based on the mortality in previous years. The unpredictable components of the mortality density are then spatially analysed in the framework of Object Oriented Spatial Statistics. Via spatial downscaling of the results obtained at the provincial level, we obtain smooth predictions at the fine scale of Italian municipalities; this also enable us to perform

Riccardo Scimone (✉)
MOX, Dipartimento di Matematica, Politecnico di Milano and Center for Analysis, Decision and Society, Human Technopole, Milano, Italy, e-mail: `riccardo.scimone@polimi.it`

Alessandra Menafoglio
MOX, Dipartimento di Matematica, Politecnico di Milano, Milano, Italy,
e-mail: `alessandra.menafoglio@polimi.it`

Laura M. Sangalli
MOX, Dipartimento di Matematica, Politecnico di Milano, Milano, Italy,
e-mail: `laura.sangalli@polimi.it`

Piercesare Secchi
MOX, Dipartimento di Matematica, Politecnico di Milano and Center for Analysis, Decision and Society, Human Technopole, Milano, Italy, e-mail: `piercesare.secchi@polimi.it`

© The Author(s) 2023
P. Brito et al. (eds.), *Classification and Data Science in the Digital Age*,
Studies in Classification, Data Analysis, and Knowledge Organization,
https://doi.org/10.1007/978-3-031-09034-9_36

anomaly detection, identifying municipalities which behave unusually with respect to the surroundings.

Keywords: COVID-19, O2S2, functional data analysis, spatial downscaling

1 Introduction and Data Presentation

At the dawn of the third year of global pandemic, we can affirm that no aspect of people's everyday life has been left untouched by the consequences of Covid-19. The virus, in addition to exacting an heavy death toll, has caused great upheavals in global economy, education systems, technological development and in countless other aspects of human life. Given this global reaching, we deem appropriate to analyse death counts from all causes, and not just those directly attributed to Covid-19, as a proxy of how Italian administrative units, be they municipalities or provinces, have been affected by the pandemic. This choice is driven by the following considerations:

- Death counts from all causes are, on many levels, high quality data: they have a very fine spatial and temporal granularity, being collected daily in each Italian municipality, they are finely stratified in many age classes, and they are not affected by errors due to incorrect attribution of the cause of death, as may happen, for example, in deciding whether or not a given death is due to Covid-19;
- They incorporate any possible shock, be it direct or indirect, which the natural death process underwent: less deaths from road accidents due to restrictive policies, more deaths from other pathologies which are left untreated because of the unnatural stress on the welfare systems, and so on;
- They are made freely available by ISTAT[1], with a substantial amounts of historical data; in particular, in the following analysis we consider data starting from the beginning of 2011 up to the end of 2020.

The purpose of the analysis of such data is twofold: (1) to study the correlation structure of the death process in Italy before and during the pandemic, assessing possible perturbations caused by its outbreak, and (2) to assess local anomalies at the municipality level (i.e., identifying municipalities which behave unusually with respect to the surrounding). This talk will entirely be devoted to presenting data and results concerning people aged between 50 and 69 years. The elderly class was the focus of [1], while analyses focusing on younger age classes can be freely examined at `https://github.com/RiccardoScimone/Mortality-densities-italy-analysis.git`.

Daily death counts for the 107 Italian provinces, in the time interval spanning from 2017 to 2020, are shown in Fig. 1: for each province, we draw death counts along the year in light blue. The black solid line is the weighted mean number of deaths, where each province has a weight proportional to its population. We also

---

[1] `https://www.istat.it/it/archivio/240401`

highlight four provinces with colours: Rome, Milan, Naples, and Bergamo. By a visual inspection, it is easy to see that, during the years 2017, 2018 and 2019, the mortality in this age class has an almost uniform behaviour, with only a very slight increase in deaths during winter, for some Provinces. Conversely, 2020 presents an abnormal behaviour in many provinces, due to the pandemic outbreak: look for example at the double peak for Milan, hit by both pandemic waves, or the single, dramatically sharp peak of Bergamo, which reached, during the first wave, higher death counts than the ones associated to provinces which are several times bigger, as Rome or Naples. By comparison with the plots in [1], on can see how all these peaks are less sharper with respect to the elderly class: this is perfectly reasonable, since people aged more than 70 years are much more susceptible to death by Covid-19.

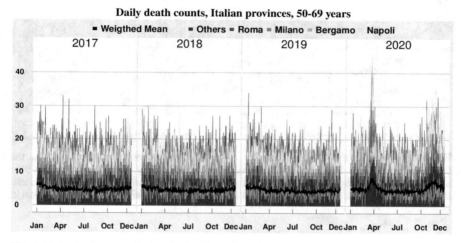

Fig. 1 Daily death counts during the last four years, for the Italian provinces. The plots refer to people aged between 50 and 69 years. For each province, death counts along the year are plotted in light blue: curves are overlaid one on top of the other to visualize their variability. The black solid line is the weighted mean number of deaths, where each province has a weight proportional to its population, while some selected provinces are highlighted in colour.

To set some notation, we denote the available death counts data as d_{iyt}, where i is a geographical index, identifying provinces or municipalities, y is the year and t is the day within year y. Moreover, we denote by T_{iy} the absolutely continuous random variable *time of death along the calendar year*, that models the instant of death of a person living in area i and passing away during year y. We hence consider the empirical discrete probability density of this random variable,

$$p_{iyt} = \frac{d_{iyt}}{\sum_t d_{iyt}} \qquad \text{for } t = 1, ..., 365$$

for each area i and year y. The family $\{p_{iy}\}_{iy}$ is the main focus of our analysis: we show these discrete densities in Fig. 2, with the same color choices of Fig. 1. It is clear that using densities provides a natural alignment of areas whose population differs significantly, providing complementary insights with respect to the absolute number of death counts: greater emphasis is given on the temporal structure of the phenomenon. For example, the astonishing behaviour of the province of Bergamo during the first pandemic wave in 2020, is now much more visible.

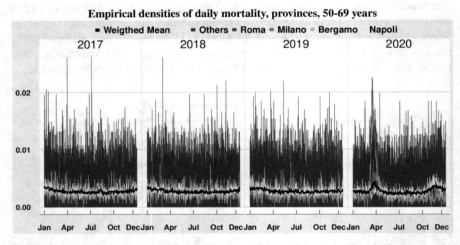

Fig. 2 Empirical densities of daily mortality, for people aged between 50 and 69 years, at the provincial scale. For each province, the empirical density of the daily mortality is plotted in light blue: densities are overlaid one on top of the other to visualize their variability. The black solid line is the weighted mean density, where the weight for each province has been set to be proportional to its population; some selected provinces are highlighted in colour.

In this talk, we will show results obtained by embedding a smoothed version of the $\{p_{iy}\}_{iy}$, i.e., an estimate $\{f_{iy}\}_{iy}$ of the continuous density functions of the $\{T_{iy}\}_{iy}$, in the Hilbert space $B^2(\Theta)$, called *Bayes space* [2, 4, 3], where Θ denotes the calendar year. This is the set (of equivalence classes) of functions

$$B^2(\Theta) = \{f : \Theta \to \mathbb{R}^+ \; s.t. \; f > 0, \; log(f) \in L^2(\Theta)\}$$

where the equivalence relation in $B^2(\Theta)$ is defined among *proportional* functions, i.e., $f =_{B^2} g$ if $f = \alpha g$ for a constant $\alpha > 0$. In [1], we also propose a preliminary exploration of the $\{p_{iy}\}_{iy}$ based on the *Wasserstein space* embedding, a very regular metric space of probability measures with a straightforward physical interpretation [5]. For the sake of brevity, we here focus on the analysis in $B^2(\Theta)$, which constitutes our main contribution.

$B^2(\Theta)$ is equipped with an Hilbert geometry, constituted by appropriate operations of sum, multiplication by a scalar, and inner product, which make it the infinite-dimensional counterpart of the Aitchison simplex used in standard compositional analysis [6, 7]: for this reason this space is considered the most suited Hilbert embedding for positive continuous density functions. The smoothed densities $\{f_{iy}\}_{iy}$

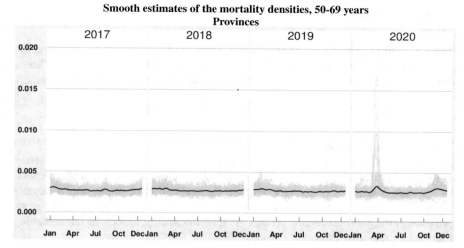

Fig. 3 Smooth estimates of the mortality densities over the 107 Italian provinces. The usual pattern of mortality is visible till 2019, while the functional process is completely different in 2020, with the two pandemic waves clearly captured by the estimated densities. The black thick lines represent the mean density, computed in B^2, with weights proportional to the population in each area.

are shown in Fig. 3: they are obtained by smoothing the $\{p_{iy}\}_{iy}$ via *compositional splines* [8, 9]. It is easy to see, by comparison with Fig. 2, how smoothing filters out a good amount of noise, much more than the case of the elderly class: this is fairly reasonable, since the death process is usually more noisy for younger age classes. From now on, the $\{f_{iy}\}_{iy}$ are analysed as a spatio-temporal functional random sample taking values in $B^2(\Theta)$. We briefly anticipate the results of such analysis:

1. The $\{f_{iy}\}_{iy}$ are decomposed, by means of a linear model formulated in $B^2(\Theta)$ [10], in a *predictable* and an *unpredictable* part, on the basis of mortality during previous years;
2. The unpredictable part is then analysed spatially in order to infer the main spatial correlation characteristics of the process; in particular, the impacts of the pandemic are investigated via functional variography [13, 14, 11, 12] and Principal Component Analysis in the B^2 space (SFPCA, [16]);
3. The results obtained at the provincial level are reduced to the municipality scale by *spatial downscaling* [15] techniques, obtaining smooth density estimates for each municipality. This provides continuous density at the municipality level, without directly smoothing the corresponding daily death process, which is quite irregular due to the reduced population of many municipalities. The spatial downscaling estimates, that are exclusively based on provincial data, are then compared with the actual measurements on municipalities, allowing for the identification of local anomalies.

Points 1 and 2 above are detailed in Section 2, while point 3 will be discussed during the talk. The reader is referred to [1] for full details on the analysis pipeline.

2 Some Results

The first step of the analysis of the random sample $\{f_{iy}\}_{iy}$, where i is indexing the 107 Italian provinces, is the formulation of a family of function-on-function linear models in $B^2(\Theta)$, extending classical models formulated in the L^2 case [17], namely

$$f_{iy}(t) = \beta_{0y}(t) + \langle \beta_y(\cdot, t), \overline{f}_{iy}\rangle_{B^2} + \epsilon_{iy}(t), \quad i = 1, \ldots 107, \quad t \in \Theta, \tag{1}$$

where $\overline{f}_{iy} = \frac{1}{4}\sum_{r=y-4}^{y-1} f_{ir}$ is the B^2 mean of the observed densities in the four years preceding year y, functional parameters $\beta_{0y}(t), \beta_y(s, t)$ are defined in the B^2 sense, as well as the residual terms $\epsilon_{iy}(t)$ and all operations of summation and multiplication by a scalar. Model (1) is trying to explain the realization of the mortality density f_{iy} for a year y in a province i as a linear function of what happened in the same province during the preceding years. It is thus interesting to look at the following functional prediction errors:

$$\delta_{iy} = f_{iy} - \hat{f}_{iy} \tag{2}$$

where

$$\hat{f}_{iy}(t) := \hat{\beta}_{0y-1}(t) + \langle \hat{\beta}_{y-1}(\cdot, t), \overline{f}_{iy}\rangle_{B^2}. \tag{3}$$

The δ_{iy} are not the estimate $\hat{\epsilon}_{iy}$ of the residual of model (1): they rather represent

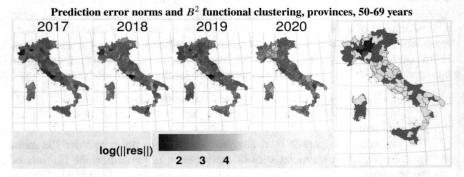

Fig. 4 First four panels, from the left: heatmaps of the B^2 norm of the prediction errors δ_{iy}, in logarithmic scale, for the elderly class. In 2020 the pandemic diffusion is clearly visible in northern Italy, while the prediction errors are generally higher on all provinces. Last panel: result of a K-mean B^2 functional clustering ($K = 3$) on the δ_{iy}, during 2020.

the error committed in forecasting f_{iy} using the model fitted at year $y - 1$. Thus, we can look at the densities δ_{iy} as the *unpredictable component* of f_{iy}, i.e., as a proxy of what happened at year y which could not be predicted by information available at the previous years, and analyze them under the spatial viewpoint. For example, we can look at the spatial heatmaps of the B^2 norms of the δ_{iy}, which are shown in Fig 4. It is clear, by looking at the magnitude of the error norms, that what happened during 2020 was to a large extent unpredictable, since almost all Italian provinces are characterized by higher errors with respect to previous years. More significantly, in 2020 a clear spatial pattern can be noticed, at least during the first wave in northern Italy: a diffusive process, having at its core the provinces most gravely hit by the first pandemic wave, seems to take place in northern Italy. This pattern is, as reasonable, slightly less evident with respect to the case of the elderly class analysed in [1]. Going in this direction, we also show in Fig 4 the result of a K-means functional clustering, set in the B^2 space, of the δ_{iy} for the year 2020. We clearly identify provinces hit by the first wave (blue cluster), while the other two clusters behave irregularly: this is a neat distinction with people aged more than 70 years, where each cluster clearly identifies different kinds of pandemic behaviour (see [1]). For a more precise investigation of the spatial correlation structure of the

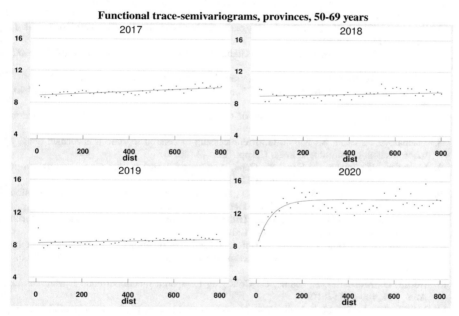

Fig. 5 Empirical trace-semivariograms for the prediction errors δ_{iy}, in people aged between 50 and 69 years. The purple lines are the corresponding fitted exponential models. Distances on the x-axes are expressed in kilometers. The last panel shows the 2020 severe perturbation of the spatial dependence structure of the process generating the prediction errors.

process across different years, from the δ_{iy} we compute a *functional trace variogram* for each year: we show them for 2017 up to 2020 in Figure 5. Without entering into the details of the mathematical definition of variograms, we can look at the fitted curves in Figure 5 as follows. Distances are on the x-axis, while on the y-axis we have a function of the spatial correlation of the process: when the curve reaches its horizontal asymptote, it has reached the total variance of the process and we are beyond the maximum correlation length. In this perspective, it is immediate to infer that not only the total variance of the functional process δ_{iy} has sharply increased in 2020, but also a significant spatial correlation has manifested, compatibly with the presence of a pandemic. In the main work [1], we further deepen the connection between the pandemic and the upheavals in the spatial structure by means of Principal Component Analysis of the δ_{iy} in the Bayes space (SFPCA, [16]).

References

1. Scimone, R., Menafoglio, A., Sangalli, L. M., Secchi, P.: A look at the spatio-temporal mortality patterns in Italy during the COVID-19 pandemic through the lens of mortality densities. Spatial Stat. (2021) doi:10.1016/j.spasta.2021.100541
2. Egozcue, J., Díaz–Barrero, J., Pawlowsky-Glahn, V.: Hilbert space of probability density functions based on Aitchison geometry. Acta Mathematica Sinica. **22**, 1175-1182 (2006)
3. Pawlowsky-Glahn, V., Egozcue, J., Boogaart, K.: Bayes Hilbert spaces. Aust. New Zeal. J. Stat. **56**, 171-194 (2014)
4. Boogaart, K., Egozcue, J., Pawlowsky-Glahn, V.: Bayes linear spaces. SORT. **34**, 201-222 (2010)
5. Villani, C.: Topics in Optimal Transportation. American Mathematical Society (2003)
6. Aitchison, J.: The statistical analysis of compositional data. J. Roy. Stat. Soc. B Stat. Meth. **44**, 139-177 (1982)
7. Aitchison, J.: The Statistical Analysis of Compositional Data. Chapman & Hall, London (1986)
8. Machalová, J., Hron, K., Monti, G.: Preprocessing of centred logratio transformed density functions using smoothing splines. J. Appl. Stat. **43** (2015)
9. Machalová, J., Talská, R., Hron, K., Gába, A.: Compositional splines for representation of density functions. Comput. Stat. **36**, 1031-1064 (2021)
10. Talská, R., Menafoglio, A., Machalová, J., Hron, K., Fišerová, E.: Compositional regression with functional response. Comput. Stat. Data Anal. **123**, 66-85 (2018)
11. Menafoglio, A., Petris, G.: Kriging for Hilbert-space valued random fields: The operatorial point of view. J. Multivariate Anal. **146** (2015)
12. Menafoglio, A., Grujic, O., Caers, J.: Universal kriging of functional data: Trace-variography vs cross-variography? Application to gas forecasting in unconventional shales. Spatial Stat. **15**, 39-55 (2016)
13. Nerini, D., Monestiez, P., Manté, C.: Cokriging for spatial functional data. J. Multivariate Anal. **101**, 409-418 (2010)
14. Menafoglio, A., Secchi, P., Dalla Rosa, M.: A universal kriging predictor for spatially dependent functional data of a Hilbert space. Electronic Journal of Statistics **7**, 2209-2240 (2013)
15. Goovaerts, P.: Kriging and semivariogram deconvolution in the presence of irregular geographical units. Mathematical Geosciences. **40**, 101-128 (2008)
16. Hron, K., Menafoglio, A., Templ, M., Hrůzová, K., Filzmoser, P.: Simplicial principal component analysis for density functions in Bayes spaces. Comput. Stat. Data Anal. **94**, 330-350 (2016)
17. Ramsay, J., Silverman, B.: Functional Data Analysis. Springer (2005)

Clustering Validation in the Context of Hierarchical Cluster Analysis: an Empirical Study

Osvaldo Silva, Áurea Sousa, and Helena Bacelar-Nicolau

Abstract The evaluation of clustering structures is a crucial step in cluster analysis. This study presents the main results of the hierarchical cluster analysis of variables concerning a real dataset in the context of Higher Education. The goal of this research is to find a typology of some relevant items taking into account both the homogeneity and the isolation of the clusters. Two similarity measures, namely the standard affinity coefficient and Spearman's correlation coefficient, were used, and combined with three probabilistic (*AVL*, *AVB* and *AV1*) aggregation criteria, from a parametric family in the scope of the *VL* (Validity Link) methodology. The best partitions were selected based on some validation indices, namely the global *STAT* levels statistics and the measures P(I2, Σ) and γ, adapted to the case of similarity coefficients. In order to evaluate the clusters and identify their most representative elements, the Mann and Whitney U statistics and the silhouette plot were also used.

Keywords: clustering validation, affinity coefficient, Spearman correlation coefficient, *VL* methodology

Osvaldo Silva (✉)
Universidade dos Açores and CICSNOVA.UAc, Rua da Mãe de Deus, 9500-321, Portugal, e-mail: osvaldo.dl.silva@uac.pt

Áurea Sousa
Universidade dos Açores and CEEAplA, Rua da Mãe de Deus, Portugal,
e-mail: aurea.st.sousa@uac.pt

Helena Bacelar-Nicolau
Universidade de Lisboa (UL) Faculdade de Psicologia and Institute of Environmental Health (ISAMB/FM-UL), Portugal, e-mail: hbacelar@psicologia.ulisboa.pt

© The Author(s) 2023
P. Brito et al. (eds.), *Classification and Data Science in the Digital Age*,
Studies in Classification, Data Analysis, and Knowledge Organization,
https://doi.org/10.1007/978-3-031-09034-9_37

1 Introduction

Cluster analysis or unsupervised classification usually concerns exploratory multi-variate data analysis methods and techniques for grouping either a set of data units or an associated set of descriptive variables in such a way that elements in the same group (cluster) are more similar to each other than elements in different clusters [6]. Therefore, it is important to validate the results obtained, bearing in mind that, in an ideal situation, the clusters should be internally homogeneous and externally well separated or isolated. Thus, according to Silva et al. ([15], p. 136), there are some important questions, such as: "i) How to compare partitions obtained using different cluster algorithms? ii) Is it possible to join information from several approaches in the decision-making process of choosing the most representative partition?"

This paper presents the main results of a hierarchical cluster analysis of variables concerning a real dataset in the field of Higher Education, in order to find a typology taking into account relevant validation measures. Two similarity measures (standard affinity coefficient and Spearman's correlation coefficient) were used, and combined with a parametric family aggregation criteria in the scope of the *VL* methodology (e.g., [10, 11, 17]).

With regard to the validation of clustering structures, some validation indices were used for the evaluation of partitions and the clusters that integrate them, which are referred to in Section 2. The main results are presented and discussed in Section 3. Section 4 contains some final remarks.

2 Data and Methods

Data were obtained from a questionnaire administered to three hundred and fifty students who were attending Higher Education in a public university, after their informed consent. The questionnaire contains, among others, eleven questions related to academic life and the respective courses.

Several algorithms of hierarchical cluster analysis of variables were applied on the data matrix. The variables (items) are: T1-Participation, T2-Interest, T3-Expectations, T4-Accomplishment, T5-Job Outlook, T6- Teachers' Professional Competence, T7-Distribution of Curricular Units, T8- Number of weekly hours of lessons, T9-Number of hours of daily study, T10-School Outcomes and T11-Assessment Methods, which were evaluated based on a Likert scale from 1 to 5 (1-Totally disagree, 2- Partially disagree, 3- Neither disagree nor agree, 4- Partially agree, 5- Totally agree).

The Ascendant Hierarchical Cluster Analysis (AHCA) was based on the standard affinity coefficient [1, 17] and Spearman's correlation coefficient. In this paper both measures of comparison were combined with three probabilistic aggregation criteria (*AVL, AVB* and *AV1*), issued from the *VL* parametric family. This methodology, in the scope of Cluster Analysis, uses probabilistic comparison functions, between pairs of elements, which correspond to random variables following a unit uniform distribution. Besides, this approach considers probabilistic aggregation criteria, which can be interpreted as distribution functions of statistics of independent random variables, that are i.i.d. uniform on [0, 1] (e.g., [17]).

Let A and B be two clusters with cardinals, respectively, α and β, and let γ_{xy} be a similarity measure between pairs of elements, $x, y \in E$ (set of elements to classify). Concerning the family I of AVL methods (e.g., SL, AVI, AVB, and AVL), the comparison functions between clusters can be summarized by the following conjoined formula:

$$\Gamma(A, B) = (p_{AB})^{g(\alpha,\beta)} \tag{1}$$

where $\alpha = Card\ A$, $\beta = Card\ B$, $p_{AB} = max[\gamma_{ab} : (a, b) \in (A \times B]$, with $1 \le g(\alpha, \beta) \le \alpha\beta$, and γ_{xy}, establishing a bridge between SL and AVL methods which have a braking effect on the formation of chains. For example, $g(\alpha, \beta) = 1$ for SL, $g(\alpha, \beta) = (\alpha + \beta)/2$ for AVI, $g(\alpha, \beta) = \sqrt{\alpha\beta}$ for AVB, and $g(\alpha, \beta) = \alpha\beta$ for AVL (see [3, 17]).

The application of the two measures of comparison between elements (Spearman correlation coefficient and standard affinity coefficient), combined with the aforementioned aggregation criteria, aims to find a typology of items corresponding to the best partition among the best partitions obtained by the several algorithms, in order to verify if there are any substantial changes in the results. Therefore, some validation indices based on the values of the corresponding proximity matrices were used, namely the global levels statistics (STAT) [1, 10, 11] and the indices P(I2mod, Σ) and γ [8], adapted to this type of matrices [16], so that the choice of the best partition is judicious and based on the desirable properties (e.g., isolation and homogeneity of the clusters). Concerning the best partitions, the respective clusters and the identification of their most representative elements were based on appropriate adaptations of the Mann and Whitney U statistics [8] and of the silhouette plots [14] to the case of similarity measures.

Each level of a dendrogram corresponds to a stage in the constitution of the partitions hierarchy. Therefore, the study of the most relevant partition(s) is strictly related to the choice of the best cut-off levels (e.g., [6, 5])

According to Bacelar Nicolau [1, 2], the global levels statistics (STAT) values must be calculated for each of the $k = 1, nivmax$ levels of the corresponding dendrograms, designating them by $STAT(k)$. At each level k, $STAT(k)$ is the global statistics that measures the total information given by the pre-order associated to the corresponding partition, in relation to the initial pre-order associated with the similarity or dissimilarity measure. A "significant" level is considered to be one that corresponds to a partition for which the global statistics undergoes a significant increase in relation to the information provided by neighbouring levels, that is, a local maximum of the differences $DIF(k) = STAT(k) - STAT(k - 1), k = 1, nivmax$.

2.1 Adaptation of the P (I2, Σ)

To evaluate the partitions, an appropriate adaptation of the index P (I2, Σ) [8] for the case of similarity measures was used, given by the following formula:

$$P(I2mod, \Sigma) = \frac{1}{c} \sum_{r=1}^{c} \frac{\sum\limits_{i \in C_r} \sum\limits_{j \notin C_r} s_{ij}}{n_r \times (N - n_r)} \tag{2}$$

where c is the number of clusters of the partition and s_{ij} is the value of the similarity measure between the element i belonging to cluster C_r and the element j belonging to another cluster. This index takes into account the number of clusters and the number of elements in each of the clusters and evaluates the isolation of clusters belonging to a given partition.

2.2 Goodman and Kruskal Index (γ)

The γ index, proposed by Goodman and Kruskal [7], has been widely used in cluster validation [9]. Comparisons are developed between all within-cluster similarities, s_{ij} and all between-cluster similarities s_{kl} [18]. A comparison is judged concordant (respectively discordant) if s_{ij} is strictly greater (respectively, smaller) than s_{kl}. The γ index is defined by:

$$\gamma = (S_+ - S_-)/(S_+ + S_-), \tag{3}$$

where S_+ (or S_-) is the number of concordant (respectively, discordant) comparisons. This index is a global stopping rule and it evaluates the fit of the partition in c clusters based on the homogeneity (high similarity between the elements within the clusters) and the isolation (low similarity of the elements between the clusters) of the clusters. Note that the higher the value of this index, the better is the adjustment of that partition.

The use of *STAT*, γ and P(I2mod, Σ) indices can help identifying the most significant levels of a dendrogram, taking into account both the homogeneity and the isolation of the clusters [15].

2.3 U Statistics (Mann and Whitney)

U statistics [12] are relevant for assessing the suitability of a cluster, combining the concepts of compactness and isolation. Thus, the "best" cluster is the one with the lowest values of global U-index, U_G, and local U-index, U_L [8]. In the present paper we used an appropriate adaptation of these indices to the case of similarity measures (for details, see [19]). Moreover, the clusters considered "ideal" are those for which U_G and U_L both take the value zero. Mann and Whitney's U statistics are useful in decision making, in situations of uncertainty, both for the evaluation of the clusters and partitions.

2.4 Silhouette Plots

We also used an appropriate adaptation of the silhouette plots [14], which allows the assessment of compactness and relative isolation of clusters. The adaptation of this measure for the case of similarity measures, $Sil(i)$, considers the average of the similarities between an element i belonging to cluster C_r, which contains $n_r (\geq 2)$ elements, and all other elements that do not belong to this cluster (see [19]). The values of this measure $\{Sil(i) : i \in C_r\}$ lie between -1 and $+1$, with "values near $+1$ indicating that element strongly belongs to the cluster in which it has been placed" ([8], p. 205). In the case of a singleton cluster, $Sil(i)$ assumes the value zero [8] in the corresponding algorithm.

3 Results and Discussion

The best partitions provided by the dendrograms are shown in Table 1.

Table 1 The best partitions concerning the dendrograms.

Coefficient	Method	The best partition	Validation indices
Affinity	*AVL*	(T1, T3, T4, T5 ,T6, T7, T8, T10, T11), (T2, T9)	STAT=5.1301 $\gamma= 0.8589$ P(I2mod,Σ)=0.2077
	AVI/AVB	(T1, T3, T4 , T5, T6, T7, T8, T10, T11), (T2), (T9)	STAT=5.3453 $\gamma= 0.8830$ P(I2mod,Σ)=0.2049
Spearman	*AVL*	(T3, T4 ,T2 , T9) (T7, T11, T8), (T6, T10), (T1), (T5)	STAT=4.0152 $\gamma= 0.8178$ P(I2mod,Σ)=0.3896
	AVI/AVB	(T3, T4 ,T2 , T9, T6) (T7, T11, T8), (T1, T10), (T5)	STAT=4.05751 $\gamma= 0.7317$ P(I2mod,Σ)=0.38177

Figure 1 shows the dendrograms obtained, respectively, by the standard affinity coefficient (left side) and Spearman's correlation coefficient (right side), both combined with the *AVL* method.

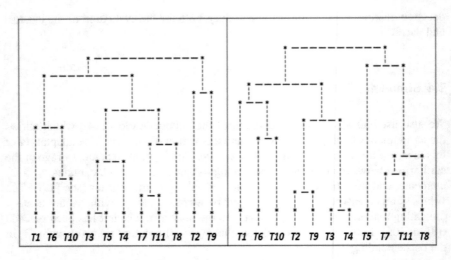

Fig. 1 Dendrograms based on standard affinity coefficient (left side) and Spearman's correlation coefficient (right side) - *AVL*.

The "best" partition obtained using the affinity coefficient and the *AVL* method is the partition into two clusters (level 9 of the aggregation process). The first cluster consists of nine items that highlight the importance of the teachers' professional competence, the structuring/content of the course and the future perspectives in relation to the career opportunities, mostly factors exogenous to the students. The second one is composed by two items (T2 and T9) which emphasize the role of interest in the study of Mathematics.

The algorithms in which the standard affinity coefficient was used are the ones that provided the best partitions and their hierarchies are the ones that remained closest to the initial pre-orders. In fact, in the case of Spearman correlation coefficient the values of *STAT* and γ indices are clearly lower than the previous ones. Moreover, the cluster {T1, T3, T4, T5, T6, T7, T8, T10, T11}, corresponding to the best partition provided by the combination of the standard affinity coefficient with the aggregation criteria *AVL*, *AV1* and *AVB*, presents (U_G =39 and U_L=4, both lower than those obtained for the cluster {T3, T4, T2, T9, T6} (U_G=65 and U_L=26) provided by the Spearman correlation coefficient combined, respectively, with *AV1* and *AVB* methods.

Focusing the attention on the two first partitions of Table 1, the only difference between them is that while the best partition provided by *AV1* and *AVB* methods contains the singletons T2 and T9, the best partition given by *AVL* joins these two singletons in the same cluster. The values of the numerical validation indices shown in Table 1 indicate that the best partition is the one provided by *AV1* and *AVB* methods. This conclusion is reinforced by the observation of the silhouette plot (see Figure 2), which indicates that the cluster joining T2 and T9, given by *AVL* method, includes the elements which have the two lowest values of *Sil* and *Sil* (T2) is negative

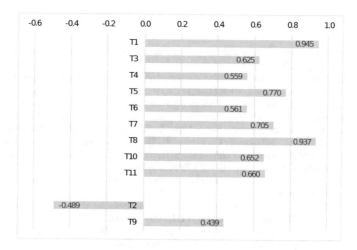

Fig. 2 Silhouette plot - standard affinity coefficient and *AVL* method.

(i.e., T2 does not fit very well in this cluster). Note that the silhouette plot cannot be used for the best partition, since it does not apply for singletons.

4 Final Remarks

This research was useful concerning the identification of relevant partitions of items in the context of Higher Education. In the cases where the affinity and the Spearman correlation coefficients were used, it was concluded that the probabilistic criteria *AV1* and *AVB* showed a higher agreement regarding the hierarchies of partitions obtained than the *AVL* method.

The validation measures *STAT*, γ and P(I2mod, Σ) help us to determine the best cut-off levels of a hierarchy of clusters, taking into account both the homogeneity and the isolation of the clusters. It should also be noted that if there is no absolute consensus between these three measures, the Mann and Whitney *U* statistics and the silhouette plot prove to be very useful, as we have seen with the application of this methodology to evaluate both the clusters and the partitions obtained.

Acknowledgements Funding. This work is financed by national funds through FCT – Foundation for Science and Technology, I.P., within the scope of the project «UIDB/04647/2020» of CICS.NOVA – Centro de Ciências Sociais da Universidade Nova de Lisboa.

References

1. Bacelar-Nicolau, H.: Analyse d'un Algorithme de Classification Automatique. Thèse de 3ème Cycle. ISUP, Paris VI (1972)
2. Bacelar-Nicolau, H.: Contributions to the Study of Comparison Coefficients in Cluster Analysis (in Portuguese). Univ. Lisbon (1980)
3. Bacelar-Nicolau, H.: On the distribution equivalence in cluster analysis. In: P. A., Devijver, & J. Kittler (eds.) Pattern Recognition Theory and Applications, NATO ASI Series, Series F. Computer and Systems Sciences, vol. 30, pp. 73-79. Springer - Verlag, New York (1987)
4. Bacelar-Nicolau, H., Nicolau, F. C., Sousa, Á., Bacelar-Nicolau, L.: Clustering of variables with a three-way approach for health sciences.Testing, Psychometrics, Methodology in Applied Psychology (TPM) (2014) doi: 10.4473/TPM21.4.56
5. Benzécri, J. P.: Analyse Factorielle des Proximités. Publication de l'Institut de Statistique de l' Universite de Paris (ISUP), XIII et XIV (1965)
6. De La Vega, W.: Techniques de la classification automatique utilisant un índice de ressemblance. Revue Française de Sociologie. **VIII**, 506–520 (1967)
7. Goodman, L. A., Kruskal, W. H.: Measures of association for cross-classifications. Journal of the American Statistical Association. **49**, 732–764 (1954)
8. Gordon, A. D.: Classification, 2nd Ed. Chapman & Hall, London (1999)
9. Hubert, L. J.: Some applications of graph theory to clustering. Psychometrika **39**(3), 283–309 (1974)) doi: 10.1007/BF02291704
10. Lerman, I. C.: Classification et Analyse Ordinale des Données. Dunod, Paris (1981)
11. Lerman, I. C.: Foundations and Methods in Combinatorial and Statistical Data Analysis and Clustering. Series: Advanced Information and Knowledge Processing. Springer-Verlag, Boston (2016)
12. Mann, H. B., Whitney, D. R.: On a test of whether one of two random variables is stochastically larger than the other. Annals of Mathematical Statistics, 50–60 (1947)
13. Nicolau, F. C., Bacelar-Nicolau, H.: Some trends in the classification of variables. In: Hayashi et al. (eds.) Data Science, Classification and Related Methods, pp. 89-98. Springer,Tokyo (1998)
14. Rousseeuw, P. J.: Silhouettes: a graphical aid to the interpretation and validation of cluster analysis. Journal of Computation and Applied Mathematics. **20**, 53–65 (1987)
15. Silva, O., Bacelar-Nicolau, H.; Nicolau, F.: A global approach to the comparison of clustering results. Biometrical Letters **49**(2), 135–147 (2013) doi: 10.2478/bile-2013-0010
16. Silva, O., Bacelar-Nicolau, H., Nicolau, F. C., Sousa, Á.: Probabilistic approach for comparing partitions. In: Manca, R., McClean, S., Skiadas, C. H.(eds.) New Trends in Stochastic Modeling and Data Analysis, pp. 113-122. ISAST (International Society for the Advancement of Science and Technology), Athens (2015)
17. Sousa, Á., Silva, O., Bacelar-Nicolau, H., Nicolau, F. C.: Distribution of the affinity coefficient between variables based on the Monte Carlo simulation method. Asian Journal of Applied Sciences. **1**(5), 236–245 (2013a)
18. Sousa, Á., Tomás, L., Silva, O., Bacelar-Nicolau, H.: Symbolic data analysis for the assessment of user satisfaction: an application to reading rooms services. European Scientific Journal (ESJ). Special/Edition **3**, 39–48 (2013b)
19. Sousa, Á., Nicolau, F., Bacelar-Nicolau, H., Silva, O.: Cluster analysis using affinity coefficient in order to identify religious beliefs profiles. European Scientific Journal (ESJ). Special/Edition **3**, 252–261 (2014)

An MML Embedded Approach for Estimating the Number of Clusters

Cláudia Silvestre, Margarida G. M. S. Cardoso, and Mário Figueiredo

Abstract Assuming that the data originate from a finite mixture of multinomial distributions, we study the performance of an integrated *Expectation Maximization* (EM) algorithm considering *Minimum Message Length* (MML) criterion to select the number of mixture components. The referred EM-MML approach, rather than selecting one among a set of pre-estimated candidate models (which requires running EM several times), seamlessly integrates estimation and model selection in a single algorithm. Comparisons are provided with EM combined with well-known information criteria – e.g. the Bayesian information Criterion. We resort to synthetic data examples and a real application. The EM-MML computation time is a clear advantage of this method; also, the real data solution it provides is more parsimonious, which reduces the risk of model order overestimation and improves interpretability.

Keywords: finite mixture model, EM algorithm, model selection, minimum message length, categorical data

1 Introduction

Clustering is a technique commonly used in several research and application areas. Most of the clustering techniques are focused on numerical data. In fact, clustering

Cláudia Silvestre (✉)
Escola Superior de Comunicação Social, Campus de Benfica do IPL 1549-014 Lisboa, Portugal,
e-mail: `csilvestre@escs.ipl.pt`

Margarida G. M. S. Cardoso
BRU-UNIDE, ISCTE-IUL, Av. das Forças Armadas, 1649-026 Lisboa, Portugal,
e-mail: `margarida.cardoso@iscte-iul.pt`

Mário Figueiredo
Instituto de Telecomunicações, Portugal, Av. Rovisco Pais 1, 1049-001 Lisboa, Portugal,
e-mail: `mario.figueiredo@tecnico.ulisboa.pt`

© The Author(s) 2023
P. Brito et al. (eds.), *Classification and Data Science in the Digital Age*,
Studies in Classification, Data Analysis, and Knowledge Organization,
https://doi.org/10.1007/978-3-031-09034-9_38

353

methods for categorical data are more challenging [12] and there are fewer techniques available [11].

In order to determine the number of clusters, model-based approaches commonly resort to information-based criteria e.g., the *Bayesian Information Criterion* (BIC) [15] or the *Akaike Information Criterion* (AIC) [1]. These criteria look for a balance between the model's fit to the data (which corresponds to maximizing the likelihood function) and parsimony (using penalties associated with measures of model complexity), thus trying to avoid over-fitting. The use of information criteria follows the estimation of candidate finite mixture models for which a predetermined number of clusters is indicated, generally resorting to an EM (*Expectation Maximization*) algorithm [7]. In this work, we focus on determining the number of clusters while clustering categorical data, using an EM embedded approach to estimate the number of clusters. This approach does not rely on selecting among a set of pre-estimated candidate models, but rather integrates estimation and model selection in a single algorithm. Our new implementation to deal with categorical variables by estimating a finite mixture of multinomials, follows a previous version described in [16]. We capitalized on the work of Figueiredo and Jain [9] for clustering continuous data and extended it for dealing with categorical data. The embedded method is thus based on a *Minimum Message Length* (MML) criterion to select the number of clusters and on an EM algorithm to estimate the model parameters.

2 Clustering with Finite Mixture Models

The literature on finite mixture models and their application is vast, including some books covering theory, geometry, and applications [8, 13, 3]. When applying finite mixture models to social sciences, the analyst is often confronted with the need to uncover sub-populations based on qualitative indicators.

2.1 Definitions and Concepts

Let $\mathbf{Y} = \{\underline{y}_i, \ i = 1, \ldots, n\}$ be a set of n independent and identically distributed (i.i.d.) sample of observations of a random vector, $\underline{Y} = [Y_1, \ldots, Y_L]'$. We assume $\underline{Y}$ follows a mixture of K components densities, $f(\underline{y}|\underline{\theta}_k)$ ($k = 1, \ldots, K$), with probabilities $\{\alpha_1, \ldots, \alpha_K\}$, where $\underline{\theta}_k$ are the distributional parameters defining the k-th component and $\Theta = \{\underline{\theta}_1, \ldots, \underline{\theta}_K, \alpha_1, \ldots, \alpha_K\}$ the set of all the parameters of the model. The α values, also called *mixing probabilities*, are subject to the usual constraints: $\sum_{k=1}^{K} \alpha_k = 1$ and $\alpha_k \geq 0, \ k = 1, \ldots, K$. The log-likelihood of the observed set of sample observations is

$$\log f(\mathbf{Y}|\Theta) = \log \prod_{i=1}^{n} f(\underline{y}_i|\Theta) = \sum_{i=1}^{n} \log \sum_{k=1}^{K} \alpha_k f(\underline{y}_i|\underline{\theta}_k). \tag{1}$$

In clustering, the identity of the component that generated each sample observation is unknown. The observed data $\mathbf{Y}$ is therefore regarded as incomplete, where the missing data is a set of indicator variables $\mathbf{Z} = \{\underline{z}_1, ..., \underline{z}_n\}$, each taking the form $\underline{z}_i = [z_{i1}, ..., z_{iK}]'$, where z_{ik} is a binary indicator: z_{ik} takes the value 1 if the observation $\underline{y}_i$ was generated by the k-th component, and 0 otherwise. It is usually assumed that the $\{\underline{z}_i, i = 1, \ldots, n\}$ are i.i.d., following a multinomial distribution of K categories, with probabilities $\{\alpha_1, \ldots, \alpha_K\}$. The log-likelihood of complete data $\{\mathbf{Y}, \mathbf{Z}\}$ is given by

$$\log f(\mathbf{Y}, \mathbf{Z}|\Theta) = \sum_{i=1}^{n} \sum_{k=1}^{K} z_{ik} \log \left[\alpha_k f(\underline{y}_i | \underline{\theta}_k) \right]. \tag{2}$$

2.2 Discrete Finite Mixture Models

Consider that each variable in $\underline{Y}$, Y_l ($l = 1, \ldots, L$) can take one of C_l categories. Conditionally on having been generated by the k-th component of the mixture, each Y_l is thus modeled by a multinomial distribution with n_l trials, C_l categories, and non-negative parameters $\underline{\theta}_{kl} = \{\theta_{klc}, c = 1, \ldots, C_l\}$, with $\sum_{c=1}^{C_l} \theta_{klc} = 1$. For a sample $y_{il}(i = 1, \ldots, n)$ of Y_l, we denote as y_{ilc} the number of outcomes in category c, which is a sufficient statistic; naturally, $\sum_{c=1}^{C_l} y_{ilc} = n_l$. Thus, with $\underline{\theta}_k = \{\underline{\theta}_{k1}, \ldots, \underline{\theta}_{kL}\}$ and $\Theta = \{\underline{\theta}_1, \ldots, \underline{\theta}_K, \alpha_1, \ldots, \alpha_k\}$, the log-likelihood function, for a set of observations corresponding to a discrete finite mixture model (mixture of multinomials). This log-likelihood can be seen as corresponding to a missing-data problem, where the missing data has exactly the same meaning and structure as above. The log-likelihood of the complete data $\{\mathbf{Y}, \mathbf{Z}\}$ is thus given by

$$\log p(\mathbf{Y}, \mathbf{Z}|\Theta) = \sum_{i=1}^{n} \sum_{k=1}^{K} z_{ik} \log \left(\alpha_k \prod_{l=1}^{L} \left[n_l! \prod_{c=1}^{C_l} \frac{(\theta_{klc})^{y_{ilc}}}{y_{ilc}!} \right] \right). \tag{3}$$

To obtain a *maximum-likelihood* (ML) or *maximum a posteriori* (MAP) estimate of the parameters of a multinomial mixture, the well-known EM algorithm is usually the tool of choice [7].

3 Model Selection for Categorical Data

Model selection is an important problem in statistical analysis [6]. In model-based clustering, the term *model selection* usually refers to the problem of determining the number of clusters, although it may also refer to the problem of selecting the structure of the clusters. Model-based clustering provides a statistical framework to solve this problem usually resorting to *information criteria*. Among the best-known information criteria we find BIC and AIC, their modifications - namely the consistent

AIC, (CAIC) and the Modified AIC (MAIC) - and also the Integrated Completed Likelihood (ICL) [14, 4]. They are all easily implemented, the final model being selected according to a compromise between its fit to data and its complexity. In this work, we use the *Minimum Message Length* (MML) criterion to choose the number of components of a mixture of multinomials. MML is based on the information-theoretic view of estimation and model selection, according to which an adequate model is one that allows a short description of the observations. MML-type criteria evaluate statistical models according to their ability to compress a message containing the data, looking for a balance between choosing a simple model and one that describes the data well. According to Shannon's information theory, if Y is some random variable with probability distribution $p(y|\Theta)$, the optimal code-length (in an expected value sense) for an outcome y is $l(y|\Theta) = -\log_2 p(y|\Theta)$, measured in bits (from the base-2 logarithm). If Θ is unknown, the total code-length function has two parts: $l(y, \Theta) = l(y|\Theta) + l(\Theta)$; the first part encodes the outcome y, while the second part encodes the parameters of the model. The first part corresponds the fit of the model to the data (better fit corresponds to higher compression), while the second part represents the complexity of the model. The message length function for a mixture of distributions (as developed in [2]) is:

$$l(y, \Theta) = -\log p(\Theta) - \log p(y|\Theta) + \frac{1}{2} \log |I(\Theta)| + \frac{C}{2} (1 - \log(12)), \quad (4)$$

where $p(\Theta)$ is a prior distribution over the parameters, $p(y|\Theta)$ is the likelihood function of mixture, $|I(\Theta)| \equiv |-E\left[\frac{\partial^2}{\partial \Theta^2} \log p(Y|\Theta)\right]|$ is the determinant of the expected Fisher information matrix, and C is the the number of parameters of the model that need to be estimated. For example, for the K mixture multinomial distributions presented in (3), $C = (K - 1) + K\left(\sum_{l=1}^{L}(C_l - 1)\right)$. The expected Fisher information matrix of a mixture leads to a complex analytical form of MML which cannot be easily computed. To overcome this difficulty, Figueiredo and Jain [9] replace the expected Fisher information matrix by its complete-data counterpart $I_c(\Theta) \equiv -E\left[\frac{\partial^2}{\partial \Theta^2} \log p(Y, Z|\Theta)\right]$. Also, they adopt independent Jeffreys' *priors* for the mixture parameters that is proportional to the square root of the determinant of the Fisher information matrix. The resulting message length function is

$$l(y, \Theta) = \frac{M}{2} \sum_{k: \alpha_k > 0} \log\left(\frac{n \alpha_k}{12}\right) + \frac{k_{nz}}{2} \log \frac{n}{12} + \frac{k_{nz}(M + 1)}{2} - \log p(y, \Theta) \quad (5)$$

where M is the number of parameters specifying each component (the dimension of each $\underline{\theta}_k$) and k_{nz} the number of components with non zero probability (for more details on the derivation of (5), see [9, 2]).

4 The MML Based EM Algorithm

In order to estimate a mixture of multinomials, we use a variant of the EM algorithm (herein termed EM-MML), which integrates both estimation and model selection, by directly minimizing (5). The algorithm results from observing that (5) contains, in addition to the log-likelihood term, an explicit penalty on the number of components (the two terms proportional to k_{nz}), and a term (the first one) that can be seen as a log-prior on the α_k parameters of Θ, that will directly affect the M-step.

E-step: The E-step of the EM-MML is precisely the same as in the case of ML or MAP estimation, since the generative model for the data is the same. Since we are dealing with a multinomial mixture, we simply have to plug the corresponding multinomial probability function yielding

$$\bar{z}_{ik}^{(t)} = \frac{\alpha_k \prod_{l=1}^{L} \left[n_l! \prod_{c=1}^{C_l} \frac{(\widehat{\theta}_{klc}^{(t)})^{y_{ilc}}}{y_{ilc}!} \right]}{\sum_{j=1}^{K} \alpha_j \prod_{l=1}^{L} \left[n_l! \prod_{c=1}^{C_l} \frac{(\widehat{\theta}_{jlc}^{(t)})^{y_{ilc}}}{y_{ilc}!} \right]}, \tag{6}$$

for $i = 1, \ldots, n$ and $k = 1, \ldots, K$.

M-step: For the M-step, noticing that the first term in (5) can be seen as the negative log-prior $-\log p(\alpha_k) = \frac{C-K+1}{2K} \log \alpha_k$ (plus a constant), and enforcing the conditions that $\alpha_k \geq 0$, for $k = 1, ..., K$ and that $\sum_{k=1}^{K} \alpha_k = 1$, yields the following updates for the estimates of the α_k parameters:

$$\widehat{\alpha}_k^{(t+1)} = \frac{\max \left\{ 0, \sum_{i=1}^{n} \bar{z}_{ik}^{(t)} - \frac{C-K+1}{2K} \right\}}{\sum_{j=1}^{K} \max \left\{ 0, \sum_{i=1}^{n} \bar{z}_{ij}^{(t)} - \frac{C-K+1}{2K} \right\}}, \tag{7}$$

for $k = 1, ..., K$. Notice that, some $\widehat{\alpha}_k^{(t+1)}$ may be zero; in that case, the k-th component is excluded from the mixture model. The multinomial parameters corresponding to components with $\widehat{\alpha}_k^{(t+1)} = 0$ need not be further calculated, since these components do not contribute to the likelihood. For the components with non-zero probability, $\widehat{\alpha}_k^{(t+1)} > 0$, the estimates of multinomial parameters are updated to their standard weighted ML estimates:

$$\widehat{\theta}_{klc}^{(t+1)} = \frac{\sum_{i=1}^{n} \bar{z}_{ik}^{(t)} y_{ilc}}{n_l \sum_{i=1}^{n} \bar{z}_{ik}^{(t)}}, \tag{8}$$

for $k = 1, \ldots, K$, $l = 1, \ldots, L$, and $c = 1, \ldots, C_l$. Notice that, in accordance with the meaning of the θ_{klc} parameters, $\sum_{c=1}^{C_l} \widehat{\theta}_{klc}^{(t+1)} = 1$.

5 Data Analysis and Results

First, we evaluate the performance of the EM-MML algorithm on 10 synthetic data sets, over 50 runs. The data sets were originated from a mixture of 3 categorical variables (with 2, 3 and 4 levels) and 2 components. The correponding Sihouette index values illustrate the structures diversity: 0.099; 0.216; 0.217; 0.230; 0.713; 0.733; 0.746; 0.778; 0.805; 0.817. The obtained results are compared with those obtained from a standard EM algorithm combined with BIC, AIC, CAIC, MAIC, and ICL criteria.

The comparison resorts to a cohesion-separation measure and a concordance measure: the Fuzzy Silhouette index [5] of the clustering structure obtained and the Adjust Rand [10] between the same clustering structure and the original one. In Table 1 we can verify there are no significant differences between the EM-MML and the other criteria, except ICL which only recovers the very well separated structures. Regarding the number of clusters, EM-MML and MAIC are tied, recovering this number correctly for all data sets.The same is not true for the other criteria: AIC identifies 3 clusters in 3 data sets and 4 clusters once; in addition, BIC and CAIC could not find any cluster structure once and ICL was unable to do it for 4 data sets. In terms of computation time, since EM-MML does not require a sequential approach, it becomes clearly faster than the other criteria (Friedman test yields $\chi^2(5)=2500$ and p-value<0.01; Post hoc tests, with Bonferroni correction, only reveal statistically significant differences between the EM-MML and the other criteria).

Table 1 Criteria performance.

Criterion	Number of data sets	Fuzzy Silhouette: 95% CI Lower ; Upper Limits[a]	Adjusted Rand: 95% CI Lower ; Upper Limits[a]
AIC	10	0.430 ; 0.741	0.545 ; 0.867
BIC	9	0.622 ; 0.935	0.728 ; 1.000
CAIC	9	0.616 ; 0.931	0.732 ; 1.000
ICL	6	0.917 ; 0.948	1.000 ; 1.000
MAIC	10	0.568 ; 0.887	0.623 ; 0.950
EM-MML	**10**	**0.561 ; 0.891**	**0.594 ; 0.955**

[a] 1000 bootstrap samples were used to estimate the Confidence Intervals (CI).

Additional insight into the performance of EM-MML is obtained by applying it to a real data set referring to the 6th European Working Conditions Survey (2015), Eurofound working conditions survey. Note that these data are the most recent.

For the purpose of our experiment, we consider the aggregate data referring to 305 European regions and the answers to the following questions: Are you able to

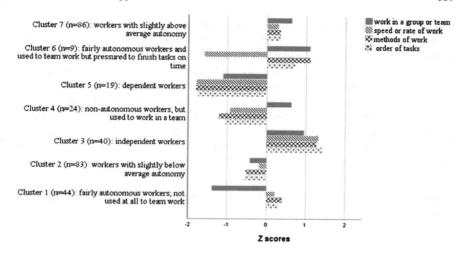

Fig. 1 Clusters' profile and their dimensions (*n*).

choose or change: a) your order of tasks; b) your methods of work; c) your speed or rate of work. Do you work in a group or team that has common tasks and can plan its work?

EM-MML selected 7 clusters, which is a smaller number than for the remaining criteria (ICL, BIC, CAIC, AIC and MAIC select 10, 12, 12, 15 and 15 respectively). This fact avoids estimation problems associated with very small segments and also improves the interpretability of the clustering solution.

The segments selected by EM-MML criterion are presented in Figure 1. Workers with slightly above average autonomy (cluster 7) live in several countries, but Ireland stands out, as well as Belgium, Germany, Netherlands, Switzerland, and the UK regions. Denmark, Estonia, Malta, and Norway are the countries where the most independent workers are found (cluster 3). The smallest cluster, 6, includes Sweden and a region of Greece and Kriti and Açores, a Greek and a Portuguese region, respectively. The cluster 5, where workers claim they have no autonomy, includes regions from many countries.

6 Discussion and Perspectives

In this work, a model selection criterion and method for finite mixture models of categorical observations was studied - EM-MML. This algorithm simultaneously performs model estimation and selects the number of components/clusters. When compared to information criteria, which are commonly associated with the use of the EM algorithm, the EM-MML method exhibits several advantages: 1) it easily recovers the true number of clusters in synthetic data sets with various degrees of separation; 2) its computations times are significantly lower than those required by standard approaches resorting to the sequential use of EM and an information criterion; 3) when applied to a real data set it produces a more parsimonious solution, thus easier to interpret. An additional advantage of this approach that stems from obtaining more parsimonious solutions is that such solutions have a higher number of observations per cluster, thus helping to overcome eventual estimation problems.

The performance of the EM-MML is encouraging for selecting the number of clusters, and the same criterion was already used for feature selection [17]. However, future research is required, namely considering data sets with different numbers of clusters and high dimensional data.

Acknowledgements This work was supported by Fundação para a Ciência e Tecnologia, grant UIDB /00315/2020.

References

1. Akaike, H.: Maximum likelihood identification of Gaussian autorregressive moving average models. Biometrika. **60**, 255–265 (1973)
2. Baxter, R. A., Olivier, J. J.: Finding overlapping components with MML. Stat. Comput. **10**(1), 5–16 (2000)
3. Bouguila, N., Fan, W.: Mixture Models and Applications (Unsupervised and Semi-Supervised Learning). Springer Nature Switzerland AG, Switzerland (2020)
4. Bozdogan, H.: Mixture-model cluster analysis using model selection criteria and a new informational measure of complexity. In: Bozdogan, H. (eds.) Proceedings of the First US/Japan Conf. Frontiers of Stat. Modeling, pp.69–113. Boston: Kluwer Academic Publishers (1994)
5. Campello, R. J., Hruschka, E. R.: A fuzzy extension of the silhouette width criterion for cluster analysis. Fuzzy Set. Syst., **157**(21), 2858–2875 (2006)
6. Celeux, G., Martin-Magniette, M. L., Maugis-Rabusseau, C., Raftery, A. E.: Comparing model selection and regularization approaches to variable selection in model-based clustering. J. Soc. Fr. Statistique. **155**(2), 57–71 (2014)
7. Dempster, A., Laird, N., Rubin, D.: Maximum likelihood estimation from incomplete data via the EM Algorithm. J. R. Stat. Soc. **39**, 1–38 (1997)
8. Everitt, B. S., Hand, D.: Finite Mixture Distributions. Chapman and Hall, New York (1981)
9. Figueiredo, M. A. T., Jain, A. K.: Unsupervised learning of finite mixture models. IEEE T. Pattern Anal. **24**, 381–396 (2002)
10. Hubert, L., Arabie, P.: Comparing partitions. J. Classif. **2**(1), 93–218 (1985).
11. Kumar, P., Kanavalli, A.: A similarity based K-means clustering technique for categorical data in data mining application. Int. J. Intell. Eng. Syst. (2021) doi: 10.22266/ijies2021.0430.05
12. Lee, C., Jung, U.: Context-based geodesic dissimilarity measure for clustering categorical data. Appl. Sciences (2021) doi: 10.3390/app11188416
13. McLachlan, G. J., Peel, D.: Finite Mixture Models. Wiley, New York (2000)
14. Novais, L., Faria, S.: Selection of the number of components for finite mixtures of linear mixed models. J. Int. Math. **24**(8), 2237–2268 (2021)
15. Schwarz, G.: Estimating the dimension of a model. Ann. Stat. **6**, 461–464 (1978)
16. Silvestre, C., Cardoso, M. G. M. S. and Figueiredo, M.: A clustering view on ESS measures of political interest: an EM-MML approach. NTTS - New Techniques and Technologies for Statistics (2017).
17. Silvestre, C., Cardoso, M. G. M. S. and Figueiredo, M.: Feature selection for clustering categorical data with an embedded modeling approach. Expert Syst. **32**(3), 444–453 (2014).

Typology of Motivation Factors for Employees in the Banking Sector: An Empirical Study Using Multivariate Data Analysis Methods

Áurea Sousa, Osvaldo Silva, M. Graça Batista, Sara Cabral, and Helena Bacelar-Nicolau

Abstract Leadership has been considerate as a competitive advantage for organizations, contributing to their success and effective and efficient performance. Motivation, on the other hand, is assumed as a basic competence of leadership. Therefore, the main purpose of this paper is to know the perceptions of bank employees on the main motivational factors in the organizational context. Data analysis was performed based on several statistical methods, among which the Categorical Principal Component Analysis (CatPCA) and some agglomerative hierarchical clustering algorithms from *VL* (*V* for Validity, *L* for Linkage) parametrical family, applied to the items that aim to assess the aspects most valued by bankers in the work context. The CatPCA allowed to extract four principal components which explain almost 70% of the total data variance. The dendrograms provided by the hierarchical clustering algorithms over the same data, exhibit four main branches, which are associated with different main motivational factors. Moreover, CatPCA and clustering results show an important correspondence concerning the main motivations in this sector.

Keywords: leadership, welfare, motivational factors, CatPCA, cluster analysis

Áurea Sousa (✉)
Universidade dos Açores and CEEAplA, Rua da Mãe de Deus, 9500-321, Portugal,
e-mail: `aurea.st.sousa@uac.pt`

Osvaldo Silva
Universidade dos Açores and CICSNOVA.UAc, Rua da Mãe de Deus, Portugal,
e-mail: `osvaldo.dl.silva@uac.pt`

M. Graça Batista
Universidade dos Açores and CEEAplA, Rua da Mãe de Deus, Portugal,
e-mail: `maria.gc.batista@uac.pt`

Sara Cabral
Universidade dos Açores, Rua da Mãe de Deus, Portugal, e-mail: `sara_crc@hotmail.com`

Helena Bacelar-Nicolau
Universidade de Lisboa (UL) Faculdade de Psicologia and Institute of Environmental Health (ISAMB/FM-UL), Portugal, e-mail: `hbacelar@psicologia.ulisboa.pt`

P. Brito et al. (eds.), *Classification and Data Science in the Digital Age*,
Studies in Classification, Data Analysis, and Knowledge Organization,
https://doi.org/10.1007/978-3-031-09034-9_39

1 Introduction

Motivation has always been subject of analysis by the scientific community, as numerous definitions have emerged. For Robbins and Judge ([21], p. 184), motivation is defined as "the processes that account for an individual's intensity, direction, and persistence of effort toward attaining a goal". These three indicators are assumed to be key-factors of motivation: intensity describes the individual's effort to achieve the proposed goals; this effort should go in a direction that benefits the organization; and, finally, the persistence with which the individual is able to maintain that effort. In this context, the individual's behavior is determined by what motivates them, which is why their performance results not only from ability and skills, but also from motivation. Moreover, motivation is complex and influenced by innumerable variables, considering the diverse needs and expectations that individuals try to satisfy in different ways [15]. Moreover, different leadership practices may lead to better or worse motivational responses from employees.

The main purpose of this paper is to analyse the perceptions of bank employees who work in the banks that operate in the Autonomous Region of the Azores on the main motivational factors in the organizational context. Our study also intends to perform a reduction of the dimensionality of the data and to find a typology of a set of items that was used to evaluate the latent variable "Motivation", regarding the most valued aspects in the work context. Thus, Section 2 concerns the materials and methods of research. Section 3 presents and discusses the main results of this study. Finally, Section 4 contains the main conclusions.

2 Materials and Methods

This study was based on a quantitative approach, using a validated questionnaire, which can be found in Cabral [7]. The sample consists of 202 bank employees (51.0 % male and 49.0 % female) of the Autonomous Region of the Azores (response rate: 6.4%). Most respondents are 36 years old or older (60.9%) and have higher education (56.7%).

The present study refers to a subset of twenty-seven items used to evaluate the latent variable "Motivation" in work context, namely: 1 - The opportunity for career advancement, 2 - Have greater responsibility, 3 - The feeling of being involved in decision making, 4 - A job that gives you prestige and status, 5 - Have an interesting and challenging job, 6 - The recognition and appreciation of others for the accomplished work, 7 - Have a good relationship with your colleagues, 8 - Have a good relationship with your superiors, 9 - A work environment where there is trust and respect, 10 - The loyalty of superiors towards the collaborators, 11 - Team spirit, 12 - Sense of belonging to the organization, 13 - An adequate discipline, 14 - There is equality of treatment and opportunities between the various employees, 15 - Earn respect and esteem of your colleagues and superiors, 16 - Professional development, 17 - Salary appropriate to the professional functions, 18 - A stable job that gives you security, 19 - Good working conditions, 20 - Balance between personal and professional life, 21 - Being able to express your opinion and ideas without fear of reprisals, 22 - Availability to solve problems/personal situations, 23 - Have a fair and adequate system of objectives and incentives, 24 - Being rewarded for overtime work, 25 - Being pressured to achieve the proposed objectives, 26 - Ability to handle pressure at work, and 27 - Appropriate training to the professional functions.

For each item, respondents could pick only one of six modalities of response according to their level of agreement or disagreement with the items that assess motivation: Totally disagree; Disagree most of the time; Slight disagree; Slight agree; Agree most of the time, and Totally agree. In this study, Categorical Principal Components Analysis (CatPCA), using the Varimax rotation method with Kaiser Normalization; and some agglomerative hierarchical clustering algorithms (AHCA) were used. Data analysis was performed using the packages IBM SPSS Statistics 26 and CLUST11 [19].

Principal Components Analysis (PCA) aims to reduce the dimensionality of the original data so that "the first few dimensions account for as much of the available information as possible" ([9], p. 83), assuming linear relationships among numeric variables. Each principal component is uncorrelated with all others, and it is expressed as a linear combination of the original variables. CatPCA optimally quantifies categorical (ordinal or nominal) variables and can handle and discover nonlinear relationships between variables (e.g., [12]). In the present study, we applied the CatPCA due to the ordinal nature of the items under analysis.

The goal of a clustering algorithm is to obtain a partition, where the elements within a cluster are similar and elements (objects/individuals/groups of individuals or variables) in different groups are dissimilar, identifying natural clustering structures in a data set (e.g., [8]). Agglomerative clustering algorithms usually start with each element to sort into its own separate cluster of size 1 (singleton). At each step, the algorithms find the two "closest" clusters, taking into account the aggregation criterion, and join them. The process continues until a cluster containing all elements to classify is obtained. The AHCA of the set of items was based on the affinity coefficient as a measure of comparison between elements, combined with two classic (Single-Linkage (SL) and Complete-Linkage (CL)) and a family of probabilistic VL (V for Validity, L for Linkage) aggregation criteria (e. g., [1, 2, 3, 10, 11, 16, 17, 18, 22]).

According to Ng et al. ([20], p. 849), "the task of finding good clusters has been the focus of considerable research in machine learning and pattern recognition". However, the identification of the best partitions using validation indices is also of crucial importance. Therefore, a pertinent question arises: "How well does the partition fit the data?" ([8], p. 505). On what validation of results is concerned, the identification of the best partitions in the present study was based on the global level statistics, $STAT$ [1, 10, 11]. The global maximum $STAT$ value indicates the best cut-off level of a dendrogram and the local maxima $STAT$ differences indicate the most significant levels.

The affinity coefficient between two distribution functions was introduced by Matusita in 1951 (e.g., [13, 14]). Bacelar-Nicolau extended it to the non-supervised classification field as a similarity measure between profiles. Let V be a set of p variables, describing a set D of N statistical data units (individuals), so that each of the $N \times p$ cells of the corresponding data table X contains one single non-negative real value x_{ik} ($i = 1,..., N$; $k = 1,..., p$) which denotes the value of the k-th variable on the i-th individual. The standard affinity coefficient $a(k, k')$ between a pair of variables, V_k and V'_k ($k, k' = 1,..., p$) is given by formula (1), where $x_{.k} = \Sigma_{i=1}^{N} x_{ik}$, $x_{.k'} = \Sigma_{i=1}^{N} x_{ik'}$.

$$a(k, k') = \Sigma_{i=1}^{N} \sqrt{\frac{x_{ik}}{x_{.k}} \frac{x_{ik'}}{x_{.k'}}} \tag{1}$$

The coefficient (1) is a symmetric similarity coefficient which takes values in [0,1] (1 for equal or proportional vectors and 0 for orthogonal vectors). Note that its mathematical formula corresponds to the inner product between the square root column profiles associated with those variables and measures a monotone tendency between column profiles. In the particular case of binary variables, the affinity coefficient coincides with the well-known Ochiai coefficient. Furthermore (e.g., [4, 6]), it is related to the Hellinger distance d by the relation $d^2 = 2(1 - a)$, which has been used in the context of spherical factor analysis by Michel Volle. Later on, the standard affinity coefficient was extended to the clustering of statistical data units or variables, mainly in a three-way approach (e.g., [3, 4, 5, 6]). The computation of the standard affinity coefficient between individuals can be performed by previously transposing the data matrix and then applying formula (1).

The probabilistic aggregation criteria on the scope of VL methodology can be interpreted as distribution functions of statistics of independent random variables, that are i.i.d. uniform on $[0, 1]$ (e.g., [3, 17]). The SL aggregation criterion can lead to very long clusters (chaining effect). On the other hand, the AVL (Aggregation Validity Link) has a tendency to form equicardinal clusters with an even number of elements. The comparison functions between a pair of clusters, A and B, concerning the family I of AVL methods can be generated by the following conjoined formula (e.g., [17, 10, 11]):

$$\Gamma(A, B) = (p_{AB})^{g(\alpha,\beta)} \tag{2}$$

with $\alpha = Card\ A$, $\beta = Card\ B$, $p_{AB} = max[\gamma_{ab} : (a, b) \in (A \times B]$, with $1 \le g(\alpha, \beta) \le \alpha\beta$, and γ_{xy} is a similarity measure between pairs of elements, x and y, of the set of elements to classify (e.g., $g(\alpha, \beta) = 1$ for SL, $g(\alpha, \beta) = \alpha\beta$ for AVL). Note that varying $g(\alpha, \beta)$ with $1 < g(\alpha, \beta) < \alpha\beta$, a sort of compromise can be built between SL and AVL methods (e.g., $g(\alpha, \beta)=(\alpha+\beta)/2$ for $AV1$). Thus, $\Gamma(A, B)$ will be "more polluted by the chain effect when $g(\alpha, \beta)$ remains near 1, and more contaminated by the symmetry effect as long as $g(\alpha, \beta)$ is in the neighbourhood of $\alpha\beta$" ([17], p. 95). Among the criteria that establish a compromise between AVL and SL methods, stands out the $AV1$ method, whose behavior is very similar to that of AVL and often provides, at its cut-off level, a partition better adjusted to the preorder than the "best" classification obtained by AVL.

3 Main Results and Discussion

Concerning the CatPCA, the best solution comprises four principal components, and the percentage of variance accounted for (PVAF) across these components is almost 70% (about 69%) of the data's total variance. All extracted components have eigenvalues above 1. Moreover, the first three main components have a very good internal consistency and the fourth component has an acceptable internal consistency, as shown by the values of the Cronbach's Alpha coefficient (see Table 1).

Table 1 Rotated component loadings of the 4-component solution - Motivational factors.

Items	PC1	PC2	PC3	PC4
M1	0.213	0.351	**0.699**	0.166
M2	0.197	0.044	**0.794**	0.211
M3	0.248	0.148	**0.763**	-0.018
M4	-0.028	0.098	**0.482**	0.442
M5	0.354	0.219	**0.674**	0.037
M6	**0.522**	0.214	0.425	0.095
M7	**0.837**	0.110	0.193	-0.114
M8	**0.774**	0.151	0.244	0.099
M9	**0.778**	0.227	0.183	-0.125
M10	**0.783**	0.269	0.227	-0.043
M11	**0.757**	0.259	0.223	-0.103
M12	**0.798**	0.155	0.227	-0.035
M13	**0.708**	0.213	0.341	0.070
M14	0.486	**0.511**	0.372	-0.257
M15	**0.775**	0.263	0.252	0.041
M16	0.432	0.364	**0.665**	0.035
M17	0.289	**0.708**	0.410	-0.046
M18	0.462	**0.641**	0.097	-0.247
M19	**0.548**	0.532	0.211	-0.034
M20	0.503	**0.609**	0.074	-0.223
M21	**0.684**	0.401	0.070	0.074
M22	**0.678**	0.399	0.019	0.054
M23	0.295	**0.770**	0.284	0.102
M24	0.174	**0.835**	0.176	-0.011
M25	0.019	-0.012	0.233	**0.864**
M26	-0.038	-0.146	0.035	**0.896**
M27	**0.543**	0.458	0.230	0.227
Eigenvalue (VAF)	7.988	4.417	4.066	2.138
Percentage accounted (PVAF)	29.59	16.36	15.06	7.92
Cronbach's Alpha	0.950	0.934	0.919	0.610

The most important items for the first dimension are items M6, M7, M8, M9, M10, M11, M12, M13, M15, M19, M21, M22, and M27, which are related to human relationships/interactions with colleagues and hierarchical superiors, so it is called

"Psychological well-being/Interpersonal relationships". This dimension explains the highest proportion of data variance (29.59%).

Concerning the second dimension, the items M14, M17, M18, M20, M23, and M24 are the most important, so this dimension was designated "Remuneration, job stability and incentive system". The most relevant items regarding the third dimension are M1, M2, M3, M4, M5, and M16; so, this dimension was called "Career progression/Professional achievement". Finally, the most important items for the fourth dimension are M25 and M26 related to "Fulfilment of the proposed objectives and the timings to achieve them".

Regarding the AHCA of the same set of items, and considering the best cut-off levels, the results of the present study are summarized in Table 2.

Table 2 The best partition - Standard affinity coefficient.

Method	The best partition	STAT	Cut-off level
SL/CL	{M1, M2, M3, M5, M8, M10, M11, M12, M13, M15, M14, M16, M18, M19, M22, M20, M6, M23, M27, M24, M21}; {M4}; {M9}; {M7}; {M25}; {M26}; {M17}	15.8858	20
AV1	{M1, M2, M3, M6, M27, M21, M5, M23, M24, M8, M15, M14, M16, M10, M13, M11, M12, M18, M19, M20, M22}; {M4, M25, M26}; {M7}; {M9}; {M17}	15.6490	22

According to the *STAT* values, the best partitions were obtained by the classic *SL/CL* and the probabilistic *AV1* methods (see Table 2). All dendrograms highlighted four main branches, which are associated with different motivational factors ("Career progression"; "Psychological well-being / Interpersonal relationships"; "Organizational environment and working conditions"; "Conformity with objectives and time to reach them"), bringing new information, and identifying some singletons, as shown in Figure 1.

4 Conclusion

Organizations and their leaders have become increasingly aware of the importance of their employees being well and that negative feelings can negatively affect productivity. Thus, it is essential to ensure the well-being of employees, taking into account the main motivational factors identified in this study. CatPCA made it possible to extract four principal components (dimensions), which explain almost 70% of the total variance of the data, which were designated, respectively, by "Psychological well-being/Interpersonal relationships"; "Remuneration, job stability and incentive system"; "Career progression/Professional achievement"; and "Fulfilment of objectives and timings to achieve them". Regarding the AHCA of the items that

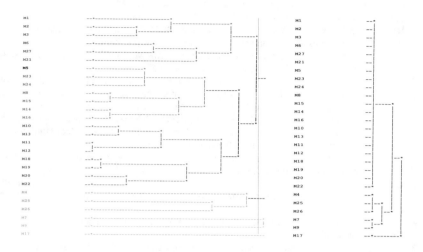

Fig. 1 Dendrogram - Standard affinity coefficient + *AV1*.

assess motivation, the dendrograms highlight four main branches, which are associated with different motivational factors called "Career progression"; "Psychological well-being / Interpersonal relationships"; "Organizational environment and working conditions"; and "Conformity with objectives and time to reach them". They carried new information and identify some singletons as well. Comparing the dendrograms, we conclude that the clusters referring to the best partitions are quite similar, with observed differences mainly concerning the few singletons. Moreover, the effective and fruitful correspondence between the AHCA and the CatPCA results may help to better understand the main types of factors identified. In fact, the four main branches of all dendrograms are related to motivational factors which corresponding interpretation are in consonance with those identified through CatPCA.

Acknowledgements This paper is financed by Portuguese national funds through FCT – Fundação para a Ciência e a Tecnologia, I.P., project number UIDB/00685/2020.

References

1. Bacelar-Nicolau, H.: Contributions to the Study of Comparison Coefficients in Cluster Analysis (in Portuguese). Univ. Lisbon (1980)
2. Bacelar-Nicolau, H.: The affinity coefficient in cluster analysis. In: Bekmann, M. J. et al. (eds.). Methods of Operations Research, pp. 507-512. Verlag Anton Hain, Munchen (1985)

3. Bacelar-Nicolau, H.: Two probabilistic models for classification of variables in frequency tables. In: Bock, H._H. (ed.) Classification and Related Methods of Data Analysis, pp. 181-186. Elsevier Sciences Publishers B.V., North Holland (1988)

4. Bacelar-Nicolau, H.: The affinity coefficient. In: Bock, H.-H. and Diday, E. (eds.) Analysis of symbolic data: Exploratory methods for extracting statistical information from complex data, Series: Studies in Classification, Data Analysis, and Knowledge Organization, pp. 160-165. SpringerVerlag, Berlin (2000) doi: 10.1007/978-3-642-57155-8

5. Bacelar-Nicolau, H., Nicolau, F.C., Sousa, Á., Bacelar-Nicolau, L.: Measuring similarity of complex and heterogeneous data in clustering of large data sets. Biocybernetics and Biomedical Engineering, PWN-Polish Scientific Publishers Warszawa. **29**(2), 9-18 (2009)

6. Bacelar-Nicolau, H., Nicolau, F. C., Sousa, Á., Bacelar-Nicolau, L.: Clustering of variables with a three-way approach for health sciences. Testing, Psychometrics, Methodology in Applied Psychology (TPM) (2014) doi: 10.4473/TPM21.4.56

7. Cabral, S.: O Impacto da Liderança na Motivação dos Colaboradores do Setor Bancário na Região Autónoma dos Açores. Universidade dos Açores, Ponta Delgada (2018)

8. Gurrutxaga, I., Muguerza, J., Arbelaitz, O., Pérez, J.M., Martín, J.I.: Towards a standard methodology to evaluate internal cluster validity indices. Pattern Recognit. Lett. **32**, 505-515 (2011)

9. Lattin, J. M., Carrol, J. D., Green, P. E.: Analyzing Multivariate Data (1st ed.). Thomson Brooks, Cole (2003)

10. Lerman, I. C.: Classification et Analyse Ordinale des Données. Dunod, Paris (1981)

11. Lerman, I. C.: Foundations and methods in combinatorial and statistical data analysis and clustering. Series: Advanced Information and Knowledge Processing. Springer-Verlag, Boston (2016)

12. Linting, M., Meulman, J. J., Groenen, P. J., van der Kooij, A. J.: Nonlinear principal components analysis: Introduction and application. Psychol. Meth. **12**(3), 336–358 (2007) doi: 10.1037/1082-989X.12.3.336

13. Matusita, K.: On the theory of statistical decision functions. Ann. Inst. Stat. Math. **3**, 1-30 (1951)

14. Matusita, K.: Decision rules, based on distance for problems of fit, two samples and estimation. Ann. Inst. Stat. Math. **26**, 631-640 (1995)

15. Mullins, L. J.: Management and Organizational Behavior. Prentice Hall, England (2005)

16. Nicolau, F. C., Bacelar-Nicolau, H.: Nouvelles méthodes d'agrégation basées sur la fonction de répartition. In: Collection Séminaires INRIA 1981, Classification Automatique et Perception par Ordinateur, pp. 45-60. Rocquencourt (1982)

17. Nicolau, F. C., Bacelar-Nicolau, H.: Some trends in the classification of variables. In: Hayashi et al. (eds.). Data Science, Classification and Related Methods, pp. 89-98. Springer, Tokyo (1998)

18. Nicolau, F. C., Bacelar-Nicolau, H.: Teaching and learning hierarchical clustering probabilistic models for categorical data. In: Proc. 54th Session of the International Statistical Institute (IASE at ISI, IPM-71). Berlin, Germany (2003) https://iase-web.org/documents/papers/isi54/3654.pdf.Cited25Jan2022

19. Nicolau, F. C., Bacelar-Nicolau, H., Sousa, F., Sousa, Á., Silva, O.: CLUST11: Cluster Analysis Software - Standard and *VL* Probabilistic Approaches. LEAD, FP-UL (2011)

20. Ng., A. Y., Jordan, M. I., Weiss, Y. In: On spectral clustering: Analysis and an algorithm (2002) https://proceedings.neurips.cc/paper/2001/file/801272ee79cfde7fa5960571fee36b9b-Paper.pdf.Cited25Jan2022

21. Robbins, S. P., Judge, T. A.: Organizational Behavior. Pearson Education, Inc., New Jersey (2015)

22. Sousa, Á., Silva, O., Bacelar-Nicolau, H., Nicolau, F. C.: Distribution of the affinity coefficient between variables based on the Monte Carlo simulation method. Asian. J. Appl. Sci. **1**(5), 236-245 (2013)

A Proposal for Formalization and Definition of Anomalies in Dynamical Systems

Jan Michael Spoor, Jens Weber, and Jivka Ovtcharova

Abstract Although many scientists strongly focus on anomaly detection in different applications and domains, there currently exists no universally accepted definition of anomalies and outliers. Using an approach based on control theory and dynamical systems, as well as a definition for anomalies as described by philosophy of science, the authors propose a generalized framework viewing anomalies as key drivers of progress for a better understanding of the dynamical systems around us. By mathematically defining anomalies and delimiting deviations within expectations from completely unforeseen instances, this paper aims to be a contribution to set up a universally accepted definition of anomalies and outliers.

Keywords: anomaly detection, outlier analysis, dynamical systems

1 Introduction

Anomalies, often interchangeably called outliers [1], are of key interest in explorative data analysis. Therefore, anomaly detection finds application in many different scientific fields, i.e., in social science, economics, engineering, and medical science [2]. In particular, research in these domains regarding databases, data mining, machine learning or statistics focuses strongly on anomaly detection [3]. Despite the wide

Jan Michael Spoor (✉)
Institut für Informationsmanagement im Ingenieurwesen (IMI), Karlsruhe Institute of Technology, Karlsruhe, Germany, e-mail: jan.spoor@kit.edu

Jens Weber
Team Digital Factory Sindelfingen, Mercedes-Benz Group AG, Sindelfingen, Germany, e-mail: jens.je.weber@mercedes-benz.com

Jivka Ovtcharova
Institut für Informationsmanagement im Ingenieurwesen (IMI), Karlsruhe Institute of Technology, Karlsruhe, Germany, e-mail: jivka.ovtcharova@kit.edu

© The Author(s) 2023
P. Brito et al. (eds.), *Classification and Data Science in the Digital Age*,
Studies in Classification, Data Analysis, and Knowledge Organization,
https://doi.org/10.1007/978-3-031-09034-9_40

range of anomaly detection, there is currently no universally accepted definition of what an outlier or anomaly is [2], and the mathematical definition depends on the selected method to find these anomalies [4].

The authors previously proposed an applied framework to formalize anomalies within the context of control theory and dynamical systems [5]. In this publication, the idea is discussed in more depth, and a generalization of the framework is proposed to extend its application area to more domains since dynamical systems are relevant in engineering and science [6] as well as in management science and economics [7]. Furthermore, the proposed definition of anomalies should also be applicable outside of the context of control theory and aims to be a contribution to set up a universally accepted definition of anomalies and outliers.

When controlling or simulating dynamical systems, a measurement and prediction process is used. Anomalies occur in this process as substantial deviations of a measured system state (an actual value) from an expected system state (a planned value) [5]. Despite simulation and planning effort, these deviations still occur. While some deviations fall within an acceptable range and within the expectations of normal system behavior, other anomalies are completely unforeseen and do not fit the set-up and expectations of the system. Three sequential questions are derived to further investigate the nature of anomalies within dynamical systems:

1. What distinguishes unforeseen system states from regular system behavior?
2. How can unforeseen system states or errors occur despite simulation?
3. How can unforeseen system states be analyzed and transferred to a standard model of a system's behavior?

2 Definition of Anomalies for Dynamical Systems

2.1 Definitions of Anomalies and Outliers

In general, it is assumed that anomalies are somehow visible within the data of the observed systems. This is also clearly stated by the definition of an outlier or anomaly as data points with a substantial deviation from the norm since this requires a normal state of the system and a measurable deviation [8]. Furthermore, the anomaly detection requires existence and knowledge of a normal state, a definition of a deviation, a metric, and a threshold measure of distance. This threshold measure of distance uses the selected metric. All distances between the norm and the data points, which are either above (in case of distance measures) or below (in case of similarity measures) the defined threshold, are assumed to be non-substantial.

Therefore, in addition, the selection of an appropriate metric becomes an important tool to accurately describe an anomaly. Some authors claim that, in a practical application, the selection of a suitable metric might be more important than the algorithm itself. For example, if clusters are clearly separated within the examined dataset in context of the selected metric, clusters will be found independently of the used method or algorithm [9]. Other authors claim that the selected method for investigating clusters is of importance [10].

To summarize, there is no trivial definition of a normal state, a deviation, and when a deviation might be substantial. Some authors therefore describe the usefulness of an analysis only within the context of the goals of the analysis [11]. Outlier detection becomes more of a technical target than an actual scientific finding of something novel since the novelty is always defined within the technical target of the analysis. Alternatively, the normal model of the data defines an anomaly [1].

This results, for example, in approaches of regression diagnostics to exclude outliers and anomalous data prior to an analysis or to conduct the analysis along the standard model in a more robust way, which is less affected by anomalies [12]. Both approaches result in the maintaining of the normal model using anomalies as if they were less adequate or not at all representative of the data set.

Since anomalies are only relevant within a context, a typology of anomalies within different dataset contexts can be created. Thus, Foorthuis [13] proposes a typology along the following dimensions: types of data (qualitative, quantitative or mixed), anomaly level (atomic or aggregated) and cardinality of relationship (univariate or multivariate). Anomalies are, within this kind of typology, always dependent on the dataset and behave differently along the measured features, which have been classified as relevant for the specific analysis. The anomaly detection becomes a detection of unfitting, surprising values while maintaining the normal model.

2.2 Definition by Philosophy of Science

If the assumptions regarding normal states, deviation, and substantiality are dropped, it is possible to discuss anomalies on a more fundamental level for understanding our surroundings and the observations of them.

To do this, anomalies have to be placed in the historic context of science and research. Since anomaly detection as a discipline of data science is placed within the scientific context [14], anomaly detection can also be analyzed as part of the scientific method and therefore a comparison with the historical understanding of anomalies in the context of science becomes relevant. By definition of Kuhn [15], anomalies play an important role in the scientific discovery of novelties:

> Discovery commences with the awareness of anomaly, i.e., with the recognition that nature has somehow violated the paradigm-induced expectations that govern normal science. It then continues with a (...) exploration of the area of anomaly. And it closes only when the paradigm theory has been adjusted so that the anomalous has become the expected.

This statement describes scientific progress as a stepwise discovery and the placement of anomalies within a normal state by science. The discussed normal state is therefore dictated by current scientific knowledge, which encompasses the predictions of the currently available and widely used models and theories. An anomaly violates the normal state by violating the predictions of these models. The steps of scientific progress are then as follows:

1. Knowledge of the anomaly.
2. Stepwise acknowledgement of observations and conceptual nature of the anomaly.
3. Change of paradigm and methods to include the anomaly in the new models, often under resistance by the scientific community itself.

Therefore, different states of an anomaly exist as follows:

1. The anomaly is completely unknown.
2. The anomaly is neither described nor modeled but was observed.
3. The anomaly is not commonly recognized and placed within the standard model.

The states of anomalies correspond to the initially defined questions in the introduction regarding the delimitation of anomalous states from normal states, the exploration of the causes for anomalies, and the modeling and planning with the now known anomalies. If the states of anomalies are used to describe practical errors in engineering, error states of systems are not anomalies. This is the case because if error states are priorly classified as such, they are therefore already known and described. This corresponds to the idea that outliers or anomalies are created by a different underlying mechanism [16] and therefore imply an unknown system behavior, which needs modeling to better describe the system. In addition, this follows the assumption of a normal state in which anomalies simply derive from a normal model [1] since they are not part of the normal model. Also, this idea relates strongly to the discussion of the relation between novelty and anomaly detection [17].

To follow the definitions by Kuhn [15], science is driven by internal progress, limited by the current methods and available resources, while external targets, defined by stakeholders, e.g., society or companies, drive technicians. This description matches the idea that the usefulness of an analysis should be evaluated within the context of its goals [11] and distinguishes two types of anomalies: "Scientific" anomalies of a novel observation and "technical" anomalies as deviations from a predefined norm using a predefined measurement of substantiality.

"Scientific" anomalies might still result in unwanted system states, which then can result in some kind of error or critical system state. Nevertheless, not every "scientific" anomaly inevitably results in an error state and not every error state is a "scientific" anomaly. An anomaly is not a "scientific" anomaly if the error state is already documented or can be described by the standard model. In this case, the anomaly becomes a "technical" anomaly.

Using the philosophy of science definition of anomalies, the normal state is the prediction by the system model, the deviation is the difference between the prediction of the system state and the measured actual state of the system, and the substantiality is defined by the noise and precision of our predictions and measurement tools.

3 Proposed Framework for a Formalization of Anomalies

To separate "scientific" and "technical" anomalies, a formerly proposed framework [5] is generalized as illustrated in Fig. 2. and mathematically defined in this section.

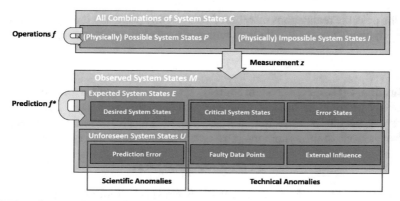

Fig. 1 Formalization of "scientific" and "technical" anomalies and system states.

Definition 1 (System State) There exists a multivariate description x_i of a state i with a finite number of features. For each feature j of state i a value x_{ij} exists, which is a realization of the feature space R_j. The value x_{ij} is the actual and precise state description of feature j at state i. Although there exists only a single true value x_{ij}, the value itself does not necessarily have to be a single data point but can be a multivariate or symbolic data value and can be of any data type.

$$\forall i \; \forall j \; \exists! \, x_{ij}, \quad x_{ij} \in R_j \tag{1}$$

The set C of all combinations of system state values with J features is given by:

$$C = \{x_i \mid \forall j \; \exists \, x_{ij} \in R_j\} = R_1 \times \dots \times R_J \tag{2}$$

Definition 2 (Operation) An operation is an analytical function f which changes the system state from state i to the following state $i + 1$. Both states belong to the set of all combinations of system states C.

$$f : C \rightarrow C, \quad f(x_i) = x_{i+1} \tag{3}$$

There exists a finite set F of functions of endogenous state transformations. This set of functions is the scope of operations that can be performed. These functions are the fundamental functionality of a system, which can be performed without any external involvement. For all functions the following expression is applied:

$$g \in F \wedge f \in F : g \circ f \in F \tag{4}$$

Using the defined function space, a restriction of reachable system states via all functions from F is defined, resulting in the set of physically possible system states.

Definition 3 (Physically Possible System States) The relation f spans the complete space of state changes of a system using the entire scope of operations. The resulting space is the set of all possible system states. The physically possible system states are the possible realizations of x_i based on a starting point and if only functions from F are applied. The set P is a group with a neutral element of operations.

$$P = \{x_i \mid \forall f \in F : f(x_i) \in P\} \subseteq C \tag{5}$$

Definition 4 (Observed System States) Of the amount J of existing features of the system state, only an amount D of features is known with $D \leq J$. Since not all system states can be measured, a function z transforms the real system states and real operations of the system into observable system states and operations.

$$z : C \to M, \quad z(x_i) = x_{i^*} \tag{6}$$

Therefore, the set $M = R_1 \times \ldots \times R_D$ is the space of all observable and known system states. Function z is the measurement process.

Definition 5 (Observed Operations) Not all functions of the whole set of function F are known or observable when planning and operating a system.

$$F' \subseteq F \tag{7}$$

Additionally, only observable system states are modeled when operating a system. The observed operations of systems are therefore projections of a subsets of known operations of F and operate within the observed and known system states.

$$F^* = z(F') \tag{8}$$

The actual conducted operations f are always from the set of operations F, but the expectation and prediction utilize, due to lack of system knowledge, only $f^* \in F^*$.

$$f^* : M \to M, \quad f^*(x_{i^*}) = x_{i+1^*} \tag{9}$$

Therefore, all states applied in operation f^* are defined as expected system states.

Definition 6 (Expected System States) The system states, which are possible if only the observed and known operations of the set F^* are applied to all system states $x_{i^*} \in E$, are the expected system behavior.

$$E = \{x_{i^*} \mid \forall f^* \in F^* : f^*(x_{i^*}) \in E\} \subseteq M \tag{10}$$

The expected system states can be further split into desired system states, where the system is running most beneficially for its usage, a critical system state, where a possible error or rare system states are measured, and error states, which are system faults with operational risks involved as defined by Basel III [18]. Applied in engineering, this definition is compatible with the definition of DIN EN 13306 since the system is at risk of being unable to perform a certain range of functions without necessarily being completely inoperable [19]. All kinds of errors, warnings and non-beneficial system states are the "technical" anomalies within the contextual analysis of the data set.

Definition 7 (Unforeseen System States) The set of unforeseen system states U are therefore all measurable system states within the realm of observable system states but not within the expected system states:

$$U = M/E \tag{11}$$

"Scientific" anomalies in unforeseen system states are measured if the real operation f differs from f^* such that a prediction error occurs:

$$f^*(x_{i^*}) \in E, \quad f^*(x_{i^*}) \neq z(f(x_i)) \notin E \tag{12}$$

"Scientific" anomalies are part of the unforeseen system states. Another reason for unforeseen system states is a measurement of an impossible system state. Anomalies originated by physically impossible system states are to be distinguished from "scientific" anomalies since the reason for their occurrence follows a different mechanism. Thus, they are assigned to the "technical" anomalies.

Definition 8 (Physically Impossible System States) Physically impossible system states I are combinations of states in set C which are not reachable using function f:

$$I = C/P \tag{13}$$

Definition 9 (External Influence) Applying changes to the system, the feature space also changes. Consequently, the space of the physically possible system states changes. Previously impossible system states become possible system states.

Definition 10 (Faulty Data Points) If a measurement is conducted incorrectly, the measured values could be within the impossible system states. Faulty data points are therefore neither measurement noise nor imprecision, but should be systematically excluded. Note that faulty data points could be within the possible system space but need to be excluded either way.

4 Conclusion

It is concluded that the anomaly concept is often loosely defined and heavily depends on assumptions of a normal state, deviation, and substantiality. These definitions are often case-specific and influenced by the conducting researchers' choice. Therefore, a rigorous definition of anomalies is capable of further streamlining the discourse and increasing a common understanding of what kind of anomaly is described.

Using "technical" and "scientific" anomalies, further research will be conducted to set up models detecting both types of anomalies separately. Differences between observed and real system states and operations are a focus of further research to more precisely analyze the hidden processes of the "scientific" anomaly generation. Also, a more fundamental discussion of the philosophical definition of anomalies within the philosophy of science and its applications to anomaly detection in general should be conducted to further gain insight into the true nature of anomalies.

The authors plan to validate the concept by using the proposed definition and framework in exemplary applications within industrial processes. Furthermore, anomaly detection methods designed for applications in dynamical systems using the proposed framework are planned to be developed.

Acknowledgements The Mercedes-Benz Group AG funds this research. The research was prepared within the framework of the doctoral program of the Institut für Informationsmanagement im Ingenieurwesen (IMI) at the Karlsruhe Institute of Technology (KIT).

References

1. Aggarwal, C. C.: Outlier Analysis. Springer Science+Business Media, New York (2013)
2. Hodge, V. J., Austin, J.A.: Survey of outlier detection methodologies. Artif. Intell. Rev. **22**, 85-126 (2004)
3. Aggarwal, C. C., Sathe, S.: Outlier Ensembles. Springer, Cham (2017)
4. Wang, X., Wang, X., Wilkes M.: New Developments in Unsupervised Outlier Detection - Algorithms and Applications. Springer, Singapore (2021)
5. Spoor, J. M., Weber, J., Ovtcharova, J.: A definition of anomalies, measurements and predictions in dynamical engineering systems for streamlined novelty detection. Accepted for the 8th International Conference on Control, Decision and Information Technologies (CoDIT), Istanbul (2022)
6. Åström, K. J., Murray, R. M.: Feedback Systems - An Introduction for Scientists and Engineers. Princeton University Press, Princeton, New Jersey (2008)
7. Sethi, S. P., Thompson, G. L.: Optimal Control Theory - Applications to Management Science and Economics. Springer Science+Business Media, Boston, MA (2000)
8. Mehrotra, K. G., Mohan, C., Huang, H.: Anomaly Detection - Principles and Algorithms. Springer International Publishing, Cham (2017)
9. Skiena, S. S.: The Data Science Design Manual. Springer International Publishing, Cham (2017)
10. James, G., Witten, D., Hastie, T., Tibshirani, R.: An Introduction to Statistical Learning. Springer Science+Business Media, New York (2013)
11. Fahrmeier, L., Hamerle, A., Tutz, G. (ed.): Multivariate Statistische Verfahren. de Gruyter, Berlin (1996)
12. Rousseeuw, P. J., Leroy, A. M.: Robust Regression and Outlier Detection. John Wiley & Sons, Inc (1987)
13. Foorthuis, R.: On the nature and types of anomalies: A review of deviations in data. Int. J. Data Sci. Anal. **12**, 297-331 (2021)
14. Cuadrado-Gallego, J. J., Demchenko, Y.: The Data Science Framework: A View from the EDISON Project. Springer Nature Switzerland AG, Cham (2020)
15. Kuhn, T.: The Structure of Scientific Revolutions. 2nd ed. The University of Chicago Press, Chicago (1970)
16. Hawkins, D.: Identification of Outliers. Chapman and Hall (1980)
17. Chandola, V., Banerjee, A., Kumar, V.: Anomaly detection: A survey. ACM Comput. Surv. **41**(3) 15 (2009)
18. Bank for International Settlements: Basel Committee on Banking Supervision: International Convergence of Capital Measurement and Capital Standards (2006)
19. DIN Deutsches Institut für Normung e. V.: DIN EN 13306: Instandhaltung - Begriffe der Instandhaltung. Beuth Verlag GmbH, Berlin (2010)

New Metrics for Classifying Phylogenetic Trees Using *K*-means and the Symmetric Difference Metric

Nadia Tahiri and Aleksandr Koshkarov

Abstract The *k*-means method can be adapted to any type of metric space and is sometimes linked to the median procedures. This is the case for symmetric difference metric (or Robinson and Foulds) distance in phylogeny, where it can lead to median trees as well as to Euclidean Embedding. We show how a specific version of the popular *k*-means clustering algorithm, based on interesting properties of the Robinson and Foulds topological distance, can be used to partition a given set of trees into one (when the data is homogeneous) or several (when the data is heterogeneous) cluster(s) of trees. We have adapted the popular cluster validity indices of Silhouette, and *Gap* to tree clustering with *k*-means. In this article, we will show results of this new approach on a real dataset (aminoacyl-tRNA synthetases). The new version of phylogenetic tree clustering makes the new method well suited for the analysis of large genomic datasets.

Keywords: clustering, symmetric difference metrics, *k*-means, phylogenetic trees, cluster validity indices

1 Introduction

In biology, one of the most significant organizing principles is the "Tree of Life" (ToL) [12]. In genetic studies, there is evidence of an enormous number of branches, but even a rough estimate of the total size of the tree remains difficult. Many recent

Nadia Tahiri (✉)
Department of Computer Science, University of Sherbrooke, Sherbrooke, QC J1K2R1, Canada,
e-mail: Nadia.Tahiri@USherbrooke.ca

Aleksandr Koshkarov
Department of Computer Science, University of Sherbrooke, Sherbrooke, QC J1K2R1, Canada;
Center of Artificial Intelligence, Astrakhan State University, Astrakhan, 414056, Russia,
e-mail: Aleksandr.Koshkarov@USherbrooke.ca

© The Author(s) 2023
P. Brito et al. (eds.), *Classification and Data Science in the Digital Age*,
Studies in Classification, Data Analysis, and Knowledge Organization,
https://doi.org/10.1007/978-3-031-09034-9_41

383

representations of ToL have emphasized either the existence of deep evolutionary relationships [7] or the knowledge of a large and diverse variety of life, with an emphasis on Eukaryotes [8]. These approaches do not consider the dramatic evolution in our understanding of the diversity of life due to genomic sampling of previously unexplored environments.

As a result, Maddison in 1991 [11] was the first to formulate the idea of multiple consensus trees when he described his phylogenetic island method. He observed that island consensus trees can differ significantly from each other and are generally better resolved than the species-wide consensus tree. The most intuitive approach to discovering and clustering genes that share similar evolutionary histories is to cluster their genetic phylogenies. In this context, Stockham et al. in 2002 [18] proposed a tree clustering algorithm based on k-means [4, 9, 10] and the Robinson and Foulds quadratic distance [15]. Their clustering algorithm aims to infer a set of strict consensus trees, minimizing information loss. They proceed by determining the consensus trees for each set of clusters in all intermediate partitioning solutions tested by k-means. This makes the Stockham et al. algorithm very expensive in terms of execution time. More recently, Tahiri et al. in 2018 [19] proposed a fast and accurate tree clustering method based on k-medoids. Finally, Silva and Wilkinson in 2021 [17] introduced a revised definition of tree islands based on any tree-to-tree metric that usefully extends this notion to any set or multiset of trees and provided an interesting discussion of biological applications of their method.

In this context, the use of a method that infers multiple supertrees (i.e., a supertree clustering method) would help discover and cluster alternative evolutionary scenarios for several ToL subtrees.

The paper is structured as follows. In the next section, we introduce a new metric for k-means algorithm based on the Robinson and Foulds distance. The section 3 presents the simulation results (on a real dataset) obtained with our algorithm compared to other clustering methods. Finally, we discuss our contributions in section 4.

2 Methods

The k-means algorithm [9, 10] is a very common algorithm for data parsing. From a set of N observations $x_i, \ldots, x_N$ each one being described by M variables, this algorithm creates a partition in k homogeneous classes or clusters. Each observation corresponds to a point in a M-dimensional space and the proximity between two points is measured by the distance between them. In the framework of k-means, the most commonly used distances are the Euclidean distance, Manhattan distance, and Minkowski distance [4]. To be precise, the objective of the algorithm is to find the partition of the N points into k clusters in such a way that the sum of the squares of the distances of the points to the center of gravity of the group to which they are assigned is minimal. To the best of our knowledge, finding an optimal partition according to the k-means least-squares criterion is known to be NP-hard [13]. Considering this fact, several polynomial-time heuristics were developed, most of which have the time complexity of $O(KNIM)$ for finding an approximate partitioning solution, where K is the maximum possible number of clusters, N is the number of objects (for example, phylogenetic trees), I is the number of iterations in the k-means algorithm, and M is the number of variables characterizing each of the N objects.

A well-known metric of comparing two tree topologies in computational biology is the Robinson-Foulds distance (RF), also known as the symmetric-difference distance [15]. Moreover, the distance RF is a topological distance, which means that it does not consider the length of the edges of the tree. The formula of RF distance can be describe as $(n_1(T_1) + n_2(T_2))$, where $n_1(T_1)$ is the number of partitions of data implied by the tree T_1, but not the tree T_2 and $n_2(T_2)$ is the number of partitions of data implied by the tree T_2 but not the tree T_1. According to Barthélemy and Monjardet [1], the majority-rule consensus tree of a set of trees is the median tree of this set. This fact makes the use of tree clustering possible.

2.1 Silhouette Index Adapted for Tree Clustering

The first popular cluster validity index we consider in our study is the Silhouette width (SH) [16]. Traditionally, the Silhouette width of the cluster k is defined as follows:

$$s(k) = \frac{1}{N_k} \left[\sum_{i=1}^{N_k} \frac{b(i) - a(i)}{max(a(i), b(i))} \right], \qquad (1)$$

where N_k is the number of objects belonging to cluster k, $a(i)$ is the average distance between object i and all other objects belonging to cluster k, and $b(i)$ is the smallest, over all clusters k' different from cluster k, of all average distances between i and all the objects of cluster k'.

We used Equations (2) and (4) for calculating $a(i)$ and $b(i)$, respectively, in our tree clustering algorithm (see also [19]). For instance, the quantity $a(i)$ can be calculated as follows:

$$a(i) = \left[\frac{\sum_{j=1}^{N_k} RF(T_{ki}, T_{kj})}{2n(T_{ki}, T_{kj}) - 6} + \xi \right] / N_k, \qquad (2)$$

where N_k is the number of trees in cluster k, T_{ki} and T_{kj} are, respectively, trees i and j in cluster k, $n(T_{ki})$ is the number of leaves in tree T_{ki}, $n(T_{kj})$ is the number of leaves in tree T_{kj}, and ξ is a penalty function which is defined as follows:

$$\xi = \alpha \times \frac{Min(n(T_{ki}), n(T_{kj})) - n(T_{ki}, T_{kj})}{Min(n(T_{kj}), n(T_{kj}))}, \qquad (3)$$

where α is the penalization (tuning) parameter, taking values between 0 and 1, used to prevent from putting to the same cluster trees having small percentages of leaves in common, and $n(T_{ki}, T_{kj})$ is the number of common leaves in trees T_{ki} and T_{kj}.

The formula for $b(i)$ is as follows:

$$b(i) = \min_{1 \leq k' \leq K, k' \neq k} \left[\frac{\sum_{j=1}^{N_{k'}} RF(T_{ki}, T_{k'j})}{2n(T_{ki}, T_{k'j}) - 6} + \xi \right] / N_{k'}, \tag{4}$$

where $T_{k'j}$ is the tree j of the cluster k', such that $k' \neq k$, and $N_{k'}$ is the number of trees in the cluster k'.

The optimal number of clusters, K, corresponds to the maximum average Silhouette width, SH, which is calculated as follows:

$$SH = \bar{s}(K) = \sum_{k=1}^{K} \left[s(k) \right] / K. \tag{5}$$

The value of the Silhouette index defined by Equation (5) ranges from -1 to +1.

2.2 *Gap* Statistic Adapted for Tree Clustering

It is worth noting that the SH cluster validity index (Equations (1) to (5)) do not allow comparing the solution consisting of a single consensus tree ($K = 1$; the calculation of SH is impossible in this case) with clustering solutions involving multiple consensus trees or supertrees ($K \geq 2$). This can be considered as an important disadvantage of the SH-based classifications because a good tree clustering method should be able to recover a single consensus tree or supertree when the input set of trees is homogeneous (e.g. for a set of gene trees that share the same evolutionary history).

The *Gap* statistic was first used by Tibshirani et al. [20] to estimate the number of clusters provided by partitioning algorithms. The formulas proposed by Tibshirani et al. were based on the properties of the Euclidean distance. In the context of tree clustering, the *Gap* statistic can be defined as follows. Consider a clustering of N trees into K non-empty clusters, where $K \geq 1$. First, we define the total intracluster distance, D_k, characterizing the cohesion between the trees belonging to the same cluster k:

$$D_k = \sum_{i=1}^{N_k} \sum_{j=1}^{N_k} \left[\frac{RF(T_{ki}, T_{kj})}{2n(T_{ki}, T_{kj}) - 6} + \xi \right]. \tag{6}$$

Then, the sum of the average total intracluster distances, V_K, can be calculated using this formula:

$$V_K = \sum_{k=1}^{K} \frac{1}{2N_k} D_k. \tag{7}$$

Finally, the *Gap* statistic, which reflects the quality of a given clustering solution including K clusters, can be defined as follows:

$$Gap_N(K) = E_N^* \{ \log(V_K) \} - \log(V_K) . \tag{8}$$

where E_N^* denotes expectation under a sample of size N from the reference distribution. The following formula [20] for the expectation of $log(V_K)$ was used in our algorithm:

$$E_N^* \{ \log(V_K) \} = \log(Nn/12) - (2/n) \log(K) , \tag{9}$$

where n is the number of tree leaves.

The largest value of the *Gap* statistic corresponds to the best clustering.

3 Results - A Biological Example

To illustrate the methods described above, we used a dataset from Woese et al. [22]. The aminoacyl-tRNA synthetases (aaRSs) are enzymes that attach the appropriate amino acid onto its cognate transfer RNA. The structure-function aspect of aaRSs has long attracted the attention of biologists [22, 6]. Moreover, the relationship of aaRSs to the genetic code is observed from the evolutionary view (the central role played by the aaRSs in translation would suggest that their histories and that of the genetic code are somehow intertwined [22]). The novel domain additions to aaRSs genes play an important role in the inference of the ToL.

We encoded 20 original aminoacyl-tRNA synthetase trees from Woese et al. [22] in Newick format and then split some of them into sub-trees to account for cases where the same species appeared more than once in the original tree. Our approach cannot handle data that includes multiple instances of the same species in the input trees. Thus, 36 aaRS trees with different numbers of leaves (including 72 species in total) were used as input of our algorithm (their Newick strings are available at: https://github.com/tahiri-lab/PhyloClust). Our approach was applied with the α parameter set to 1.

First, we implemented our new approach with the *Gap* statistic cluster validity index which suggested the presence of 7 clusters of trees in the data, thus suggesting a heterogeneous scenario of their evolution. Then, we conducted the computation using the *SH* cluster validity index and obtained 2 clusters of trees each of which could be represented by its own supertree. The first cluster obtained using *SH* included 19 trees for a total of 56 organisms, whereas the second cluster included 17 trees for a total of 61 organisms. The supertrees (see Figure 1) for the two obtained clusters of trees were inferred using the CLANN program [5]. Further, we decided to infer the most common horizontal gene transfers which characterized the evolution of gene trees included in the two obtained tree clusters. The method of [3], reconciling the species and gene phylogenies to infer transfers, was used for this purpose. The species phylogenies followed the NCBI taxonomic classification. These phylogenies were not fully resolved (the species phylogeny in Figure 1a contains 9 internal nodes

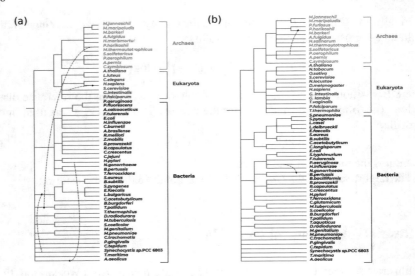

Fig. 1 Nonbinary species tree corresponding to the NCBI taxonomic classification are represented with (a) 56 species for cluster 1. The 4 HGTs (indicated by arrows) were found by the SH index for the first cluster; (b) 61 species with α equal 1 for cluster 2. The 2 HGTs (indicated by arrows) were found by the SH index with α equal 1 for the second cluster. We applied Most Similar Supertree Method ($dfit$) [5] implemented in CLANN Software with mrp criterion. This criterion is a matrix representation employing parsimony criterion.

with a degree higher than 3 and the species phylogeny in Figure 1b contains 10 internal nodes with a degree higher than 3).

We used the version of the HGT (Horizontal Gene Transfer) algorithm available on the T-Rex web site [2] to identify the scenarios of HGT events that reconcile the species tree and each of the supertrees. We choose the same root between species trees and supertrees: the root which split Bacteria to the clade of Eukayota and Archaea.

For the first cluster composed of 56 species, we obtained 40 transfers with 22 regular and 18 trivial HGTs. Trivial HGTs are necessary to transform a non-binary tree into a binary tree. We removed the trivial HGTs and selected between regular HGTs. The non-trivial HGTs with low representation are most likely due to the tree reconstruction artefacts. In Figure 1a, we illustrated only those HGTs that are most represented in the dataset.

We followed the same procedure for the second cluster composed of 61 species and obtained 42 transfers with 28 regular and 14 trivial HGTs that are not represented here. We selected only the most popular HGTs in the dataset. All other transfers are represented in Figure 1b.

The transfers link of *P. horikoshii* to the clade of *spirochetes* (i.e. *B. burgdorferi* and *T. pallidum*) was found by [3, 14]. The transfers of *P. horikoshii* to *P. aerophilum* were also found by [14]. These results confirmed the existing HGT of [3, 14].

4 Discussion

Many research groups are estimating trees containing several thousands to hundreds of thousands of species, toward the eventual goal of the estimation of the Tree of Life, containing perhaps several million leaves. These phylogenetic estimations present enormous computational challenges, and current computational methods are likely to fail to run even with datasets on the low end of this range. One approach to estimate a large species tree is to use phylogenetic estimation methods (such as maximum likelihood) on a supermatrix produced by concatenating multiple sequence alignments for a collection of markers; however, the most accurate of these phylogenetic estimation methods are extremely computationally intensive for datasets with more than a few thousand sequences. Supertree methods, which assemble phylogenetic trees from a collection of trees on subsets of the taxa, are important tools for phylogeny estimation where phylogenetic analyses based upon maximum likelihood (ML) are infeasible.

In this article, we described a new algorithm for partitioning a set of phylogenetic trees in several clusters in order to infer multiple supertrees, for which the input trees have different, but mutually overlapping sets of leaves. We presented new formulas that allow the use of the popular Silhouette and *Gap* statistic cluster validity indices along with the Robinson and Foulds topological distance in the framework of tree clustering based on the popular k-means algorithm. The new algorithm can be used to address a number of important issues in bioinformatics, such as the identification of genes having similar evolutionary histories, e.g. those that underwent the same horizontal gene transfers or those that were affected by the same ancient duplication events. It can also be used for the inference of multiple subtrees of the Tree of Life. In order to compute the Robinson and Foulds topological distance between such pairs of trees, we can first reduce them to a common set of leaves. After this reduction, the Robinson and Foulds distance is normalized by its maximum value, which is equal to $2n - 6$ for two binary trees with n leaves. Overall, the good performance achieved by the new algorithm in both clustering quality and running time makes it well suited for analyzing large genomic and phylogenetic datasets. A C++ program, called PhyloClust (Phylogenetic trees Clustering), implementing the discussed tree partitioning algorithm is freely available at `https://github.com/tahiri-lab/PhyloClust`.

Acknowledgements We would like to thank Andrey Veriga and Boris Morozov for helping us with the analysis of Aminoacyl-tRNA synthetases data. We also thank Compute Canada for providing access to high-performance computing facilities. This work was supported by Fonds de Recherche sur la Santé of Québec and University of Sherbrooke grant.

References

1. Barthelemy, J., Monjardet, B.: The median procedure in cluster analysis and social choice theory. Math. Soc. Sci. **1**, 235-267 (1981)
2. Boc, A., Legendre, P., Makarenkov, V.: An efficient algorithm for the detection and classification of horizontal gene transfer events and identification of mosaic genes. Algorithms From And For Nature And Life. pp. 253-260 (2013)
3. Boc, A., Philippe, H., Makarenkov, V.: Inferring and validating horizontal gene transfer events using bipartition dissimilarity. Syst. Biol. **59**, 195-211 (2010)
4. Bock, H.: Clustering methods: a history of k-means algorithms. Selected Contributions In Data Analysis And Classification. pp. 161-172 (2007)
5. Creevey, C., Fitzpatrick, D., Philip, G., Kinsella, R., O'Connell, M., Pentony, M., Travers, S., Wilkinson, M., McInerney, J.: Does a tree–like phylogeny only exist at the tips in the prokaryotes?. Proc. Roy. Soc. Lond. B Biol. Sci. **271**, 2551-2558 (2004)
6. Godwin, R., Macnamara, L., Alexander, R., Salsbury Jr, F.: Structure and dynamics of tRNA-met containing core substitutions. ACS Omega. **3**, 10668-10678 (2018)
7. Gouy, R., Baurain, D., Philippe, H.: Rooting the tree of life: the phylogenetic jury is still out. Phil. Trans. Biol. Sci. **370**, 20140329 (2015)
8. Hinchliff, C., Smith, S., Allman, J., Burleigh, J., Chaudhary, R., Coghill, L., Crandall, K., Deng, J., Drew, B., Gazis, R. et al.: Synthesis of phylogeny and taxonomy into a comprehensive tree of life. Proc. Natl. Acad. Sci. Unit. States Am. **112**, 12764-12769 (2015)
9. Lloyd, S.: Least squares quantization in PCM. IEEE Trans. Inform. Theor. **28**, 129-137 (1982)
10. MacQueen, J. et al.: Some methods for classification and analysis of multivariate observations. Proceedings of the Fifth Berkeley Symposium On Mathematical Statistics and Probability. **1**, 281-297 (1967)
11. Maddison, D.: The discovery and importance of multiple islands of most-parsimonious trees. Syst. Biol. **40**, 315-328 (1991)
12. Maddison, D., Schulz, K., Maddison, W. et al.: The tree of life web project. Zootaxa. **1668**, 19-40 (2007)
13. Mahajan, M., Nimbhorkar, P., Varadarajan, K.: The planar k-means problem is NP-hard. International Workshop On Algorithms And Computation. pp. 274-285 (2009)
14. Makarenkov, V., Boc, A., Delwiche, C., Philippe, H. et al.: New efficient algorithm for modeling partial and complete gene transfer scenarios. Data Science And Classification. 341-349 (2006)
15. Robinson, D., Foulds, L.: Comparison of phylogenetic trees. Math. Biosci. **53**, 131-147 (1981)
16. Rousseeuw, P.: Silhouettes: a graphical aid to the interpretation and validation of cluster analysis. J. Comput. Appl. Math. **20**, 53-65 (1987)
17. Silva, A., Wilkinson, M.: On defining and finding islands of trees and mitigating large island bias. Syst. Biol. **706**, 1282-1294 (2021)
18. Stockham, C., Wang, L., Warnow, T.: Statistically based postprocessing of phylogenetic analysis by clustering. Bioinformatics. **18**, S285-S293 (2002)
19. Tahiri, N., Willems, M., Makarenkov, V.: A new fast method for inferring multiple consensus trees using k-medoids. BMC Evol. Biol. **18**, 1-12 (2018)
20. Tibshirani, R., Walther, G., Hastie, T.: Estimating the number of clusters in a dataset via the gap statistic. J. Roy. Stat. Soc. B Stat. Meth. **63**, 411-423 (2001)
21. Whidden, C., Zeh, N., Beiko, R.: Supertrees based on the subtree prune-and-regraft distance. Syst. Biol. **63**, 566-581 (2014)
22. Woese, C., Olsen, G., Ibba, M., Soll, D.: Aminoacyl-tRNA synthetases, the genetic code, and the evolutionary process. Microbiol. Mol. Biol. Rev. **64**, 202-236 (2000)

On Parsimonious Modelling via Matrix-variate t Mixtures

Salvatore D. Tomarchio

Abstract Mixture models for matrix-variate data have becoming more and more popular in the most recent years. One issue of these models is the potentially high number of parameters. To address this concern, parsimonious mixtures of matrix-variate normal distributions have been recently introduced in the literature. However, when data contains groups of observations with longer-than-normal tails or atypical observations, the use of the matrix-variate normal distribution for the mixture components may affect the fitting of the resulting model. Therefore, we consider a more robust approach based on the matrix-variate t distribution for modeling the mixture components. To introduce parsimony, we use the eigen-decomposition of the components scale matrices and we allow the degrees of freedom to be equal across groups. This produces a family of 196 parsimonious matrix-variate t mixture models. Parameter estimation is obtained by using an AECM algorithm. The use of our parsimonious models is illustrated via a real data application, where parsimonious matrix-variate normal mixtures are also fitted for comparison purposes.

Keywords: matrix-variate, mixture models, clustering, parsimonious models

1 Introduction

The matrix-variate model-based clustering literature is expanding more and more over the last few years, as confirmed by the high number of contributions using finite mixture models for the modelization of matrix-variate data [1, 2, 3, 4, 5, 6, 7, 8]. This kind of data is arranged in three-dimensional arrays, and depending on the entities indexed in each of the three layers, different data examples might be considered [9]. In many of these applications, we observe a $p \times r$ matrix for each statistical

Salvatore D. Tomarchio (✉)
University of Catania, Department of Economics and Business, Catania, Italy,
e-mail: daniele.tomarchio@unict.it

© The Author(s) 2023
P. Brito et al. (eds.), *Classification and Data Science in the Digital Age*,
Studies in Classification, Data Analysis, and Knowledge Organization,
https://doi.org/10.1007/978-3-031-09034-9_42

observation. Thus, from a model-based clustering perspective, the challenge is to suitably cluster realization coming from random matrices.

One problem of matrix-variate mixture models is the potentially high number of parameters. To cope with this issue, [5] have recently proposed a family of parsimonious mixtures based on the matrix-variate normal (MVN) distribution. Nevertheless, for many datasets, the tails of the MVN distribution are often shorter than required. This has several consequences on parameter estimation as well as in the proper data classification [4, 7]. Therefore, in this paper we relax the normality assumption of the mixture components by using (in a parsimonious setting) the matrix-variate t (MVT) distribution. The MVT distribution has been used within the finite mixture model paradigm by [10] in an unconstrained framework. Here, to introduce parsimony in this model, (i) we use the eigen-decomposition of the two scale matrices of each mixture component and (ii) we allow the degrees of freedom to be tied across the groups. This produces the family of 196 parsimonious matrix-variate MVT mixture models (MVT-Ms) discussed in Section 2. Parameter estimation is implemented by using an alternating expectation-conditional maximization (AECM) algorithm [12]. In Section 3, our parsimonious MVT-Ms, along with parsimonious matrix-variate MVN mixture models (MVN-Ms) for comparison purposes, are fitted to a Swedish municipalities expenditure dataset. The differences in terms of fitting among the two families of models are illustrated. The estimated parameters and the data partition of the overall best fitting model are also commented. Finally, some conclusions are drawn in Section 4.

2 Methodology

2.1 Parsimonious Mixtures of Matrix-variate t Distributions

The probability distribution function (pdf) of a $p \times r$ random matrix $\mathcal{X}$ coming from a finite mixture model is

$$f_{\text{MIXT}}(\mathbf{X}; \mathbf{\Omega}) = \sum_{g=1}^{G} \pi_g f(\mathbf{X}; \mathbf{\Theta}_g), \tag{1}$$

where π_g is the gth mixing proportion, such that $\pi_g > 0$ and $\sum_{g=1}^{G} \pi_g = 1$, $f(\mathbf{X}; \mathbf{\Theta}_g)$ is the gth component pdf with parameter $\mathbf{\Theta}_g$, and $\mathbf{\Omega}$ contains all of the parameters of the mixture. In this paper, for the gth component of model (1), we adopt the MVT distribution having pdf

$$f_{\text{MVT}}(\mathbf{X}; \mathbf{\Theta}_g) = \frac{|\mathbf{\Sigma}_g|^{-\frac{r}{2}} |\mathbf{\Psi}_g|^{-\frac{p}{2}} \Gamma\left(\frac{pr+\nu_g}{2}\right)}{(\pi \nu_g)^{\frac{pr}{2}} \Gamma\left(\frac{\nu_g}{2}\right)} \left[1 + \frac{\delta_g\left(\mathbf{X}; \mathbf{M}_g, \mathbf{\Sigma}_g, \mathbf{\Psi}_g\right)}{\nu_g} \right]^{-\frac{pr+\nu_g}{2}}, \tag{2}$$

where $\delta_g\left(\mathbf{X}; \mathbf{M}_g, \boldsymbol{\Sigma}_g, \boldsymbol{\Psi}_g\right) = \text{tr}\left[\boldsymbol{\Sigma}_g^{-1}(\mathbf{X} - \mathbf{M}_g)\boldsymbol{\Psi}_g^{-1}(\mathbf{X} - \mathbf{M}_g)'\right]$, $\mathbf{M}_g$ is the $p \times r$ component mean matrix, $\boldsymbol{\Sigma}_g$ is the $p \times p$ component row scale matrix, $\boldsymbol{\Psi}_g$ is the $r \times r$ component column scale matrix and $v_g > 0$ is the component degree of freedom. It is interesting to recall that the pdf in (2) can be hierarchically obtained via the matrix-variate normal scale mixture model when the mixing random variable W is a gamma distribution with scale and rate parameters set to $v_g/2$ [10]. Specifically, a hierarchical representation of MVT distribution can be given as follows

1. $W \sim \mathcal{G}\left(v_g/2, v_g/2\right)$,
2. $\mathbf{X}|W = w \sim \mathcal{N}(\mathbf{M}_g, \boldsymbol{\Sigma}_g/w, \boldsymbol{\Psi}_g)$,

where $\mathcal{G}\left(\cdot\right)$ is a gamma distribution and $\mathcal{N}(\cdot)$ denotes the MVN distribution. This representation will be convenient for parameter estimation presented in Section 2.2.

As discussed in Section 1, the mixture model in (1) may be characterized by a potentially high number of parameters. To address this concern, we firstly use the eigen-decomposition of the components scale matrices $\boldsymbol{\Sigma}_g$ and $\boldsymbol{\Psi}_g$. In detail, we recall that a generic $q \times q$ scale matrix $\boldsymbol{\Phi}_g$ can be decomposed as [11]

$$\boldsymbol{\Phi}_g = \lambda_g \boldsymbol{\Gamma}_g \boldsymbol{\Delta}_g \boldsymbol{\Gamma}_g', \tag{3}$$

where $\lambda_g = |\boldsymbol{\Phi}_g|^{1/q}$, $\boldsymbol{\Gamma}_g$ is a $q \times q$ orthogonal matrix whose columns are the normalized eigenvectors of $\boldsymbol{\Phi}_g$, and $\boldsymbol{\Delta}_g$ is the scaled ($|\boldsymbol{\Delta}_g| = 1$) diagonal matrix of the eigenvalues of $\boldsymbol{\Phi}_g$. By constraining the three components in (3), the following family of 14 parsimonious structures is obtained: EII, VII, EEI, VEI, EVI, VVI, EEE, VEE, EVE, VVE, EEV, VEV, EVV, VVV, where "E" stands for equal, "V" means varying and "I" denotes the identity matrix.

If we apply the decomposition in (3) to $\boldsymbol{\Sigma}_g$ and $\boldsymbol{\Psi}_g$, we obtain $14 \times 14 = 196$ parsimonious structures. However, to solve a well-known identifiability issue related to the scale matrices of matrix-variate distributions [1, 3, 5], we impose the restriction $|\boldsymbol{\Psi}_g| = 1$, which makes the parameter λ_g unnecessary, and reduces the number of parsimonious structures related to $\boldsymbol{\Psi}_g$ from 14 to 7: II, EI, VI, EE, VE, EV, VV. Thus, we have $14 \times 7 = 98$ parsimonious structures for the component scale matrices.

To further increase the parsimony of model (1), we also consider the option of constraining the component degrees of freedom v_g. The nomenclature used is the same to that adopted for the scale matrices. This option, combined with that discussed above for the scale matrices, allows us to produce a total of $98 \times 2 = 196$ parsimonious MVT-Ms.

2.2 An AECM Algorithm for Parameter Estimation

To estimate the parameters of our family of mixture models, we implement an AECM algorithm. By using the hierarchical representation of Section 2.1, our complete data are $\mathbf{S}_c = \{\mathbf{X}_i, \mathbf{z}_i, w_i\}_{i=1}^N$, where $\mathbf{z}_i = (z_{i1}, \ldots, z_{iG})'$, such that $z_{ig} = 1$ if observation i belongs to group g and $z_{ig} = 0$ otherwise, and w_i is the realization of W. Therefore, the complete-data log-likelihood can be written as

$$\ell_c\left(\boldsymbol{\Omega};\mathbf{S}_c\right) = \ell_{1c}\left(\boldsymbol{\pi};\mathbf{S}_c\right) + \ell_{2c}\left(\boldsymbol{\Xi};\mathbf{S}_c\right) + \ell_{3c}\left(\boldsymbol{\vartheta};\mathbf{S}_c\right), \tag{4}$$

where

$$\ell_{1c}\left(\boldsymbol{\pi};\mathbf{S}_c\right) = \sum_{i=1}^{N}\sum_{g=1}^{G} z_{ig}\ln\left(\pi_g\right),$$

$$\ell_{2c}\left(\boldsymbol{\Xi};\mathbf{S}_c\right) = \sum_{i=1}^{N}\sum_{g=1}^{G} z_{ig}\left[-\frac{pr}{2}\ln\left(2\pi\right) + \frac{pr}{2}\ln\left(w_{ig}\right) - \frac{r}{2}\ln|\boldsymbol{\Sigma}_g| - \frac{p}{2}\ln|\boldsymbol{\Psi}_g| - \frac{w_{ig}\delta_g\left(\mathbf{X};\mathbf{M}_g,\boldsymbol{\Sigma}_g,\boldsymbol{\Psi}_g\right)}{2}\right], \tag{5}$$

$$\ell_{3c}\left(\boldsymbol{\vartheta};\mathbf{S}_c\right) = \sum_{i=1}^{N}\sum_{g=1}^{G} z_{ig}\left\{\frac{\nu_g}{2}\ln\left(\frac{\nu_g}{2}\right) - \ln\left[\Gamma\left(\frac{\nu_g}{2}\right)\right] + \left(\frac{\nu_g}{2}-1\right)\ln\left(w_{ig}\right) - \frac{\nu_g}{2}w_{ig}\right\},$$

with $\boldsymbol{\pi} = \left\{\pi_g\right\}_{g=1}^{G}$, $\boldsymbol{\Xi} = \left\{\mathbf{M}_g,\boldsymbol{\Sigma}_g,\boldsymbol{\Psi}_g\right\}_{g=1}^{G}$ and $\boldsymbol{\vartheta} = \left\{\nu_g\right\}_{g=1}^{G}$.

Our AECM algorithm then proceeds as follows (notice that, the parameters marked with one dot are the updates of the previous iteration, while those marked with two dots are the updates at the current iteration):

E-step At the E-step we have to compute the following quantities

$$\ddot{z}_{ig} = \frac{\dot{\pi}_g f_{\mathrm{MVT}}\left(\mathbf{X}_i;\dot{\boldsymbol{\Theta}}_g\right)}{\sum_{h=1}^{G}\dot{\pi}_h f_{\mathrm{MVT}}\left(\mathbf{X}_i;\dot{\boldsymbol{\Theta}}_h\right)} \quad\text{and}\quad \ddot{w}_{ig} = \frac{pr + \dot{\nu}_g}{\dot{\nu}_g + \dot{\delta}_g\left(\mathbf{X}_i;\dot{\mathbf{M}}_g,\dot{\boldsymbol{\Sigma}}_g,\dot{\boldsymbol{\Psi}}_g\right)}. \tag{6}$$

There is no need to compute the expected value of $\ln\left(W_{ig}\right)$, given that we do not use this quantity to update ν_g.

CM-step 1 At the first CM-step, we have the following updates

$$\ddot{\pi}_g = \frac{\sum_{i=1}^{N}\ddot{z}_{ig}}{N} \quad\text{and}\quad \ddot{\mathbf{M}}_g = \frac{\sum_{i=1}^{N}\ddot{z}_{ig}\ddot{w}_{ig}\mathbf{X}_i}{\sum_{i=1}^{N}\ddot{z}_{ig}\ddot{w}_{ig}}.$$

Because of space constraints, we cannot report here the updates of each parsimonious structure related to $\boldsymbol{\Sigma}_g$ and $\boldsymbol{\Psi}_g$. However, they can be obtained by generalizing the results in [5]. The only differences consist in the updates of the row and column scatter matrices of the gth component, that are here defined as

$$\ddot{\mathbf{W}}_g^R = \sum_{i=1}^{N}\ddot{z}_{ig}\ddot{w}_{ig}\left(\mathbf{X}_i - \ddot{\mathbf{M}}_g\right)\boldsymbol{\Psi}_g^{-1}\left(\mathbf{X}_i - \ddot{\mathbf{M}}_g\right)',$$

$$\ddot{\mathbf{W}}_g^C = \sum_{i=1}^{N}\ddot{z}_{ig}\ddot{w}_{ig}\left(\mathbf{X}_i - \ddot{\mathbf{M}}_g\right)'\ddot{\boldsymbol{\Sigma}}_g^{-1}\left(\mathbf{X}_i - \ddot{\mathbf{M}}_g\right).$$

CM-step 2   At the second CM-step, we firstly define the "partial" complete-data log-likelihood function according to the following specification

$$\ell_{pc}\left(\mathbf{\Omega};\mathbf{S}_{pc}\right) = \ell_{1c}\left(\boldsymbol{\pi};\mathbf{S}_{pc}\right) + \sum_{i=1}^{N}\sum_{g=1}^{G} z_{ig}\ln f_{\mathrm{MVT}}(\mathbf{X}_i;\mathbf{\Theta}_g), \tag{7}$$

where "partial" refers to fact that the complete data are now defined as $\mathbf{S}_{pc} = \{\mathbf{X}_i, \mathbf{z}_i\}_{i=1}^{N}$. Then, $\ddot{v}_g$ is determined by maximizing

$$\sum_{i=1}^{N} \ddot{z}_{ig}\ln f_{\mathrm{MVT}}(\mathbf{X}_i;\ddot{\mathbf{\Theta}}_g) \quad \text{or} \quad \sum_{i=1}^{N}\sum_{g=1}^{G} \ddot{z}_{ig}\ln f_{\mathrm{MVT}}(\mathbf{X}_i;\ddot{\mathbf{\Theta}}_g),$$

over $v_g \in (0, 100)$, depending on the parsimonious structure selected, i.e. V or E, respectively. Notice that, an higher upper bound could also have been selected for the maximization problem but, with the already chosen value, the differences between an estimated MVT distribution and the nested MVN distribution would be negligible. Furthermore, when a heavy-tailed distribution approaches to normality, the precision of the estimated tailedness parameters is unreliable [4].

3 Real Data Application

Here, we analyze the Municipalities dataset contained in the **AER** package [13] for the R statistical software. It consists of expenditure information for $N = 265$ Swedish municipalities over $r = 9$ years (1979–1987). For each municipality, we measure the following $p = 3$ variables: (i) total expenditures, (ii) total own-source revenues and (iii) intergovernmental grants received.

We fitted parsimonious MVT-Ms and MVN-Ms for $G \in \{1, 2, 3, 4, 5\}$ to the data, and for each family of models the Bayesian information criterion (BIC) [14] is used to select the best fitting model. According to our results, we found that the best among MVN-Ms has a BIC of -82362.61, a VVV-EE structure and $G = 4$ groups, while the best among MVT-Ms has a BIC of -82701.59, a VVE-EE-V structure and $G = 3$ groups. Thus, the overall best fitting model is that selected for MVT-Ms. The MVN-Ms seem to overfit the data, given that an additional group is detected. This is not an unusual behavior, given that the tails of normal mixture models cannot adequately accommodate deviations from normality, and additional groups are consequently found in the data [4, 7, 15]. Anyway, the best fitting models of the two families agree in finding varying volumes and shapes in the components row scale matrices and equal shapes and orientations in the components column scale matrices.

Figure 1 illustrates the parallel coordinate plots of the data partition detected by the VVE-EE-V MVT-Ms. The dashed lines correspond to the estimated mean for that variable, across the time, in that group. We notice that the first group contains municipalities having, on average, slightly higher expenditures, an intermediate

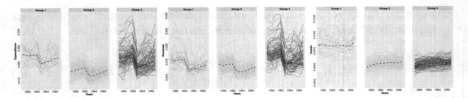

Fig. 1 Parallel coordinate plots of the data partition obtained by the VVE-EE-V MVT-Ms. The dashed lines correspond to the estimated means.

level of revenues and higher levels of intergovernmental grants than the other two groups. Furthermore, it seems to cluster several outlying observations, as confirmed by the estimated degree of freedom $v_1 = 3.75$, which implies quite heavy tails for this mixture component. The second group shows the lowest average levels of expenditures and revenues, but a similar amount of received grants to that of the third group. Interestingly, this group does not presents many outlying observations, as also supported by the estimated degree of freedom $v_2 = 10.95$. Lastly, the third group has the highest levels of revenues but, as already said, it is similar to the other two groups in the other variables. Also in this case, we have a moderately heavy tail behavior given that the estimated degree of freedom is $v_3 = 6.05$.

To evaluate the correlations of the variables with each other and over time, for the three groups, we now report the correlation matrices $\mathbf{R}_{(\cdot)}$ related to the covariance matrices associated to Σ_g and Ψ_g:

$$\mathbf{R}_{\Sigma_1} = \begin{bmatrix} 1.00 & 0.48 & 0.14 \\ 0.48 & 1.00 & -0.06 \\ 0.14 & -0.06 & 1.00 \end{bmatrix}, \mathbf{R}_{\Sigma_2} = \begin{bmatrix} 1.00 & 0.55 & 0.18 \\ 0.55 & 1.00 & -0.07 \\ 0.18 & -0.07 & 1.00 \end{bmatrix}, \mathbf{R}_{\Sigma_3} = \begin{bmatrix} 1.00 & 0.73 & 0.22 \\ 0.73 & 1.00 & -0.02 \\ 0.22 & -0.02 & 1.00 \end{bmatrix},$$

$$\mathbf{R}_{\Psi_1} = \mathbf{R}_{\Psi_2} = \mathbf{R}_{\Psi_3} = \begin{bmatrix} 1.00 & 0.80 & 0.72 & 0.67 & 0.65 & 0.59 & 0.58 & 0.55 & 0.52 \\ 0.80 & 1.00 & 0.79 & 0.73 & 0.69 & 0.62 & 0.62 & 0.57 & 0.54 \\ 0.72 & 0.79 & 1.00 & 0.80 & 0.73 & 0.69 & 0.66 & 0.63 & 0.60 \\ 0.67 & 0.73 & 0.80 & 1.00 & 0.79 & 0.73 & 0.71 & 0.67 & 0.64 \\ 0.65 & 0.69 & 0.73 & 0.79 & 1.00 & 0.83 & 0.80 & 0.73 & 0.71 \\ 0.59 & 0.62 & 0.69 & 0.73 & 0.83 & 1.00 & 0.80 & 0.76 & 0.73 \\ 0.58 & 0.62 & 0.66 & 0.71 & 0.80 & 0.80 & 1.00 & 0.81 & 0.78 \\ 0.55 & 0.57 & 0.63 & 0.67 & 0.73 & 0.76 & 0.81 & 1.00 & 0.79 \\ 0.52 & 0.54 & 0.60 & 0.64 & 0.71 & 0.73 & 0.78 & 0.79 & 1.00 \end{bmatrix}.$$

When $\mathbf{R}_{\Sigma_1}$, $\mathbf{R}_{\Sigma_2}$ and $\mathbf{R}_{\Sigma_3}$ are considered, we notice that, as it might be reasonable to expect, the correlations between total-expenditures and total-own source revenues or intergovernmental grants received are positive, despite they increase as we move from the first to the third group. Conversely, there exists a slightly negative correlation between total-own source revenues and intergovernmental grants received. However, there would be no great differences among the groups in this case. As concerns $\mathbf{R}_{\Psi_1}$, $\mathbf{R}_{\Psi_2}$ and $\mathbf{R}_{\Psi_3}$, we observe that the correlation among the columns, i.e. between time points, decreases as the temporal distance increases. Furthermore, considering the dimensionality of these column matrices, it is readily understandable the benefit, in terms of number of parameters to be estimated, of an EE parsimonious structure with respect to a fully unconstrained model.

Finally, we analyze the uncertainty of the detected classification. This can be computed, for each observation, by subtracting the probability z_{ig} of the most likely group from 1 [16]. The lower the uncertainty is, the stronger the assignment becomes. The quantiles of the obtained uncertainties can be used to measure the quality of the classification. In this regard, we noticed that 75% of the observations have an uncertainty equal or lower than 0.05. However, we observed a maximum value of 0.50. This happens when groups intersect, since uncertain classifications are expected in the overlapping regions [17]. Relatedly, a more detailed information can be gained by looking at the uncertainty plot illustrated in Figure 2, which reports the (sorted) uncertainty values of all the municipalities. We see that the municipalities clustered

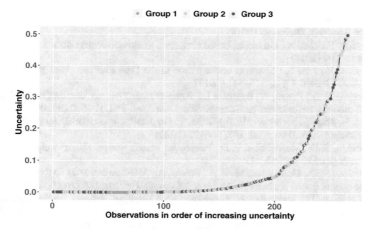

Fig. 2 Uncertainty plot for the `Municipalities` dataset.

in the first group, excluding a couple of cases, have practically null uncertainties. This applies to a lesser extent to the municipalities in the other two groups, given the slightly higher number of exceptions. For example, there are 15 observations (approximately 5% of the total sample size) that have uncertainty values greater than 0.3. However, and as said above, this is due to the closeness between the groups, which can be confirmed by looking at the parallel plots in Figure 1.

4 Conclusions

One serious concern of matrix-variate mixture models is the potentially high number of parameters. Furthermore, many real data requires models having heavier-than-normal tails. To address both aspects, in this paper a family of 196 parsimonious mixture models, based on the matrix-variate t distribution, is introduced. The eigen-decomposition of the components scale matrices, as well as constraints on the components degrees of freedom, are used to attain parsimony. An AECM algorithm for parameter estimation has been presented. Our family of models have been fitted to a real dataset along with parsimonious mixtures of matrix-variate normal distributions. The results demonstrate the best fitting results of our models, and the overfitting tendency of matrix-variate normal mixtures. Lastly, the estimated parameters and data partition for the best of our models have been reported and commented.

Acknowledgements This work was supported by the University of Catania grant PIACERI/CRASI (2020).

References

1. Gallaugher, M. P. B., McNicholas P. D.: Finite mixtures of skewed matrix variate distributions. Pattern Recognit. **80**, 83–93 (2018)
2. Melnykov, V., Zhu, X.: On model-based clustering of skewed matrix data. J. Multivar. Anal. **167**, 181–194 (2018)
3. Melnykov, V., Zhu, X.: Studying crime trends in the USA over the years 2000–2012. Adv. Data Anal. Classif. **13**(1), 325–341 (2019)
4. Tomarchio, S. D., Punzo, A., Bagnato, L.: Two new matrix-variate distributions with application in model-based clustering. Comput. Stat. Data Anal. **152**, 107050 (2020)
5. Sarkar, S., Zhu, X., Melnykov, V., Ingrassia, S.: On parsimonious models for modeling matrix data. Comput. Stat. Data Anal. **142**, 106822 (2020)
6. Tomarchio, S. D., McNicholas, P. D., Punzo, A.: Matrix normal cluster-weighted models. J. Classif. **38**(3), 556–575 (2021)
7. Tomarchio, S. D., Gallaugher, M. P. B., Punzo, A., McNicholas, P. D.: Mixtures of matrix-variate contaminated normal distributions. J. Comput. Gr. Stat. 1–9 (2022)
8. Tomarchio, S. D., Ingrassia, S., Melnykov, V.: Modelling students' career indicators via mixtures of parsimonious matrix-normal distributions. Aust. N. Z. J. Stat. 1–16 (2022)
9. Viroli, C.: Model based clustering for three-way data structures. Bayesian Anal. **6**(4), 573–602 (2011)
10. Doğru, F. Z., Bulut, Y. M., Arslan, O.: Finite mixtures of matrix variate t distributions. Gazi Univ. J. Sci. **29**(2), 335–341 (2016)
11. Celeux, G., Govaert, G.: Gaussian parsimonious clustering models. Pattern Recognit. **28**(5), 781–793 (1995)
12. Meng, X. L., Van Dyk, D.: The EM algorithm-an old folk-song sung to a fast new tune. J. Royal Stat. Soc. B. **59**(3), 511–567 (1997)
13. Kleiber, C., Zeileis, A.: Applied Econometrics with R. Springer-Verlag, New York (2008)
14. Schwarz, G.: Estimating the dimension of a model. Ann. Stat. **6**(2), 461–464 (1978)
15. Gallaugher, M. P. B., Tomarchio, S. D., McNicholas, P. D., Punzo, A.: Multivariate cluster weighted models using skewed distributions. Adv. Data Anal. Classif. 1–32 (2021)
16. Fraley, C., Raftery, A. E.: Enhanced model-based clustering, density estimation, and discriminant analysis software: MCLUST. J. Classif., **20**(2), 263–286 (2003)
17. Tomarchio, S. D., Punzo, A.: Dichotomous unimodal compound models: application to the distribution of insurance losses. J. Appl. Stat. **47**(13-15), 2328–2353 (2020)

Evolution of Media Coverage on Climate Change and Environmental Awareness: an Analysis of Tweets from UK and US Newspapers

Gianpaolo Zammarchi, Maurizio Romano, and Claudio Conversano

Abstract Climate change represents one of the biggest challenges of our time. Newspapers might play an important role in raising awareness on this problem and its consequences. We collected all tweets posted by six UK and US newspapers in the last decade to assess whether 1) the space given to this topic has grown, 2) any breakpoint can be identified in the time series of tweets on climate change, and 3) any main topic can be identified in these tweets. Overall, the number of tweets posted on climate change increased for all newspapers during the last decade. Although a sharp decrease in 2020 was observed due to the pandemic, for most newspapers climate change coverage started to rise again in 2021. While different breakpoints were observed, for most newspapers 2019 was identified as a key year, which is plausible based on the coverage received by activities organized by the Fridays for Future movement. Finally, using different topic modeling approaches, we observed that, while unsupervised models partly capture relevant topics for climate change, such as the ones related to politics, consequences for health or pollution, semi-supervised models might be of help to reach higher informativeness of words assigned to the topics.

Keywords: climate change, Twitter, environment, time series, topic modeling

Gianpaolo Zammarchi (✉)
University of Cagliari, Viale Sant'Ignazio 17, 09123, Cagliari, Italy,
e-mail: gp.zammarchi@unica.it

Maurizio Romano
University of Cagliari, Viale Sant'Ignazio 17, 09123, Cagliari, Italy,
e-mail: romano.maurizio@unica.it

Claudio Conversano
University of Cagliari, Viale Sant'Ignazio 17, 09123, Cagliari, Italy, e-mail: conversa@unica.it

P. Brito et al. (eds.), *Classification and Data Science in the Digital Age*,
Studies in Classification, Data Analysis, and Knowledge Organization,
https://doi.org/10.1007/978-3-031-09034-9_43

403

1 Introduction

Climate change is one of the biggest challenges for our society. Its consequences which include, among others, glaciers melting, warming oceans, rising sea levels, and shifting weather or rainfall patterns, are already impacting our health and imposing costs on society. Without drastic action aimed at reducing or preventing human-induced emissions of greenhouse gasses, these consequences are expected to intensify in the next years. Despite its global and severe impacts, individuals may perceive climate change as an abstract problem [1]. It is also a well-known fact that the level of information plays a crucial role in the awareness about a topic (e.g. healthy food [2] and smoking [3]) . Media represent a crucial source of information and can exert substantial effects on public opinion, thus helping to raise the awareness on climate change. For instance, media can explain climate change consequences as well as portraying actions that governments, communities and single individuals can take. For this reason, it is important to distinguish themes that might have gained popularity from those that may have seen a decrease of interest. Nowadays, social media have become a reliable and popular source of information for people from all around the world. Twitter is one of the most popular microblogging services and is used by many traditional newspapers on a daily basis. While we can hypothesize that in the last few years the media coverage on climate change might have risen, due for instance to international climate strike movements, the recent emergence of the coronavirus disease 2019 (COVID-19) pandemic might have led to a decrease of attention on other relevant topics.

Aims of this work were to: (1) assess trends in media coverage on climate change using tweets posted by main international newspapers based in United Kingdom (UK) and United States (US), and (2) identify the main topics discussed in these tweets using topic modeling.

2 Dataset and Methods

We downloaded all tweets posted from 2012 January 1st to 2021 December 31st from the official Twitter account of six widely known newspapers based in UK (The Guardian, The Independent and The Mirror) or US (The New York Times, The Washington Post and The Wall Street Journal) leading to a collection of 3,275,499 tweets. Next, we determined which tweets were related to climate change and environmental awareness based on the presence of at least one of the following keywords: "climate change", "sustainability", "earth day", "plastic free", "global warming", "pollution", "environmentally friendly" or "renewable energy". We plotted the number of tweets on climate change posted by each newspaper during each year using R v. 4.1.2 [4].

We analyzed the association between the number of tweets on climate change and the whole number of tweets posted by each newspaper using Spearman's correlation analysis. For each year and for each newspaper, we computed and plotted the differences in the number of posted tweets compared to the previous year, for either (a) tweets related to climate change and (b) all tweets. Finally, we used the changepoint R package [5] to conduct an analysis aimed at identifying structural breaks, i.e. unexpected changes in a time series. In many applications, it is reasonable to believe that there might be m breakpoints (especially if some exogenous event occurs) in which a shift in mean value is observed. The changepoint package estimates the breakpoints using several penalty criteria such as the Bayesian Information Criterion (BIC) or the Akaike Information Criterion (AIC). We estimated the breakpoints using the Binary Segmentation (BinSeg) method [6] implemented in the package.

Lastly, we used tweets posted by The Guardian to perform topic modeling, a method for classification of text into topics. Preprocessing (including lemmatization, removal of stopwords and creation of the document term matrix) was conducted with tm [7] and quanteda [8] in R. We used two different approaches: 1) Latent Dirichlet Allocation (LDA) implemented in the textmineR R package [9]; and 2) Correlation Explanation (CorEx), an approach alternative to LDA that allows both unsupervised as well as semi-supervised topic modeling [10].

3 Results

3.1 Analysis of Tweet Trends and Breakpoints

Among 3,275,499 collected tweets, we identified 11,155 tweets related to climate change and environmental awareness. Figure 1A shows the number of tweets on climate change posted by each of the analyzed newspapers from 2012 to 2021, while Figure 1B the total number of tweets posted by each newspaper.

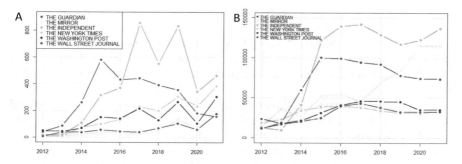

Fig. 1 Number of tweets on climate change (A) or total number of tweets (B) posted by the six newspapers from 2012 to 2021.

For the majority of newspapers, the number of tweets on climate change increased from 2014 to 2019, saw a sharp decrease in 2020, in correspondence of the emergence of the COVID-19 pandemic, and a subsequent rise in 2021. On the other hand, the

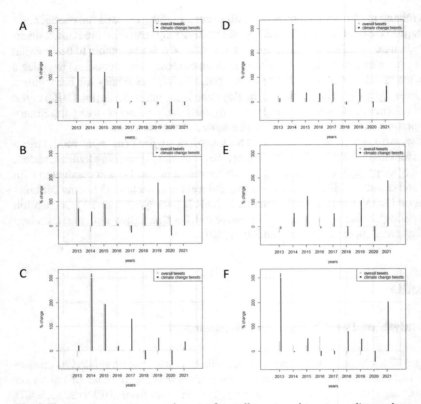

Fig. 2 Year-over-year percentage changes of overall tweets and tweets on climate change. A: The Guardian, B: The Mirror, C: The Independent, D: The New York Times, E: The Washington Post, F, The Wall Street Journal.

number of tweets on climate change posted by The Guardian showed a peak during 2015 and a subsequent decrease. However, it must be noted that The Guardian is also the newspaper that showed a more pronounced decrease in the overall number of tweets.

The number of tweets on climate change was significantly positively correlated with the overall number of tweets posted from 2012 to 2021 for four newspapers (The Guardian, Spearman's rho = 0.95, $p < 0.001$; The Mirror, Spearman's rho = 0.95, $p < 0.001$; The Independent, Spearman's rho = 0.76, $p = 0.016$; The Washington Post, Spearman's rho = 0.70, $p = 0.031$) but not for The New York Times (Spearman's rho = 0.18, $p = 0.63$) or The Wall Street Journal (Spearman's rho = 0.49, $p = 0.15$). Year-over-year percentage changes among either tweets related to climate change or all posted tweets can be observed in Figure 2.

Looking at Figure 2, we can observe a great variability in the posted number of tweets during the years, both for the total number of tweets and for the number of tweets on climate change. While the analysis aimed at identifying structural changes

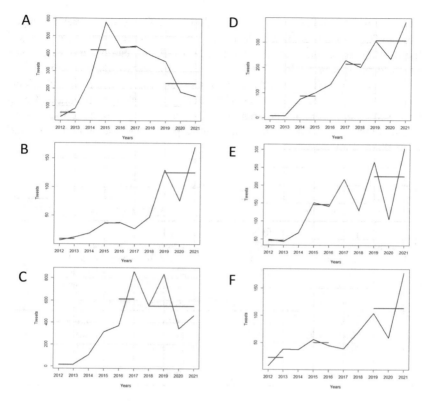

Fig. 3 Structural changes in the time series of tweets related to climate change. A: The Guardian, B: The Mirror, C: The Independent, D: The New York Times, E: The Washington Post, F, The Wall Street Journal. The red line represents the years between two breakpoints.

in the time series comprising tweets on climate change identified three or four breakpoints for all newspapers, wide variability was observed regarding the specific year in which these structural changes were identified (Figure 3). Despite the great variability, Figure 3 shows that even if a common breakpoint cannot be identified, 2019 was a key year for five out of six newspapers (except for The Independent).

3.2 Topic Modeling

Finally, we exploited the topic modeling approach to identify and analyze the main topics discussed by newspapers in their tweets. Due to space limitations, we focus only on The Guardian since this newspaper showed a trend in contrast with the others. Data comes from 2,916 tweets posted by The Guardian analyzed using LDA and CorEx. For LDA, a range of 5-20 unsupervised topics was tested, with the most interpretable results obtained with 10 topics (Table 1). The topic coherence ranged from 0.01 to 0.34 (mean: 0.13). For each topic, bi-gram topic labels were assigned with the labeling algorithm implemented in textmineR. We can observe that topics are related to politics or leaders (Topics 3, 7 and 10), environmental scientists or climate journalists (Topics 1 and 5), energy sources (Topics 4 and 8) and effects of climate change (Topics 2, 6 and 9). The intertopic distance map obtained with LDAvis is shown in Figure 4. The area of each circle is proportional to the relative prevalence of that topic in the corpus, while inter-topic distances are computed based on Jensen-Shannon divergence.

Table 1 Top terms for the ten topics identified with LDA.

dana_nuccitelli	air_pollution	barack_obama	renewable_energy	john_abraham
dana	pollution	fight	energy	john
dana_ nuccitelli	air	obama	renewable	trump
nuccitelli	air_pollution	trump	renewable_energy	australia
live	study	plan	uk	tackle
trump	finds	battle	sustainability	abraham
air_pollution	donald_trump	fossil_fuel	extreme_weather	pope_francis
pollution	trump	report	world	pollution
air	schoolstrike	fossil	paris	study
air_pollution	school	ipcc	leaders	tackling
uk	great	warns	talks	pope
tackle	donald	stop	deal	scientists

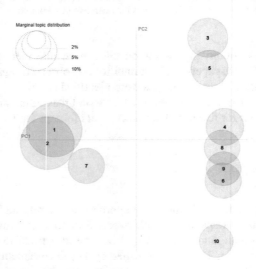

Fig. 4 Intertopic distance map.

Finally, we conducted a semi-supervised topic modeling analysis based on anchored words using CorEx. When anchoring a word to a topic, CorEx maximizes the mutual information between that word and the topic, thus guiding the topic model towards specific subsets of words. A model with 5 topics and three anchored words for each topic (Table 2) showed a total correlation (i.e. the measure maximized by CorEx when constructing the topic model) of 4.36. This value was higher compared to the one observed with an unsupervised CorEx analysis with the same number of topics (total correlation = 0.97, topics not shown due to space limits). Topics related to politics (Topic 3) and science (Topic 5) were found to be the most informative in our dataset based on the total correlation metric.

Table 2 Topics with anchored words and examples of tweets.

Topic	Topic words	Examples of tweets per topic
1	**school, strike, march,** schoolstrik, climatestrikeuk, ukschoolstrik, schoolstrikeclim, climatemarch, arabia, saudi	EPA wipes its climate change site day before march on Washington
2	**ocean, ice, environment,** john, dana, nuccitelli, air, abraham, sea, reed	Chasing Ice filmmakers plumb the 'bottomless' depths of climate change - new clip from @GuardianEco
3	**trump, obama, lead,** donald, barack, ivanka, brighton, repli, administr, pick	Trump administration pollution rule strikes final blow against environment
4	**plastic, fuel, oil,** fossil, compani, pictur, wast, big, bay, photo	Engaging with oil companies on climate change is futile
5	**studi, scientist, research,** find, link, say, show, death, prematur, speci	Microplastic pollution revealed 'absolutely everywhere' by new research

The anchored words are reported in bold.

4 Discussion

The present study aims to evaluate how some of the most relevant British and American newspapers have given space to the topic of climate change on their Twitter page in the last decade. Apart from The Guardian, which shows a decreasing trend in the number of tweets related to climate change, all the other newspapers showed an overall growing trend, except during 2020. During this year, the number of tweets related to climate change declined for all six newspapers. This was most probably due to the COVID-19 outbreak that was massively covered by all media. By analyzing the breakpoints in Figure 3, it is possible to observe that 2019 was a relevant year for climate change. This is plausible considering that, starting from the end of 2018, the strikes launched by the Fridays for Future movement to raise awareness on the issue of climate change, gained high media coverage.

Our topic modeling analysis showed that the main topics defined using unsupervised models such as LDA are mostly related to politics, environmental scientists, energy sources and effects of climate change. While unsupervised models capture relevant topics, using CorEx we found a semi-supervised model to be able to reach a higher total correlation, which is a measure of informativeness of the topics, compared to an unsupervised model with the same number of topics.

As future developments, we plan to extend our analyses to newspapers from other countries. We believe our work to be useful to gain more knowledge and awareness about the climate change topic and on how much space relevant newspapers have given to this issue on social media. Increasing the knowledge about the nature of the topics covered by newspapers will lay the basis for future studies aimed at evaluating public awareness on this highly relevant challenge.

References

1. Van Lange, P. A. M., Huckelba, A. L.: Psychological distance: How to make climate change less abstract and closer to the self. Curr. Opin. Psychol. **42**, 49–53 (2021)
2. Wakefield, M., Flay B., Nichter M., Giovino G.: Role of the media in influencing trajectories of youth smoking. Addiction, **98**, 79-103 (2003)
3. Dumanovsky, T., Huang, C. Y., Bassett, M. T., Silver, L. D.: Consumer awareness of fast-food calorie information in New York City after implementation of a menu labeling regulation. American Journal of Public Health, **12**, 2520-2525 (2010)
4. R Core Team. R: A language and environment for statistical computing. R Foundation for Statistical Computing, Vienna, Austria (2020). Available via `http://www.R-project.org`
5. Killick, R., Eckley, I. A.: changepoint: An R Package for changepoint analysis. J. Stat. Softw. **58**, 1–19 (2014)
6. Scott, A.J., Knott, M.: A cluster analysis method for grouping means in the analysis of variance, Biometrics, **30**, 507-512 (1974)
7. Feinerer, I., Hornik, K., Meyer, D.: Text mining infrastructure in R. J. Stat. Softw. **25**, 1–54 (2008)
8. Benoit, K., Watanabe, K., Wang, H., Nulty, P., Obeng, A., Müller, S., Matsuo, A.: quanteda: An R package for the quantitative analysis of textual data. J. Open Source Softw. **3**, 774 (2018)
9. Jones, T., Doane, W.: Package 'textmineR'. Functions for Text Mining and Topic Modeling (2021). Retrieved from `https://cran.r-project.org/web/packages/textmineR/textmineR.pdf`
10. Gallagher, R., Reing, K., Kale, D., Ver Steeg, G.: Anchored correlation explanation: Topic modeling with minimal domain knowledge. Trans. Assoc. Comput. **5**, 529-542 (2017)

Index

k-means, 181
3-way network, 125

Abdesselam, R., 1
adjacency matrix, 1
affinity coefficient, 285
anomaly detection, 309
Anton, C., 9
Antonazzo, F., 17
Arnone, E., 25
Ascari, R., 31
Aschenbruck, R., 39
Ashofteh, A., 47
association measures, 197
AUC, 150
automated planning, 87

Bacelar-Nicolau, H., 285, 301
Batagelj, V., 55
Batista, M. G., 301
Bayesian inference, 31
Bayesian methodology, 189
Beaudry, É., 86
bi-stochastic matrix, 181
blockmodeling, 55
bootstrap method, 150
Bouaoune, M. A., 71
Bouchard, K., 213
Boutalbi, R., 63
brain-computer interface, 269

Cabral, S., 301
Campos, J., 47
Cappé, O., 221
Cardoso, M. G. M. S., 293
categorical data, 293
categorical time series, 197

CatPCA, 301
Chabane, N., 71
Chadjipadelis, T., 79, 237
Champagne Gareau, J., 86
classification of textual documents, 103
climate change, 333
cluster analysis, 229, 301
cluster stability, 39, 157
cluster validation, 39
cluster validity indices, 317
clustering, 47, 71, 173, 197, 205, 317, 325
clustering validation, 285
clustering with relational constraints, 55
clusterwise, 63
co-clustering, 63
community detection, 125
complex network, 125
constraints, 117
contaminated normal distribution, 9, 253
Conversano, C., 333
correspondence analysis, 79
count data, 31
COVID-19, 79, 278
Cunial, E., 25

D'Urso, P., 197
data analysis, 237
data mining, 63
decision boundaries, 17
democracy, 237
dependancy chains, 87
Di Nuzzo, C., 95
dimensionality reduction, 71
Dobša, J., 103
Dvořák, J., 245
dynamical systems, 309

© The Author(s) 2023
P. Brito et al. (eds.), *Classification and Data Science in the Digital Age*,
Studies in Classification, Data Analysis, and Knowledge Organization,
https://doi.org/10.1007/978-3-031-09034-9

ECM algorithm, 253
EEG, 269
EM algorithm, 9, 293
emotions, 269
environment, 333
evidence-based policy making, 79
expectation-maximization, 221

factorial k-means, 181
Faria, B. M., 269
Figueiredo, M., 293
finite mixture model, 293
Fontanella, S., 261
Forbes, F., 221
Fort, G., 221
fraud detection, 111
functional data, 9, 245
functional data analysis, 25, 261, 278
fuzzy sets, 205

Gama, J., 111
García-Escudero, L. A., 117
Gaussian mixture model, 157
Gaussian process, 213
Genova, V. G., 124
Giordano, G., 124
Giubilei, R., 133
graph clustering, 133
graphical LASSO, 261
grocery shopping recommendation, 71
Górecki, T., 141

Hayashi, K., 149
Hennig, C., 157
hierarchical cluster analysis, 79
hierarchical clustering, 1
Hoshino, E., 149
hyperparameter tuning, 111
hyperquadrics, 17

Ievoli, R., 229
Ignaccolo, R., 261
image processing, 165
indicator processes, 197
Ingrassia, S., 17, 95
intelligent shopping list, 71
Ippoliti, L., 261
item classification, 47

Janáček, P., 165

k-means, 317
Kalina, J., 165
Karafiátová, I., 245

kernel density estimation, 133
kernel function, 95
Kiers, H. A. L., 103
Koshkarov, A., 317

L_1-penalty, 261
López-Oriona, Á., 197
Labiod, L., 63, 173, 181
LaLonde, A., 189
leadership, 301
learning from data streams, 111
leave-one-out cross-validation, 150
Lee, H. K. H., 213
Love, T., 189
low-energy replacements, 165
LSA, 103

machine and deep learning, 269
machine learning, 71
Magopoulou, S., 79
Makarenkov, V., 71, 86
Markov chain Monte Carlo, 189
Markov decision process, 87
Masís, D., 205
matrix-variate, 325
Mayo-Iscar, A., 117
Mazoure, B., 71
measurement error, 47
Menafoglio, A., 277
Meng, R., 213
Migliorati, S., 31
minimum message length, 293
minorization-maximization, 221
mixed-mode official surveys, 47
mixed-type data, 39
mixture model, 31, 189, 261
mixture modelling, 117
mixture models, 325
mixture of regression models, 253
mixtures of regressions, 17
mobility data, 125
mode-based clustering, 133
model based clustering, 117
model selection, 293
model-based cluster analysis, 253
model-based clustering, 9, 17
Morelli, G., 117
motivational factors, 301
multidimensional scaling, 229
multivariate data analysis, 1
multivariate methods, 237
multivariate regression, 31
multivariate time series, 213

Nadif, M., 63, 173, 181
Nakanishi, E., 149
neighborhood graph, 1
network analysis, 55
networked data, 173
networks, 133
neural networks, 245
Nguyen, H. D., 221
noise component, 157
nonparametric statistics, 133
number of clusters, 157
numerical smoothing, 205

O2S2, 278
Obatake, M., 149
online algorithms, 221
optimized centroids, 165
outlier analysis, 309
Ovtcharova, J., 309

pair correlation function, 245
Palazzo, L., 229
Panagiotidou, G., 237
parallel computing, 87
parameter estimation, 221
parsimonious models, 325
Pawlasová, K., 245
Perrone, G., 253
phylogenetic trees, 317
Piasecki, P., 141
political behavior, 237
projection matrix, 150
Pronello, N., 261
proximity measure, 1

Ragozini, G., 124
random forest, 141
rare disease, 150
recommender systems, 71
reduced k-means, 103, 181
regional healthcare, 229
Reis, L. P., 269
religion, 237
representation learning, 173
reversible jump, 189
Riani, M., 117
robustness, 165
Rodrigues, D., 269
Romano, M., 333

Sakai, K., 149
Sangalli, L. M., 25, 277
Scimone, R., 277
Secchi, P., 277

seemingly unrelated regression, 253
Segura, E., 205
semiparametric regression with roughness
    penalty, 25
sensitivity and specificity, 150
silhouette width, 261
Silva, O., 285, 301
Silvestre, C., 293
similarity forest, 141
Smith, I., 9
social networks, 55
Soffritti, G., 253
Sousa, Á., 285, 301
sparsity, 165
spatial data analysis, 25
spatial downscaling, 278
spatial point patterns, 245
Spearman correlation coefficient, 285
spectral clustering, 95
spectral rotation, 173
split-merge procedures, 189
Spoor, J. M., 309
stochastic approximation, 221
stochastic optimization, 213
strongly connected components, 87
supervised classification, 245
supervised learning, 269
Suzuki, M., 149
symbolic data analysis, 55
symmetric difference metrics, 317
Szepannek, G., 39

Tahiri, N., 71, 317
tensor, 63
tertiary education, 125
three-way data, 95
Tighilt, R. A. S., 71
time series, 141, 229, 333
time series classification, 141
time-varying correlation, 213
Tomarchio, S. D., 325
topic modeling, 333
Trejos, J., 205
trimmed k-means, 229
trimming, 157
Twitter, 333

user tuning, 157

variational inference, 213
Vilar. J. A., 197
Vitale, M. P., 124
VL methodology, 285

Weber, J., 309
weighting methods, 47
welfare, 301
Wilhelm, A. F. X., 39
Wu, T., 189

Xavier, A., 205

Young, D. R., 189

Zammarchi, G., 333

Łuczak, T., 141

Printed in the United States
by Baker & Taylor Publisher Services